NEW HORIZONS IN ASTRONOMY
Frank N. Bash Symposium 2007

COVER ILLUSTRATION:

Graphic from the conference poster designed by S. Stone.

Images courtesy of NASA/JPL-Caltech.

ASTRONOMICAL SOCIETY OF THE PACIFIC CONFERENCE SERIES

A SERIES OF BOOKS ON RECENT DEVELOPMENTS IN ASTRONOMY AND ASTROPHYSICS

Volume 393

EDITORIAL STAFF

Managing Editor: J. W. Moody
Assistant Managing Editor: Jonathan Barnes
Publication Manager: Lisa B. Roper
Editorial Assistant: Amy Schuff
E-book Specialist: Jeremy Roper
Web Developer/Technical Consultant: Jared Bellows
L^{A}T$_E$X Consultant: T. J. Mahoney (Spain) – tjm@iac.es

PO Box 4666, Room C454 – ESC, Brigham Young University, Provo, Utah, 84602-4666
Phone: 801-422-2111   Fax: 801-422-0553
E-mail: aspcs@aspbooks.org   E-book site: http://www.aspbooks.org

PUBLICATION COMMITTEE

Lynne Hillenbrand (2009), Chair
California Institute of Technology

Liz Bryson (2009)
Canada-France Hawaii Telescope

Daniela Calzetti (2010)
University of Massachusetts

Gary J. Ferland (2008)
University of Kentucky

Ed Guinan (2010)
Villanova University

Luis Ho (2010)
The Observatories of the Carnegie
Institution of Washington

Scott J. Kenyon (2008)
Smithsonian Astrophysical Observatory

Joe Patterson (2009)
Columbia University

Catherine A. Pilachowski (2009)
Indiana University

René Racine (2008)
Université de Montréal

ASPCS Volumes may be found as e-books with color images at
http://www.aspbooks.org

A listing of recently published volumes may be found at the back of this volume.

For a complete listing of ASPCS and IAU Volumes published by the ASP see
http://www.astrosociety.org/pubs.html

ASTRONOMICAL SOCIETY OF THE PACIFIC
CONFERENCE SERIES

Volume 393

NEW HORIZONS IN ASTRONOMY
Frank N. Bash Symposium 2007

Proceedings of a workshop held at
The University of Texas, Austin, Texas, USA
14–16 October 2007

Edited by

Anna Frebel
McDonald Observatory, University of Texas, Austin, Texas, USA

Justyn R. Maund
Department of Astronomy, University of Texas, Austin, Texas, USA

Juntai Shen
McDonald Observatory, University of Texas, Austin, Texas, USA

and

Michael H. Siegel
McDonald Observatory, University of Texas, Austin, Texas, USA

SAN FRANCISCO

ASTRONOMICAL SOCIETY OF THE PACIFIC
390 Ashton Avenue
San Francisco, California, 94112-1722, USA

Phone: 415-337-1100
Fax: 415-337-5205
E-mail: service@astrosociety.org
Web Site: www.astrosociety.org
E-books: www.aspbooks.org

All Rights Reserved
© 2008 by Astronomical Society of the Pacific.
ASP Conference Series - First Edition

No part of the material protected by this copyright notice may be reproduced or utilized in any form or by any means—graphic, electronic, or mechanical, including photocopying, taping, recording, or by any information storage and retrieval system—without written permission from the Astronomical Society of the Pacific.

ISBN: 978-1-58381-656-1
e-Book ISBN: 978-1-58381-657-8

Library of Congress (LOC) Cataloging in Publication (CIP) Data:
Main entry under title
Library of Congress Control Number (LCCN): 2008926839

Printed in the United States of America by Sheridan Books, Ann Arbor, Michigan

Frank N. Bash

Contents

Preface xi

Participants xiii

Conference Photograph xxi

Part 1. Reviews

Physical Properties of Small Solar System Bodies and Implications for
    Formation and Evolution of Planetary Systems 3
       J. P. Emery
Expanding and Improving the Search for Habitable Worlds 19
       A. M. Mandell
The Evolution of Primordial Circumstellar Disks 35
       L. A. Cieza
Connecting the Dots: Low-Mass Stars, Brown Dwarfs, and Planets 51
       K. Cruz
Metal-Poor Stars 63
       A. Frebel
Instrumentation in the ELT Era 79
       S. Barnes
Globular Clusters in the Local Universe 95
       D. Macky
Galaxy Formation 111
       T. Okamoto
Exploring Galaxy Formation and Reionization Towards the Cosmic Dawn 127
       M. Ouchi
Supernovae 141
       R. M. Quimby
Recent Progress on the Evolution of Rapidly Star-forming Galaxies and
    Active Galactic Nuclei 155
       J. R. Rigby

Part 2. Research Highlights

Distances to the High Velocity Clouds: A Forty-Year Mystery on the Way to Solution 179
    J. C. Barentine, B. P. Wakker, D. G. York, J. C. Howk, R. Wilhelm, H. van Woerden, R. F. Peletier, T. C. Beers, P. Richter, Ž. Ivezić, and U. J. Schwarz

Hydrodynamic Instabilities in Jet-Induced Supernovae: Results of 2D Simulations 183
    S. M. Couch, J. Craig Wheeler, and M. Milosavljević

The Hobby-Eberly Telescope Chemical Abundances of Stars in the Halo (CASH) Project III. Abundance Analysis of Three Bright Hamburg/ESO Survey Stars 187
    L. A. Davies, A. Frebel, J. J. Cowan, C. Allende Prieto, and C. Sneden

Using Stellar Photospheres as Chronometers for Studies of Disk Evolution   191
    C. P. Deen and D. T. Jaffe

Through the Thick and Thin: Electron Density Measurements of the Local Interstellar Medium 195
    R. E. Falcon and S. Redfield

Mass-to-light Ratio of Ly α Emitters: Implications of Lyα Surveys at Redshifts $z = 5.7$, 6.5, 7, and 8.8 199
    E. Fernandez and E. Komatsu

The Hobby-Eberly Telescope Chemical Abundances of Stars in the Halo (CASH) Project I. Observations of the First Year 203
    A. Frebel, C. Allende Prieto, I. U. Roederer, M. Shetrone, J. Rhee, C. Sneden, T. C. Beers, and J. J. Cowan

The Hobby-Eberly Telescope Chemical Abundances of Stars in the Halo (CASH) Project II. The Li-, r- and s-Enhanced Metal-Poor Giant HK-II 17435-00532 207
    A. Frebel, I. U. Roederer, M. Shetrone, C. Allende Prieto, J. Rhee, R. Gallino, S. Bisterrzo, C. Sneden, T. C. Beers, and J. J. Cowan

Morphological Transformations of Galaxies in the A901/02 Supercluster from STAGES 211
    A. Heiderman, S. Jogee, D. Bacon, M. Balogh, M. Barden, F. D. Barazza, E. F. Bell, A. Böhm, J. A. R. Caldwell, M. E. Gray, B. Häußler, C. Heymans, K. Jahnke, E. van Kampen, S. Koposov, K. Lane, I. Marinova, D. McIntosh, K. Meisenheimer, C. Y. Peng, H.-W. Rix, S. F. Sánchez, R. Somerville, A. Taylor, L. Wisotzki, C. Wolf, and X. Zheng

Radiative Feedback in the Formation of the First Protogalaxies 215
    J. L. Johnson, T. H. Greif, and V. Bromm

Radioactive ^{30}S Beam to Study X-Ray Bursts 219
    D. Kahl, A. A. Chen, J. Chen, S. Hayakawa, A. Kim, S. Kubono, S. Michimasa, K. Setoodehnia, Y. Wakabayashi, and H. Yamaguchi

The Cluster-Merger Shock in 1E 0657-56: Faster than a Speeding Bullet? 223
  J. Koda, M. Milosavljević, P. R. Shapiro, D. Nagai, and E. Nakar

Beryllium Abundances in Solar Mass Stars 227
  J. A. Krugler and A. M. Boesgaard

Characterizing Barred Galaxies in the Abell 901/902 Supercluster from
  STAGES 231
  I. Marinova, S. Jogee, D. Bacon, M. Balogh, M. Barden, F. D. Barazza,
  E. F. Bell, A. Böhm, J. A. R. Caldwell, M. E. Gray, B. Häußler,
  C. Heymans, K. Jahnke, E. van Kampen, S. Koposov, K. Lane,
  D. H. McIntosh, K. Meisenheimer, C. Y. Peng, H.-W. Rix, S. F. Sánchez,
  A. Taylor, L. Wisotzki, C. Wolf, and X. Zheng

Exploring the Impact of Galaxy Interactions over Seven Billion Years with
  CAS 235
  S. H. Miller, K. Penner, E. Bell, S. Jogee, M. Barden, K. Jahnke,
  R. Skelton, F. Barazza, C. Heymans, A. Robaina, S. Sanchez, X. Zheng,
  A. Borch, C. Peng, C. Papovich, S. Beckwith, D. McIntosh, J. Caldwell,
  C. Conselice, B. Haeussler, H. Rix, C. Wolf, R. Somerville,
  K. Meisenheimer, and L. Wisotzki

Amplitude Limitation in Multi-Periodic Pulsating White Dwarfs 239
  M. H. Montgomery

The Bolocam 1.1mm Galactic Plane Survey 243
  M. K. Nordhaus, N. J. Evans, J. Aguirre, J. Bally, C. Burnett,
  M. Drosback, J. Glenn, G. Laurent, R. Chamberlin, E. Rosolowsky,
  J. Vaillancourt, J. Walawender, and J. Williams

The Brightest Serendipitous X-ray Sources in ChaMPlane 247
  K. Penner, M. van den Berg, J. Hong, S. Laycock, P. Zhao, and
  J. Grindlay

An Investigation of the Canis Major Over-density 251
  W. Lee Powell Jr, R. J. Wilhelm, and K. Carrell

Signatures of Granulation in the Spectra of K-Dwarfs 255
  I. Ramirez, C. Allende Prieto, D. L. Lambert, and M. Asplund

the Hobby-Eberly Telescope (HET) Delivers the First Ground-Based
  Detection of an Exoplanetary Atmosphere 259
  S. Redfield, M. Endl, W. D. Cochran, and L. Koesterke

Europium, Samarium, and Neodymium Isotopic Fractions in Metal-Poor
  Stars 263
  I. U. Roederer, J. E. Lawler, C. Sneden, J. J. Cowan, J. Sobeck, and
  C. A. Pilachowski

Coordinated Galaxy Growth on Inner and Outer Disk Scales: Analysis of
  an Unusually Resonant S0 Galaxy and Its Companion 267
  T. Scarborough and S. Kannappan

The HST/ACS Survey of Galactic Globular Clusters: First Results 271
  M. Siegel, A. Sarajedini, B. Chaboyer, A. Dotter, S. Majewski, and
  D. Nidever

The Impact of Cosmic Rays on Population III Star Formation 275
    A. Stacy and V. Bromm

Constraining Galaxy Evolution With Bulge-Disk-Bar Decomposition . . . 279
    T. Weinzirl, S. Jogee, and F. D. Barazza

The Relics of Structure Formation: Extra-Planar Gas and High-Velocity Clouds Around M31 283
    T. Westmeier, C. Brüns, and J. Kerp

Embers of the Dark Ages: Recombination Radiation from the First Stars   287
    A. Wilson, V. Bromm, and J. L. Johnson

Author Index 303

Preface

The novel suggestion of an astronomy symposium dedicated entirely to the work and ideas of young researchers first arose in 2003 as a way to honor Frank N. Bash, retiring Director of McDonald Observatory. Recognizing Frank's longtime advocacy of education and outreach, and interest in rising talent at both the Observatory and the University of Texas at Austin, his colleagues Dan Jaffe and Neal Evans decided to celebrate his career by inviting a set of outstanding postdoctoral researchers from various subfields of astronomy to give review talks at UT Austin, creating a symposium that would emphasize "a look ahead at what will be the big issues and opportunities for progress in their fields over the next decade." Local UT postdocs served as panelists and moderators for extended discussion sessions after the talks, and the entire symposium brilliantly showcased both the current research and the research *vision* of the next generation of astronomers.

The 2003 symposium was so successful that it inspired several UT astronomy professors, including Frank Bash, to pledge funds for a biennial series of similar conferences, this time organized by the UT postdocs themselves (with invaluable assistance from faculty and staff). The result was a second successful symposium, the "Frank N. Bash Symposium 2005: New Horizons in Astronomy," colloquially known as "BashFest '05." This incarnation of the symposium set ground rules for all future BashFests, including inviting UT postdocs to give some of the review talks, and the addition of a poster session open to graduate students and postdocs from across the country. In the preface to the BashFest '05 proceedings, the editors expressed the hope that future symposia could continue to expand upon their success.

It was with those lofty expectations that the SOC and LOC of the 2007 symposium began their work in the fall of 2006. As imitation is the sincerest form of flattery, we did not tinker with the successful formula of the previous BashFests. And so, the "Frank N. Bash Symposium 2007: New Horizons in Astronomy" was held on the campus of UT Austin October 14 – 16, 2007. Thirteen postdocs, including two from UT, two former UT graduate students, and two overseas participants, provided spectacular reviews of a wide range of astronomy, from the state of minor planets in our solar system out to the fringes of cosmology to the future instrumentation needed to continue our quests for knowledge of our Universe.

In addition to our speakers, a total of 34 posters were contributed by UT postdocs, graduate students, and participants from far-flung institutions such as Texas Tech, the University of Oklahoma, Michigan State, and McMaster University. Extended moderated sessions and informal break-time conversations fertilized new ideas and planted seeds for future collaboration. Graduate students participated in two packed lunchtime Q&A sessions with the speakers, with conversations that drifted from science to the job market to the gorgeous Texas weather. The LOC went above and beyond the call of duty, arranging a marvelous opening reception at the Star of Texas Inn, a scrumptious banquet at Fonda San Miguel, and a night on the town in "The Live Music Capital of the World" that lasted until past sunrise the following morning. And the t-shirts were once again a hot commodity, thanks in large part to stirring efforts by Monica Kidd and Shelley Stone, who went to great lengths to convince local artists that there are *no* photographs of Saturn in front of a non-black background (Saturn being very obstinate about not posing in a studio, but only out in space).

All in all, BashFest '07 was yet another tremendous success. If there was one shortcoming, it was the small number of non-UT attendees at the conference. We worked hard at advertising the conference outside of UT, so that it might become a truly national event. Interest among non-UT postdocs was high, but many supervisors balked at having to pay travel expenses for their postdocs and/or graduate students to attend a "non-scientific" conference. Perhaps if money were available for partial travel scholarships for non-UT attendees, this hurdle could be overcome.

The success of BashFest '07 would not have been possible without the contributions of numerous people. We especially thank all of the contributors in this volume, whose presentations and posters were the single most important reason for the symposium's success. Dean Mary Ann Rankin, Astronomy Department Chair Neal Evans, McDonald Observatory Director David Lambert, and Professor Frank Bash provided insightful opening and closing remarks. The session moderators, postdocs Mike Endl, Mike Siegel, Juntai Shen, and Erin Bonning, kept the symposium running on time and helped to encourage lively discussions. Monica Kidd, Shelley Stone, and Gordon Orris manned the registration tables and policed the opening reception and symposium dinner, in addition to their crucial assistance in every aspect of symposium planning. Lara Eakins excelled as the audio-visual guru before and during the symposium. The conference website and fabulous graphics were designed and maintained by Jim Umbarger. Photography is courtesy of Martin Harris Photography. Welsh Corgi judging was generously provided by Queen Elizabeth II. We are also grateful for the support and guidance of Seth Redfield and Sheila Kannappan, whose expertise in planning the 2005 symposium proved invaluable. Finally, the conference was organized by a dedicated and imaginative team of postdocs (starred below), grad students (double starred), staff, and faculty. Serving on the LOC were *Stuart Barnes (Chair), *Anna Frebel, **Mike Dunham, Monica Kidd, Mary Lindholm, Shelley Stone, and Jim Umbarger. The SOC consisted of *Justyn Maund (Co-Chair), *Kurtis Williams (Co-Chair), *Mike Montgomery, *Kyungjin Ahn, and Eiichiro Komatsu.

Finally, we wish to recognize the generous financial support of the McDonald Observatory Board of Visitors. The Bash Symposium series is fully funded by a generous annual contribution from the Board of Visitors; without their support, BashFest '07 would not have been possible.

Justyn Maund and Kurtis Williams, co-chairs, SOC

Participants

J. J. ADAMS, Department of Astronomy, University of Texas at Austin, 1 University Station C1400, Austin, TX, 78712-0259, USA ⟨jjadams@astro.as.utexas.edu⟩

A. M. ADKINS, Austin Astronomical Society, P.O. Box 12831, Austin, TX, 78711-2831, USA ⟨anne@hadkins.com⟩

R. D. AKSAMIT, McDonald Observatory Board of Visitors, Suite 2300, 711 Louisiana Street, Houston, TX, 77002-2770, USA

A. S. AL MUHAMMAD, Department of Physics, University of Texas at Austin, 1 University Station C1600, Austin, TX, 78712-0264, USA ⟨anwar_s_a@hotmail.com⟩

J. C. BARENTINE, Department of Astronomy, University of Texas at Austin, 1 University Station C1400, Austin, TX, 78712-0259, USA ⟨jcb@astro.as.utexas.edu⟩

S. I. BARNES, McDonald Observatory, University of Texas at Austin, 1 University Station C1402, Austin, TX, 78712-0259, USA ⟨stuart@astro.as.utexas.edu⟩

R. J. BARNIOL DURAN, Department of Astronomy, University of Texas at Austin, 1 University Station C1400, Austin, TX, 78712-0259, USA ⟨rbarniol@mail.utexas.edu⟩

F. N. BASH, Department of Astronomy, University of Texas at Austin, 1 University Station C1400, Austin , TX, 78712-0259, USA ⟨fnb@astro.as.utexas.edu⟩

A. E. BAUER, Department of Astronomy, University of Texas at Austin, 1 University Station C1400, Austin, TX, 78712-0259, USA ⟨amanda@astro.as.utexas.edu⟩

A. J. BAYLESS, Department of Astronomy, University of Texas at Austin, 1 University Station C1400, Austin, TX, 78712-0259, USA ⟨baylessa@astro.as.utexas..edu⟩

S. BICKERSTAFF, McDonald Observatory Board of Visitors, 7705 Merrybrook Circle, Austin, TX, 78731, USA ⟨SBickerstaff@law.utexas.edu⟩

G. A. BLANC, Department of Astronomy, University of Texas at Austin, 1 University Station C1400, Austin, TX, 78712-0259, USA ⟨gblancm@astro.as.utexas.edu⟩

E. BONNING, Department of Astronomy, University of Texas at Austin, 1 University Station C1400, Austin, TX, 78712-0259, USA ⟨bonning@astro.as.utexas.edu⟩

V. BROMM, Department of Astronomy, University of Texas at Austin, 1 University Station C1400, Austin, TX, 78712-0259, USA ⟨vbromm@astro.as.utexas.edu⟩

K. W. CARRELL, Physics Department, Texas Tech University, Box 41051, Lubbock, TX, 79409, USA ⟨kenneth.w.carrell@ttu.edu⟩

D. W. CHANDLER, Central Texas Astronomical Society, 3409 Whispering Oaks, Temple, TX, 76504, USA ⟨chandler@vvm.com⟩

A. M. CHANDLER, Austin Astronomical Society, 10026 Michael Dale, Austin, TX, 78736-7834, USA ⟨anachandler@isp.com⟩

J. CHANDLER, Austin Astronomical Society, 10026 Michael Dale, Austin, TX, 78736-7834, USA ⟨chanjj@isp.com⟩

J. M. CHAVEZ, Department of Astronomy, University of Texas at Austin, 1 University Station C1400, Austin, TX, 78712-0259, USA ⟨jchavez@astro.as.utexas.edu⟩

J. CHEN, Department of Astronomy, University of Texas at Austin, 1 University Station C1400, Austin, TX, 78712-0259, USA ⟨jhchen@astro.as.utexas.edu⟩

L. A. CIEZA, Institute for Astronomy, University of Hawaii at Manoa, 2680 Woodlawn Drive, Honolulu, HI, 96822, USA ⟨lcieza@ifa.hawaii.edu⟩

A. L. COCHRAN, McDonald Observatory, University of Texas at Austin, 1 University Station C1402, Austin, TX, 78712-0259, USA ⟨anita@barolo.as.utexas.edu⟩

W. D. COCHRAN, McDonald Observatory, University of Texas at Austin, 1 University Station C1402, Austin, TX, 78712-0259, USA ⟨wdc@astro.as.utexas.edu⟩

S. M. COUCH, Department of Astronomy, University of Texas at Austin, 1 University Station C1400, Austin, TX, 78712-0259, USA ⟨smc@astro.as.utexas.edu⟩

K. CRUZ, Department of Astronomy, California Institute of Technology, 1200 California Blvd. MS 105-24, Pasadena, CA, 91125, USA ⟨kelle@astro.caltech.edu⟩

L. A. DAVIES, Homer L. Dodge Department of Physics and Astronomy, University of Oklahoma, 440 W. Brooks St., Norman, OK, 73019, USA ⟨lauradavies@ou.edu⟩

W. DAVIS, McDonald Observatory Board of Visitors, P.O. Box 7397, Austin, TX, 78713, USA ⟨woody.davis@austin.utexas.edu⟩

C. P. DEEN, Department of Astronomy, University of Texas at Austin, 1 University Station C1400, Austin, TX, 78712-0259, USA ⟨deen@astro.as.utexas.edu⟩

T. DILLE, McDonald Observatory Board of Visitors, 7917 Brightman, Austin, TX, 78733, USA ⟨tdille@earthlink.net⟩

H. L. DINERSTEIN, Department of Astronomy, University of Texas at Austin, 1 University Station C1400, Austin , TX, 78712-0259, USA ⟨harriet@astro.as.utexas.edu⟩

M. M. DUNHAM, Department of Astronomy, University of Texas at Austin,
1 University Station C1400, Austin, TX, 78712-0259, USA
⟨ mdunham@astro.as.utexas.edu ⟩

L. E. EAKINS, Department of Astronomy, University of Texas at Austin,
1 University Station C1400, Austin, TX, 78712-0259, USA
⟨ lara@astro.as.utexas.edu ⟩

J. P. EMERY, NASA Ames Research Center , MS 245-6, Moffett Field, CA,
94035, USA ⟨ jemery@mail.arc.nasa.gov ⟩

M. ENDL, McDonald Observatory, University of Texas at Austin,
1 University Station C1402, Austin, TX, 78712-0259, USA
⟨ mike@astro.as.utexas.edu ⟩

N. J. EVANS, Department of Astronomy, University of Texas at Austin,
1 University Station C1400, Austin, TX, 78712-0259, USA
⟨ nje@astro.as.utexas.edu ⟩

R. E. FALCON, Department of Astronomy, University of Texas at Austin,
1 University Station C1400, Austin, TX, 78712-0259, USA
⟨ cylver@astro.as.utexas.edu ⟩

E. R. FERNANDEZ, Department of Astronomy, University of Texas at Austin,
1 University Station C1400, Austin, TX, 78712-0259, USA
⟨ beth@astro.as.utexas.edu ⟩

B. FOR, Department of Astronomy, University of Texas at Austin,
1 University Station C1400, Austin, TX, 78712-0259, USA
⟨ biqing@astro.as.utexas.edu ⟩

A. D. FORESTELL, Department of Astronomy, University of Texas at Austin,
1 University Station C1400, Austin, TX, 78712-0259, USA
⟨ amydove@astro.as.utexas.edu ⟩

A. FREBEL, McDonald Observatory, University of Texas at Austin,
1 University Station C1402, Austin, TX, 78712-0259, USA
⟨ anna@astro.as.utexas.edu ⟩

M. GASKELL, Department of Astronomy, University of Texas at Austin,
1 University Station C1400, Austin, TX, 78712-0259, USA
⟨ gaskell@astro.as.utexas.edu ⟩

C. L. GRAY, Department of Astronomy, University of Texas at Austin,
1 University Station C1400, Austin, TX, 78712-0259, USA
⟨ candaceg@astro.as.utexas.edu ⟩

W. F. GUEST, McDonald Observatory Board of Visitors, 2243 Stanmore,
Houston, TX, 77019, USA ⟨ williamguest@gmail.com ⟩

L. HAO, Radiophysics & Space Research, Cornell University, 108 Space
Science Building, Ithaca, NY, 14850, USA ⟨ haol@isc.astro.cornell.edu ⟩

M. HARRIS, Symposium Photographer, 201 West Alpine, Austin, TX, 78704,
USA

P. M.HARVEY, Department of Astronomy, University of Texas at Austin,
1 University Station C1400, Austin, TX, 78712-0259, USA
⟨ pmh@astro.as.utexas.edu ⟩

J. HEASLEY, McDonald Observatory Board of Visitors, Texas Bankers Association, 209 W. 10th St., Austin, TX, 78701, USA

A. HEIDERMAN, Department of Astronomy, University of Texas at Austin, 1 University Station C1400, Austin, TX, 78712-0259, USA
⟨ alh@astro.as.utexas.edu ⟩

M. K.HEMENWAY, Department of Astronomy, University of Texas at Austin, 1 University Station C1400, Austin, TX, 78712-0259, USA
⟨ marykay@astro.as.utexas.edu ⟩

W. P.HOBBY, McDonald Observatory Board of Visitors, P. O. Box 326, Houston, TX, 77001-0326, USA ⟨ dchambers@hobbycomm.com ⟩

J. C.HUTH, Austin Astronomical Society, 320 Ranch House Rd, Wimberley, TX, 78676, USA ⟨ fresheggs91@yahoo.com ⟩

S. HWANG, Department of Astronomy, University of Texas at Austin, 1 University Station C1400, Austin, TX, 78712-0259, USA
⟨ shwang@astro.as.utexas.edu ⟩

E. JEFFERY, Department of Astronomy, University of Texas at Austin, 1 University Station C1400, Austin, TX, 78712-0259, USA
⟨ ejeffery@astro.as.utexas.edu ⟩

D. JEONG, Department of Astronomy, University of Texas at Austin, 1 University Station C1400, Austin, TX, 78712-0259, USA
⟨ djeong@astro.as.utexas.edu ⟩

S. JOGEE, Department of Astronomy, University of Texas at Austin, 1 University Station C1400, Austin , TX, 78712-0259, USA
⟨ sj@astro.as.utexas.edu ⟩

J. L. JOHNSON, Department of Astronomy, University of Texas at Austin, 1 University Station C1400, Austin, TX, 78712-0259, USA
⟨ jljohnson@astro.as.utexas.edu ⟩

D. M. KAHL, Department of Physics and Astronomy, McMaster University, 1280 Main Street West ABB 241, Hamilton, ON, L8S 4M1, Canada
⟨ kahldm@mcmaster.ca ⟩

S. KANG, Department of Physics, University of Texas at Austin, 1 University Station C1600, Austin, TX, 78712-0264, USA
⟨ sjkang@mail.utexas.edu ⟩

M. M. KIDD, Department of Astronomy, University of Texas at Austin, 1 University Station C1400, Austin, TX, 78712-0259, USA
⟨ monicak@astro.as.utexas.edu ⟩

A. KIM, Department of Astronomy, University of Texas at Austin, 1 University Station C1400, Austin, TX, 78712-0259, USA
⟨ agnes@astro.as.utexas.edu ⟩

H. J. KIM, Department of Astronomy, University of Texas at Austin, 1 University Station C1400, Austin, TX, 78712-0259, USA
⟨ hyojeong@astro.as.utexas.edu ⟩

J. KODA, Department of Astronomy, University of Texas at Austin,
1 University Station C1400, Austin, TX, 78712-0259, USA
⟨junkoda@physics.utexas.edu⟩

J. KODOSKY, McDonald Observatory Board of Visitors, 22 Cousteau Ln.,
Austin, TX, 78746, USA ⟨jkodosky@austin.rr.com⟩

E. KOMATSU, Department of Astronomy, University of Texas at Austin,
1 University Station C1400, Austin, TX, 78712-0259, USA
⟨komatsu@astro.as.utexas.edu⟩

M. R. KRAUSE, McDonald Observatory, University of Texas at Austin,
1 University Station C1402, Austin, TX, 78712-0259, USA
⟨mkrause@astro.as.utexas.edu⟩

J. A. KRUGLER, Department of Physics and Astronomy, Michigan State
University, East Lansing, MI, 48824, USA ⟨kruglerj@msu.edu⟩

D. L. LAMBERT, McDonald Observatory, University of Texas at Austin,
1 University Station C1402, Austin, TX, 78712-0259, USA
⟨director@astro.as.utexas.edu⟩

M. LINDHOLM, Department of Astronomy, University of Texas at Austin,
1 University Station C1400, Austin, TX, 78712-0259, USA
⟨mary@astro.as.utexas.edu⟩

D. MACKEY, Institute for Astronomy, Royal Observatory, Blackford Hill,
Edinburgh, Midlothian, EH9 3HJ, UK ⟨dmy@roe.ac.uk⟩

P. J. MACQUEEN, McDonald Observatory, University of Texas at Austin,
1 University Station C1402, Austin, TX, 78712-0259, USA
⟨pjm@wairau.as.utexas.edu⟩

E. E. MAMAJEK, Harvard-Smithsonian Center for Astrophysics, 60 Garden
St., MS-51, Cambridge, MA, 2138, USA ⟨emamajek@cfa.harvard.edu⟩

A. M. MANDELL, NASA Goddard Space Flight Center, MS 690.3, Greenbelt,
MD, 20771, USA ⟨Avi.Mandell@gsfc.nasa.gov⟩

I. S. MARINOVA, Department of Astronomy, University of Texas at Austin,
1 University Station C1400, Austin, TX, 78712-0259, USA
⟨marinova@astro.as.utexas.edu⟩

H. MARK, Department of Aerospace Engineering and Engineering Mechanics,
University of Texas at Austin,
1 University Station C0600, Austin, TX, 78712, USA
⟨hmark@mail.utexas.edu⟩

E. MARTIOLI, Department of Astronomy, University of Texas at Austin,
1 University Station C1400, Austin, TX, 78712-0259, USA
⟨eder@astro.as.utexas.edu⟩

J. R. MAUND, McDonald Observatory, University of Texas at Austin,
1 University Station C1402, Austin, TX, 78712-0259, USA
⟨jrm@astro.as.utexas.edu⟩

S. H. MILLER, Department of Astronomy, University of Texas at Austin,
1 University Station C1400, Austin, TX, 78712-0259, USA
⟨sarahmiller@mail.utexas.edu⟩

M. MONTGOMERY, Department of Astronomy, University of Texas at Austin, 1 University Station C1400, Austin, TX, 78712-0259, USA ⟨mikemon@astro.as.utexas.edu⟩

M. K. NORDHAUS, Department of Astronomy, University of Texas at Austin, 1 University Station C1400, Austin, TX, 78712-0259, USA ⟨nordhaus@astro.as.utexas.edu⟩

T. OKAMOTO, Department of Physics, University of Durham, South Road, Durham, County Durham, DH1 3LE, UK ⟨takashi.okamoto@durham.ac.uk⟩

J. ORR, San Antonio Astronomical Association , 13235 Trentwood, San Antonio, TX, 78231-2256, USA ⟨JosephOrr@aol.com⟩

G. P. ORRIS, Department of Astronomy, University of Texas at Austin, 1 University Station C1400, Austin, TX, 78712-0259, USA ⟨argus@astro.as.utexas.edu⟩

M. OUCHI, Carnegie Observatories, 813 Santa Barbara St., Pasadena, CA, 91101, USA ⟨ouchi@ociw.edu⟩

D. J. PATNAUDE, Harvard-Smithsonian Center for Astrophysics, 60 Garden St, MS-51, Cambridge, MA, 2138, USA ⟨dpatnaude@cfa.harvard.edu⟩

K. D. PENNER, Department of Astronomy, University of Texas at Austin, 1 University Station C1400, Austin, TX, 78712-0259, USA ⟨kpenner@mail.utexas.edu⟩

D. F. PITRE, , 16301 Decker Lake Rd, Manor, TX, 78653, USA ⟨farmerpittre@lycos.com⟩

W. L. POWELL, Physics Department, Texas Tech University, P.O. Box 41051, Lubbock, TX, 79409, USA ⟨william.l.powell@ttu.edu⟩

R. QUIMBY, Department of Astronomy, California Institute of Technology, 1200 California Blvd. MS 105-24, Pasadena, CA, 91125, USA ⟨quimby@phobos.caltech.edu⟩

I. P. RAMIREZ, Department of Astronomy, University of Texas at Austin, 1 University Station C1400, Austin, TX, 78712-0259, USA ⟨ivan@astro.as.utexas.edu⟩

E. RAMIREZ-RUIZ, Department of Astronomy, University of California at Santa Cruz, 1156 High Street, Santa Cruz, CA, 95060, USA ⟨enrico@ucolick.org⟩

M. A. RANKIN, College of Natural Sciences, University of Texas at Austin, 1 University Station G2500, Austin, TX, 78712, USA ⟨rankin@mail.utexas.edu⟩

S. REDFIELD, Department of Astronomy, University of Texas at Austin, 1 University Station C1400, Austin, TX, 78712-0259, USA ⟨sredfield@astro.as.utexas.edu⟩

J. G. RIES, Department of Astronomy, University of Texas at Austin, 1 University Station C1400, Austin, TX, 78712-0259, USA ⟨moon@astro.as.utexas.edu⟩

J. R. RIGBY, Carnegie Observatories, 813 Santa Barbara St, Pasadena, CA, 91101, USA ⟨jrigby@ociw.edu⟩

V. W. ROBINSON, 105 Hidden Valley, McDonald Observatory Board of Visitors, P.O. Box 1234, Ft. Davis, TX, 79734, USA ⟨starsend@wildblue.net⟩

I. U. ROEDERER, Department of Astronomy, University of Texas at Austin, 1 University Station C1400, Austin, TX, 78712-0259, USA ⟨iur@astro.as.utexas.edu⟩

P. M. ROSE, Texas House of Representatives, P.O. Box 1053, Dripping Springs, TX, 78620, USA

S. T. SALVIANDER, Department of Astronomy, University of Texas at Austin, 1 University Station C1400, Austin, TX, 78712-0259, USA ⟨triples@astro.as.utexas.edu⟩

J. M. SCALO, Department of Astronomy, University of Texas at Austin, 1 University Station C1400, Austin, TX, 78712-0259, USA ⟨parrot@astro.as.utexas.edu⟩

T. A. SCARBOROUGH, Department of Astronomy, University of Texas at Austin, 1 University Station C1400, Austin, TX, 78712-0259, USA ⟨tara.scarborough@mail.utexas.edu⟩

R. SHEN, Department of Astronomy, University of Texas at Austin, 1 University Station C1400, Austin, TX, 78712-0259, USA ⟨rfshen@astro.as.utexas.edu⟩

J. SHEN, Department of Astronomy, University of Texas at Austin, 1 University Station C1400, Austin, TX, 78712-0259, USA ⟨shen@astro.as.utexas.edu⟩

G. A. SHIELDS, Department of Astronomy, University of Texas at Austin, 1 University Station C1400, Austin, TX, 78712-0259, USA ⟨shields@astro.as.utexas.edu⟩

M. SHOJI, Department of Astronomy, University of Texas at Austin, 1 University Station C1400, Austin, TX, 78712-0259, USA ⟨mshoji@astro.as.utexas.edu⟩

M. H. SIEGEL, Department of Astronomy, University of Texas at Austin, 1 University Station C1400, Austin, TX, 78712-0259, USA ⟨siegel@astro.as.utexas.edu⟩

C. SNEDEN, Department of Astronomy, University of Texas at Austin, 1 University Station C1400, Austin, TX, 78712-0259, USA ⟨chris@verdi.as.utexas.edu⟩

A. R. STACY, Department of Astronomy, University of Texas at Austin, 1 University Station C1400, Austin, TX, 78712-0259, USA ⟨minerva@astro.as.utexas.edu⟩

S. D. STONE, Department of Astronomy, University of Texas at Austin, 1 University Station C1400, Austin , TX, 78712-0259, USA ⟨s.stone@astro.as.utexas.edu⟩

S. B. STRONG, Department of Astronomy, University of Texas at Austin, 1 University Station C1400, Austin, TX, 78712-0259, USA ⟨sholmes@astro.as.utexas.edu⟩

N. B. SUNTZEFF, Department of Physics, Texas A&M, College Station, TX, 77845, USA ⟨nsuntzeff@tamu.edu⟩

K. M. TAUTE, Center for Nonlinear Dynamics, University of Texas at Austin, 1 University Station C1610, Austin, TX, 78712-0259, USA ⟨ktaute@chaos.ph.utexas.edu⟩

R. G. TULL, McDonald Observatory, University of Texas at Austin, 1 University Station C1402, Austin, TX, 78712-0259, USA ⟨rgt@mail.utexas.edu⟩

J. F. UMBARGER, Department of Astronomy, University of Texas at Austin, 1 University Station C1400, Austin, TX, 78712-0259, USA ⟨jimum@astro.as.utexas.edu⟩

A. URBAN, Department of Astronomy, University of Texas at Austin, 1 University Station C1400, Austin, TX, 78712-0259, USA ⟨aurban@asto.as.utexas.edu⟩

W. WANG, Department of Astronomy, University of Texas at Austin, 1 University Station C1400, Austin, TX, 78712-0259, USA ⟨weisong@astro.as.utexas.edu⟩

Y. WATANABE, Department of Astronomy, University of Texas at Austin, 1 University Station C1400, Austin, TX, 78712-0259, USA ⟨yuki@astro.as.utexas.edu⟩

T. M. WEINZIRL, Department of Astronomy, University of Texas at Austin, 1 University Station C1400, Austin, TX, 78712-0259, USA ⟨timw@astro.as.utexas.edu⟩

T. WESTMEIER, Australia Telescope National Facility , P.O. Box 76, Epping, NSW, 1710, Australia ⟨tobias.westmeier@csiro.au⟩

J. C. WHEELER, Department of Astronomy, University of Texas at Austin, 1 University Station C1400, Austin, TX, 78712-0259, USA ⟨wheel@astro.as.utexas.edu⟩

K. A. WILLIAMS, Department of Astronomy, University of Texas at Austin, 1 University Station C1400, Austin, TX, 78712-0259, USA ⟨kurtis@astro.as.utexas.edu⟩

A. C. WILSON, Department of Astronomy, University of Texas at Austin, 1 University Station C1400, Austin, TX, 78712-0259, USA ⟨amywaymy@mail.utexas.edu⟩

N. WOOD, McDonald Observatory Board of Visitors, 20595 Highland Lake Drive, Lago Vista, TX, 78646, USA ⟨nhwood33@juno.com⟩

J. WOOD, McDonald Observatory Board of Visitors, 20595 Highland Lake Drive, Lago Vista, TX, 78646, USA ⟨nhwood33@juno.com⟩

P. YOACHIM, McDonald Observatory, University of Texas at Austin, 1 University Station C1402, Austin, TX, 78712-0259, USA ⟨yoachim@astro.as.utexas.edu⟩

Part I

Reviews

Physical Properties of Small Solar System Bodies and Implications for Formation and Evolution of Planetary Systems

Joshua P. Emery

Carl Sagan Center, SETI Institute, Mountain View, CA, USA

Abstract. In the past decade and a half, planetary astronomy has produced an unprecedented increase in the number of known small Solar System bodies, including the discovery of an entirely new population – the Kuiper Belt. With the prospect of several upcoming large surveys, this explosion of numbers will continue, setting the stage for dynamical studies of ever increasing sophistication. The true potential of these databases of objects for understanding the formation and evolution of planetary systems is realized when combined with physical studies. In this way, trends, groups, or other correlations between physical and dynamical/orbital properties can be identified from which the history of the Solar System can be pieced together. Compositional analysis of the Main Belt of asteroids unveiled a correlation between heliocentric distance and composition that led to a relatively quiescent model for at least the inner Solar System. Dynamical structure of the Kuiper Belt, on the other hand, points to a more active system. Kuiper Belt objects trapped in mean motion resonances with Neptune constitute strong evidence for orbital migration of the giant planets. The details of this dynamical evolution and potential effects on the rest of the Solar System are still debated. Physical properties of Jupiter Trojan asteroids and Kuiper Belt objects have particular importance for distinguishing among dynamical models. Ongoing and planned spacecraft missions and continued telescopic observations, aided by upcoming advancements in observing capabilities, promise a tremendous improvement in our understanding of the physical properties of small Solar System bodies.

1. Introduction

Of all locations throughout the universe, the Solar System has the most direct influence on us and is also the most accessible. Its eight planets contain most of the mass in the present-day disk of the Solar System, all except Uranus and Neptune have been known since antiquity, and their properties are fascinatingly diverse and complex. As a result, these planets have received, perhaps appropriately, the bulk of focus – in terms of funding and work effort – of planetary astronomy and exploration. When it comes to looking for clues to the origin and evolution of the Solar System, however, the small bodies (asteroids, comets, Kuiper Belt objects), may have as much or possibly more to reveal than their larger and more famous kin.

The known population of small bodies has grown dramatically since the discovery of 1 Ceres in 1801. Ninety years later, 323 asteroids were known, and by 1979 there were just over 2100 numbered asteroids, which grew to about 3400 by 1989. The 1990s saw the advent of dedicated large-scale discovery surveys, and the IAU Minor Planet Center now lists more than 170,000 numbered small bodies. Larger-scale surveys that are currently planned promise to raise this number to well over 10^6 in the coming years.

These objects are real test particles that record the gravitational conditions during the evolution of the Solar System. As additional objects have been discovered and their orbits mapped, more and more dynamical structure has become apparent. It is often very difficult to untangle from current orbits alone the web of potential dynamical mechanisms that could have led to the observed structure. Combining this information with physical studies (e.g., compositional types) offers a very powerful means to identify source regions and thereby strongly constrain potential evolutionary paths.

Many small bodies remain relatively unaltered since formation, and therefore are reservoirs of the original material from which they formed. These objects provide our most direct look into the make-up of the solar nebula. This unaltered state is in contrast to the planets and other larger bodies that have undergone differentiation and so do not provide nearly as clear of a window to the past. Distant objects, both large and small, are very cold, and therefore even if they have undergone some internal alteration, they still contain signatures of the solar nebula in the region where some of the ices, silicates, and organics were inherited from the interstellar medium.

The intent of this paper is to illustrate the benefits of linking physical and dynamical studies of small bodies. An overview of small body populations in the Solar System is followed by a review of the compositional structure of the Main Belt, as an example of successful synthesis of physical and dynamical properties. An overview of the outer Solar System sets the stage for discussing the pivotal nature of the Jupiter Trojan asteroids and the ongoing work to understand the various populations in the Kuiper Belt. The paper concludes with a short summary of future prospects for physical characterization of small bodies.

2. Overview of the Solar System

Fig. 1 illustrates the distribution of small bodies throughout the Solar System. The largest concentrations of small bodies are found in regions where the planets have minimal gravitational influence (e.g., the Main Belt and the Kuiper Belt). Nevertheless, some objects do reside in orbits that cross those of the planets (e.g., near-Earth objects and Centaurs). Such orbits are not generally stable over the lifetime of the Solar System, so those populations are most likely transient. Close inspection reveals further dynamical substructure that also has implications for current and past dynamical evolution. Each group labelled in Fig. 1 is described briefly in this section, and the Main Belt, Trojan asteroids, and Kuiper Belt are discussed in more detail below.

Near-Earth objects: For the purposes of Fig. 1, near-Earth objects (NEOs) are defined as having perihelia less than 1.3 AU. In such orbits, gravitational interactions with the terrestrial planets are likely, and average NEO lifetimes are $\sim 10^7$ yrs (Gladman et al. 2000). These objects must therefore be continually resupplied. Several source regions have been identified in the Main Belt (resonances with Jupiter) based on dynamical arguments (Gladman et al. 1997). The size (absolute magnitude) distribution of NEOs is distinct from that of Main

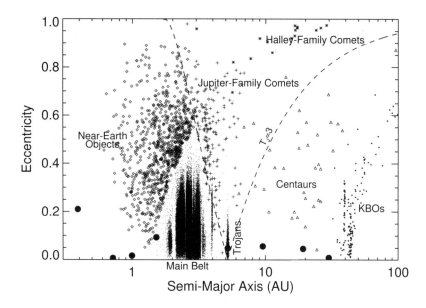

Figure 1. The dynamical structure of the Solar System as revealed by a plot of orbital eccentricity and semi-major axis. The various groups discussed in the text readily separate themselves. The dashed line marks the boundary where the Tisserand parameter with respect to Jupiter equals three. T_J remains approximately constant when an object has a gravitational interaction with Jupiter, despite changes in orbital elements. Comets typically have $T_J < 3$ and asteroids typically have $T_J > 3$. The eight planets are plotted as large, filled circles.

Belt asteroids, with a relative excess of smaller bodies. It is now thought that radiation forces, which more strongly affect smaller objects, can drive asteroids from near a resonance into that resonance, thereby feeding them into near Earth space (e.g., Bottke et al. 2002). The proximity of NEOs to Earth and the undeniable likelihood that some will eventually impact Earth add an extra impetus to not only discovering and tracking as many of them as possible, but also to determining their physical properties. The effectiveness of various hazard mitigation scenarios can depend on composition (e.g., metallic, silicate, comet-like) and internal structure (rubble-pile or monolith).

Main Belt asteroids: The Main Belt is the largest repository of asteroids in the Solar System. Most Main Belt asteroids are concentrated between about 2.1 and 3.3 AU, though some orbit outside this range, most notably the Hungaria group at ∼1.9 AU and the Hilda group at ∼3.9 AU. Gravitational forces from Jupiter, or possibly large planetesimals scattered inward by Jupiter, stirred up the early asteroid belt, keeping the young asteroids from further accreting into one or more larger bodies. Resonances with Jupiter create gaps (Kirkwood gaps) at several locations (e.g., ∼2.5 AU) and confine objects to isolated groups in other locations (e.g., Hildas at ~3.9 AU). The Main Belt is estimated to contain ~ 7×10^5 asteroids larger than 1 km (Ivezić et al. 2001).

Trojan asteroids: Each planet has gravitationally stable regions 60° in front of and 60° behind the planet in its orbit, the L4 and L5 Lagrange points. Objects that orbit within these regions are termed Trojan objects, from the tradition of naming Jupiter Trojan asteroids after heroes from the Trojan war. Only three planets support Trojan populations. There are four asteroids known in the Lagrange regions of Mars, the only rocky planet with known Trojans (Connors et al. 2005; Rivkin et al. 2007), and six objects have been discovered in Neptune's L4 region (Sheppard & Trujillo 2006). By far the largest population of Trojans belongs to Jupiter. Nearly 2300 Jupiter Trojans have been discovered, and a total population of ~6 $\times 10^5$ larger than 1 km is expected (Jewitt et al. 2000; Yoshida & Nakamura 2005), comparable to the estimated population of the Main Belt. For the remainder of this paper, the term Trojan will refer only to Jupiter Trojan asteroids.

Comets: Comets are defined morphologically as objects that exhibit comae and/or tails, which are formed as a result of sublimation of ices. Jupiter family comets have relatively short orbits (<20 yrs) and high eccentricities that bring them close to the Sun frequently. Halley family comets have much longer periods that bring them into the inner Solar System very infrequently. The immediate source of these, the Oort Cloud, encircles the Solar System at *very* large distances and was originally populated by objects scattered from among the giant planets (Gladman & Duncan 1990; Dones et al. 2004). Single apparition comets are also from the Oort Cloud. Jupiter family comets, by contrast, are derived from the Kuiper Belt (Levison & Duncan 1997).

Centaurs: Objects that orbit the Sun among the giant planets are called Centaurs. Such orbits are only stable for on the order of 10^7 yrs, so Centaurs, like NEOs, must be constantly replenished. It is generally well accepted that they come from the Kuiper Belt and, after migrating through the giant planet region, may end up as Jupiter family comets (Duncan et al. 1995; Levison & Duncan 1997). Several Centaurs exhibit cometary activity at distances that are far too great to be explained by H_2O ice sublimation. Possible explanations include sublimation of CO (e.g., Fanale & Salvail 1997; Brown & Luu 1998) or amorphous to crystalline phase transition of H_2O ice (e.g., Prialnik et al. 1995).

Kuiper Belt objects: The Kuiper Belt is a disk of material that circles the Sun beyond the orbit of Neptune, now officially including Pluto as its second largest member. The first Kuiper Belt object (KBO) was discovered in 1992 (Jewitt & Luu 1993), and 1140 had been detected as of 7 Nov 2007 (157 numbered + 983 unnumbered). As discovery of KBOs has progressed, several dynamical subpopulations have been recognized.

3. Main Belt Composition

Combined analysis of physical and dynamical properties is certainly not a new idea. While we are at a point when such a synthesis for small bodies of the outer Solar System is about to produce significant advances in our understanding of the formation and evolution of planetary systems, it is useful to look back at similar paradigm setting progressions. Compositional analysis of Main Belt asteroids offers an excellent illustration.

Remote examinations of asteroids with the purpose of determining physical properties began in earnest in the 1950s, and really took off in the 1970s (see Bowell & Lumme 1979). UBV photometry and the onset of spectrophotometry (McCord et al. 1970) uncovered variations in physical properties (e.g., colors, albedos) among the asteroids. These differences were parametrized by a taxonomy system analogous to stellar classifications (Chapman et al. 1975; Bowell et al. 1978). The major groups were assigned letter designations: C-types exhibit flat to slightly red (increasing reflectance with increasing wavelength) visible spectral slopes with a possible absorption at $\lambda < 0.4\,\mu$m, S-types show a moderate to strong absorption at $\lambda > 0.7\,\mu$m and a moderate absorption at $\lambda < 0.7\,\mu$m, E-, M-, and P-types all have featureless, flat to moderately red visible spectra and were originally divided by albedo, and D-types exhibit very red, featureless visible spectra. Several sub-classes were also defined. Analysis of the radial positions of asteroids found that the taxonomic types are not distributed randomly in the Main Belt. There is a systematic variation of asteroid type with heliocentric distance (Gradie & Tedesco 1982). Recently, Bus & Binzel (2002a,b) conducted a systematic survey of visible spectra of nearly 1500 Main Belt asteroids. This high-quality, internally consistent database led to some slight restructuring of the original classes (e.g., the E, M, and P-types are recombined in to a singe X class) and identification of additional sub-classes, but generally supports and extends the previous efforts. The heliocentric distribution of classes found by Bus & Binzel is "fully consistent with the original findings of Gradie & Tedesco (1982), but show[s] more detailed structure" (Fig. 2).

Remote observations (photometry, spectroscopy) of asteroids and laboratory studies of meteorites have led to compositional interpretations of the different asteroid types. From the radial distribution of taxonomic types and inferred compositions, a view of condensation conditions in different parts of the Main Belt developed in which the inner belt E-type asteroids are due to high temperature spectrally neutral siliceous condensates which grade into the S-types (moderate temperature siliceous condensates), then C-types (low temperature carbonaceous-rich siliceous condensates), and the outer belt P- and D-types are from lower temperature more carbonaceous-rich siliceous condensates (see Gradie et al. 1989). Continued observations add details to possible mechanisms of formation and conditions in the solar nebula, but the overall model of accretion conditions proposed in the early 1980s is still the accepted view.

The compositional data of Main Belt asteroids presented in this section, along with the generally accepted interpretation in terms of accretion conditions, points to a relatively quiescent solar nebula, at least in the inner region. The asteroids are built of material that condensed from a nebula in thermodynamic equilibrium, and they have remained relatively static since that time so that their compositions are indicative of the early conditions in the regions in which they currently reside. The largest uncertainty with this paradigm comes from the outer belt P- and D-type asteroids. The meteorite collection contains no good spectral analog to these asteroids, and their spectra are consistently featureless. Interpretations of their compositions are therefore somewhat circularly influenced by assumptions that the model from the inner and mid Main Belt extends to the outer belt (i.e., essentially C-type asteroids with more organic material). The sections below address this point in more detail.

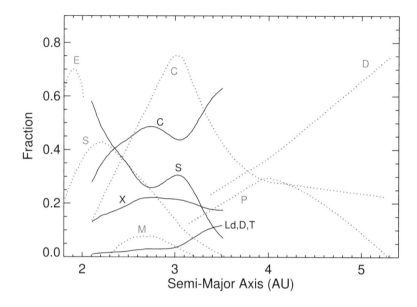

Figure 2. The fractions of the total asteroid population made up of each taxonomic type is plotted as a function of semi-major axis in the Main Belt and Trojan region. The light gray, dotted curves are from Gradie & Tedesco (1982). The dark, solid curves are from Bus & Binzel (2002b) and represent a more recent analysis using a larger set of visible spectra.

4. The Outer Solar System and Delivery of Material Inward

4.1. Significant "Recent" Events in Planetary Astronomy

Over the past 15 years or so, a number of discoveries and advances have occurred that give a new perspective on Solar System evolution. It helps to look at data from the outer Main Belt outward in the context of these advances.

Dedicated asteroid searches have lead to an explosion in the number of known small bodies in the Solar System. In the last decade alone, the known asteroid population has increased by over two orders of magnitude. A more complete inventory brings into focus details of dynamical structure that could not previously be seen. Upcoming surveys (PanSTARRS, LSST) promise to continue this trend and take it to new heights of asteroid discovery.

Improvements in detectors and telescopes is hopefully a more continuous and ongoing process. Two particular developments that have helped expand the bounds of asteroid science are high-quality infrared detectors and adaptive optics. Near-infrared (NIR) spectroscopy ($0.8 - 4.0\,\mu$m) has extended the compositional investigations of asteroids described above into a new spectral region, enabling reliable assessment of silicate mineralogy and searches for signatures of ices, hydrated silicates, and organic materials. Adaptive optics now allows direct imaging of asteroids and has led to the discovery of a large number of asteroid binaries and even several multiple component systems (Merline et al. 2002; Richardson & Walsh 2006). Astrometry of binary orbits is used to derive mass and, if the size is known, density, which is a critical parameter for understanding bulk composition.

Advances in numerical techniques and computer processors have combined to enable significant progress in dynamical simulations. Symplectic N-body integrators that have recently been applied to planetary dynamical problems are much faster than traditional algorithms, are very accurate, and have been successfully modified to handle close encounters (Duncan et al. 1998; Chambers 1999). The steady increase in computer processor speeds and the ease with which processors can be clustered together permits much more sophisticated simulations than were possible even ten years ago.

The discovery of extra-solar planets has caused tremendous excitement, both in public and scientific circles, which only continues to grow as smaller and potentially habitable planets are discovered (see Chapter Mandell). Ideas of planetary system formation and evolution have been improved significantly by the unexpectedly wide range of system morphologies observed. In particular, whereas planet migration had been thought about before discovery of exo-planet systems (Fernández & Ip 1984), understanding of this process flowered with the need to explain some of these systems.

The discovery of the Kuiper Belt similarly opened a new frontier in the Solar System. Compositional analyses of KBOs promise to deepen our understanding of the conditions in the outer solar nebula the way similar studies of Main Belt asteroids did for the inner solar nebula. Dynamical structure within the Kuiper Belt has also made it clear that the giant planets of the Solar System have migrated since their formation. The degree of migration and the potential effects it had on the Solar System are areas of active study, which will be best addressed with a synthesis of physical and dynamical investigations.

4.2. Overview of Kuiper Belt Structure and Models

The Kuiper Belt is not a uniform disk of material, but rather has dynamical structure that is in some ways more distinct than the Kirkwood gaps and Hilda-like resonance groups in the Main Belt. Several dynamical subpopulations, which have undergone varying degrees of orbital scattering, are currently recognized (Fig. 3). A large number of KBOs are trapped in resonances with Neptune (3:2, 2:1, 5:2, etc) and are hence termed *resonant* objects. *Scattered disk* objects have endured gravitational interactions with Neptune that scattered them into orbits that are typically characterized by high eccentricity and large semi-major axis. *Centaurs*, discussed above, have also been scattered by Neptune, though in this case into orbits that cross those of the giant planets. A small number of objects have been detected with extremely distant perihelia that are not consistent with scattering by Neptune, or that at least require an additional mechanism to produce the large perihelia. These are called *extended scattered* or *detached* objects, Sedna (perihelion of 76 AU and semi-major axis of 485 AU) being the most prominent example. The "classical" Kuiper Belt itself is generally divided into two groups. *Hot classical* objects have relatively circular orbits, but somewhat high orbital inclinations ($\gtrsim 5°$), presumably from gravitational interactions with Neptune. The *cold classical* population, on the other hand, has similarly low eccentricity orbits, but also with low inclinations ($\lesssim 5°$). Cold classical KBOs are the only group hypothesized to remain in or close to their original orbits (Chiang et al. 2007; Lykawka & Mukai 2007).

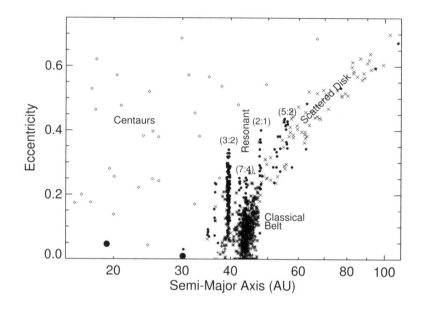

Figure 3. Dynamical structure of the Kuiper Belt illustrated with a plot of orbital semi-major axis and eccentricity. A few, but not all, of the more prominent mean motion resonances with Neptune are labeled. The "cold" and "hot" classical populations are defined by inclination and are not distinguished on this plot. Uranus and Neptune are shown as two large, filled circles.

Even before the Kuiper Belt was recognized as a population, Pluto's orbit provided the first clue to the structure to be discovered. Pluto resides in the 3:2 mean motion resonance with Neptune. Not long before the discovery of the Kuiper Belt, it was suggested that Pluto was trapped in this resonance as Neptune migrated outward (Malhotra 1993). The later recognition of a large population of resonant KBOs, and subsequent focus on improved dynamical simulations, have led to widespread agreement that Uranus and Neptune have migrated outward since their formation, though most of the details of migration are targets of vigorous research.

A review of the multitude of models to explain the structure of the outer Solar System is beyond the scope of this paper. These models range from simple, relatively quiescent outward migration of Uranus and Neptune (Malhotra 1995) to those with more early activity in the outer nebula (Chiang et al. 2007). These models often require additional mechanisms besides migration (e.g., passing stars, additional Neptune-sized planetesimals, galactic tides) to explain some of the details of KBO orbits, but generally leave the inner Solar System unaffected. Each model has its own implications for the evolutionary details of the outer Solar System (regions of origin of different groups, amount of mixing, etc.).

As observations of KBOs continue (see Section 4.4 below) physical signatures will be used to distinguish these various models from one another.

Recently, a somewhat more violent model has been proposed whose effects stretch all the way to the inner Solar System (Tsiganis et al. 2005). This model (call the "Nice" model as it was developed at the Nice Observatory), posits a compact early Solar System with Neptune and Uranus just outside the orbit of Saturn and a massive disk ($\sim$35 $M_\oplus$) beyond that cuts off at $\sim$30 AU. The planets slowly migrate until Jupiter and Saturn cross their 1:2 mean-motion resonance. This resonance crossing increases the eccentricities of Jupiter and Saturn, which in turn initially destabilizes the orbits of Uranus and Neptune. The ice giants are thrust into the massive disk and settle back down into stable orbits. The Nice model seems to reproduce a number of characteristics of the Solar System that are difficult to explain with some of the other models. See Levison et al. (2007) and references therein for more details. Several predictions related to various small body populations are very distinct from more quiescent models and testable with physical observations.

4.3. Trojan Asteroids

Orbits in the Jupiter Trojan swarms are stable over >4.5 Gyr, though the region of stability is decreasing and the overall diffusion of objects is currently out of rather than into stable librating orbits (Levison et al. 1997). Marzari & Scholl (1998) showed that a growing Jupiter would naturally capture objects into the Lagrange regions, though their test particles did not achieve the high inclinations observed for some current Trojans. Gas drag in the early nebula could also have aided in capture of Trojans. In either of these cases, capture of objects already orbiting near Jupiter is most likely. The Nice model, on the other hand, predicts that as Jupiter and Saturn pass through their 2:1 mean motion resonance, the Trojan swarms are first emptied of their initial residents, then repopulated (as Jupiter and Saturn exit the resonance state) with material primarily originating in the Kuiper Belt (Morbidelli et al. 2005). The compositions of Trojan asteroids therefore offer a critical test between very different models of Solar System dynamical evolution. These different models, in turn, have far broader implications concerning the formative stages and evolution of the Solar System (and, by extension, other planetary systems), including the structure of the Kuiper Belt, the properties of outer planet systems, and the impact history of the inner Solar System (i.e., late-heavy bombardment).

Large Trojan asteroids (D $\gtrsim$ 50 km) have uniformly low albedos ($p_v \sim 0.03$ to 0.06) as measured radiometrically (Cruikshank 1977; Tedesco et al. 2002; Fernández et al. 2003). Recent observations from the Spitzer Space Telescope suggest that smaller Trojans may have systematically higher albedos (Fernández et al. 2006). Reflectance spectroscopy at visible wavelengths had failed to discover any absorption features, but reveal red spectral slopes, comparable to outer belt D-type asteroids (e.g., Fornasier et al. 2007). Whereas there is no correlation between size (or any other physical parameters) and spectral slope, redder Trojans do appear to extend to higher inclinations (Szabó et al. 2007). No ultrared slopes comparable to many Centaurs and KBOs have been detected among the Trojans (Fig. 4). The low albedo and red slope were originally modeled by mixtures of (hydrated) silicates, carbon black, and complex organics (Gradie &

Veverka 1980). This result was incorporated into the solar nebula condensation sequence described above in which increasing organic content is responsible for red slopes in the outer belt and Trojan swarms (Gradie et al. 1989).

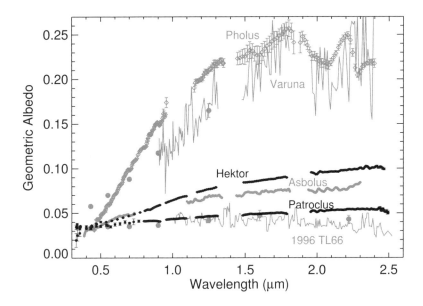

Figure 4. Visible and near-infrared spectra of representative Trojan asteroids (Hektor and Patroclus), Centaurs (Asbolus and Pholus), and KBOs (1996 TL66 and Varuna). Centaurs and KBOs exhibit a very large range of spectral slopes from neutral to ultra-red. Some Centaur and KBO spectra are featureless, but many contain absorptions due to ices (e.g., H_2O near $2.0\,\mu m$ and CH_3OH near $2.3\,\mu m$ for Pholus). Spectra of Trojan asteroids, on the other hand, are uniformly featureless and do not exhibit any ultra-red spectral slopes.

Near-infrared spectroscopy (0.8 to $4.0\,\mu m$) has also failed to detect any clear absorption features, including no evidence for H_2O, no 1- and 2-μm silicate bands, and no absorptions from organics or hydrated minerals (Luu et al. 1994; Dumas et al. 1998; Emery & Brown 2003; Dotto et al. 2006; Yang & Jewitt 2007). Cruikshank et al. (2001) and Emery & Brown (2004) demonstrated that vis-NIR spectra can be modeled without the use of organics (just silicates and amorphous carbon). Emery & Brown (2004) further note that the absence of absorptions in the 3–4 μm range strongly limits the type and abundance of organics possible on these surfaces. Emery et al. (2007) describe two distinct spectral groups among the Trojans, rather than a continuum between neutral and red spectral slopes. Discrete mineralogical features attributed to fine-grained (∼few μm), anhydrous silicates were recently detected in mid-IR thermal emission spectra of three Trojans using the Spitzer Space Telescope. The mineralogy may resemble that of cometary silicates, and the spectral shape indicates that the surfaces are probably either very porous or that the grains are embedded in a matrix that is relatively transparent in the mid-IR (Emery et al. 2006). Some Trojans display more muted 10-μm bands, indicative of differences in composition and/or grain size. Some Main Belt D-type asteroids exhibit similar emissivity features, but others are different. Two Centaurs have somewhat similar emissivity spectra, but differences in positions and shapes of peaks suggest different silicate mineralogy between Centaurs and Trojans (Fig. 5).

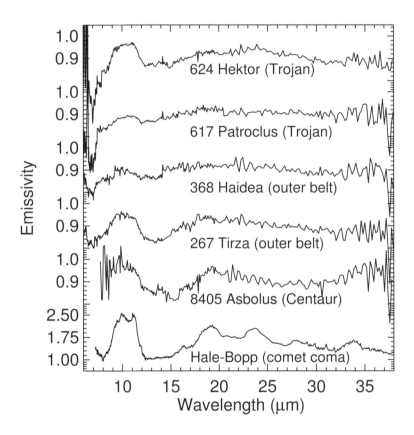

Figure 5. Emissivity spectra of two Trojan asteroids along with several potentially related objects. The plateau near 10 μm is indicative of fine-grained silicates. The Trojans with large 10-μm plateaus also have redder vis-NIR slopes. All of the asteroid spectra are from the Spitzer Space Telescope, and the Hale-Bopp spectrum is from ISO.

The measurements with the most direct implication for the internal composition of Trojan asteroids are densities of 617 Patroclus and 624 Hektor. The discovery that Patroclus is a binary, combined with follow-up astrometry of the two components, has yielded a density of 0.8 ± 0.2 g/cm^3 (Marchis et al. 2006a). The most straightforward interpretation of this low density includes both significant bulk porosity and a relatively significant ice fraction in the interior. A higher density of ∼2.5 g/cm^3 was derived for Hektor, both from the orbit of a recently discovered moonlet (Marchis et al. 2006b) and from detailed models of the rotational light curve (Lacerda & Jewitt 2007). Such a high density for Hektor does not require an ice component.

The most obvious inference to draw from the data amassed thus far of Trojans is that these asteroids are even more circumspect about providing a glimpse behind their gates than their Homerian namesakes. Nevertheless, a few points are worth noting. The current spectral (color) distribution among Trojan asteroids does not match that among KBOs. Most notably, the reddest spectral slopes of Centaurs and KBOs are absent among the Trojans, as are signatures of ices. One scenario that has been put forth posits Trojan formation as KBOs, but on migrating inward the surface is altered in such a way as to mask the interior, primordial, ultra-red material. Two suggested alteration mechanisms are ice sublimation creating a dust mantle (e.g., Jewitt 2002) and irradiation carbonizing the organics (e.g., Moroz et al. 2004; Fornasier et al. 2007). In either scenario, ultra-red material and ices would be present below the surface and could be exposed by a sufficiently large impact. Whether such an exposure should have been detected in one of the more than 50 Trojans for which vis-NIR spectra and/or colors have been published depends on the masking timescale and mantle thickness, neither of which have been rigorously modeled.

Another potential explanation for the differences between Trojans and KBOs is that they are not derived from the same population. Differences in silicate mineralogy, as derived from emissivity spectra, between Trojans and two Centaurs that have very similar vis-NIR spectra as Trojans may support this possibility. However, with only two Centaur emissivity spectra to compare, it is far too soon to firmly draw such a conclusion. Densities do not provide much insight in this case. The two known Trojan densities are very different from one another and span a range that is encompassed in both the Main Belt and among Centaurs and KBOs. The presence of two compositional classes among the Trojans, measured both with NIR colors and emissivity spectra, suggests that there may have been two source regions for the Trojans. Perhaps the Trojans sample both the local (∼5 AU) solar nebula and more distant regions.

4.4. Kuiper Belt Composition

The dynamical complexity of the Kuiper Belt described above is matched by the complexity in physical properties of KBOs. Visible spectra of KBOs and Centaurs are generally featureless with spectral slopes that range from neutral to much redder than those of Trojan asteroids (Fig. 4). Barucci et al. (2005), using colors out to J-band (1.25 μm), have devised a taxonomic classification system that divides KBOs and Centaurs into four groups, BB, BR, IR, RR in order of increasing spectral slope. For comparison, all Trojan asteroids would fall in either the BB or BR classes. NIR spectra reveal H_2O absorption bands for many KBOs and Centaurs. The ice bands do not correlate with the color groups, with the possible exception that the BB class seems to show a slightly higher ice content (Barucci et al. 2006; Barkume et al. 2008). Spectra of two RR objects show an absorption near 2.35 μm that has been attributed to methanol (Cruikshank et al. 1998; Barucci et al. 2006), lending support to the hypothesis that the ultra-red color is due to organic material. CH_4 and N_2 have been detected on the surfaces of several of the largest (Pluto-sized) KBOs. Presumably their large size (and cold temperatures) have led to the retention of these volatile ices (Schaller & Brown 2007). Nearly all of the KBOs with the deepest H_2O absorption bands follow relatively similar orbits. It has been suggested that these objects constitute a dynamical family created by a large impact into the object (136108) 2003 EL_{61} (Brown et al. 2007). Albedos of most Centaurs and KBOs range from about 0.03 to 0.20, though the Pluto-sized objects have albedos of 0.6 to 0.8 (Stansberry et al. 2008).

Of the Kuiper Belt dynamical groups described above, only the cold classical population appears to exhibit distinct physical properties. An early hint of a color-inclination correlation among KBOs is more properly described as the cold classical objects having redder spectral slopes than the other classes (Gulbis et al. 2006). Cold classical objects also have higher average albedos. In general, binary systems are approximately as common among Centaurs and KBOs as in the Main Belt, but Stephens & Noll (2006) note a much higher fraction among the cold classical KBOs. Taken together, the physical and dynamical properties suggest that the cold classical KBOs did not suffer the same degree of mixing as their kin, but rather formed in place, at the edge of the primordial Kuiper Belt. The redder colors then further suggest that either the ultra-red (organic?) material was more abundant in this region, or if it is an irradiation product, that the precursor material was more abundant in this region.

Continuing observations target the detection and characterization of ices besides H_2O (NH_3, CH_4, C_2H_6, CO_2, N_2, etc.) and the phase of H_2O itself (i.e., amorphous or crystalline). These searches will provide insight into condensation of volatiles, possible geologic activity (e.g., Cook et al. 2007), and the irradiation environment (Mastrapa & Brown 2006). Characterization of the dark material is also of utmost importance. Tying these physical observations to dynamical models is the critical next step for unraveling source regions and putting the observational results in proper context regarding the formation and evolution of the Solar System.

5. Prospects for the Future

Small bodies conceal within their physical and dynamical properties key clues to untangling formation and evolutionary processes of planetary systems. Planetary astronomers are in the middle of what will most probably be a paradigm setting step in revealing those processes. That step is progressing through a comprehensive approach combining detailed analysis enabled by spacecraft missions with the "big picture" view put into focus by intercomparisons of telescopic data of the various dynamical populations.

Several missions planned or already underway will lead to a deeper understanding of the physical properties of small bodies from in-situ and sample return investigations. The Dawn mission is currently on its way to orbit, in turn, two of the largest (but very different from one another) Main Belt asteroids, Ceres and Vesta. The New Horizons spacecraft, after a spectacularly successful Jupiter system flyby, is barreling toward Pluto, and possibly one or two KBOs beyond. These up-close views provide important ground-truth context for interpretation of telescopic data. Several potential asteroid and comet sample return missions are under study to allow laboratories on Earth to bring their full arsenals to bear. As these missions move forward, continued remote characterization from ground- and space-based telescopes will provide the mechanism for applying that detailed knowledge of a few objects to the vast array of less accessible targets. Upcoming large-scale surveys (e.g., LSST and PanSTARRS) promise to increase the number of known small bodies by orders of magnitude. The Spitzer Space Telescope has a substantial legacy to offer in terms of sizes and albedos of probably a few tens of thousands of asteroids and mineralogy of one or two hundred. SOFIA will enable observations at wavelengths not easily visible from the ground, really opening the 3–4 μm region, which is important for studying ices, organics, and hydrated minerals, and the mid-IR to asteroid science. In the farther future, the James Webb Space Telescope offers the potential for physical characterization of a very large number of objects of all dynamical classes. Larger telescopes, improved instrumentation, and ingenious astronomers are also on the verge of detecting distinct small body populations in other planetary systems. Comparisons between systems can be expected to lead to unforeseen insights. The prospects for the future of studying physical properties of small bodies therefore look very bright, and the prospects for understanding the formation and evolution of planetary systems even brighter.

References

Barkume, K.M., Brown, M.E., & Schaller, E.L. 2008, AJ, in press
Barucci, M.A., Belskaya, I.N., Fulchignoni, M., & Birlan, M. 2005, AJ, 130, 1291
Barucci, M.A., Merlin, F., Dotto, E., et al. 2006, A&A, 455, 725
Bottke, W., Vokrouhlický, D., Rubincam, D., & Brož, M. 2002, in Asteroids III, ed. W. Bottke, P. Paolicchi, R. Binzel, & A. Cellina (Univ. Ariz. Press, Tucson), 395
Bowell, E. & Lumme, K. 1979, in Asteroids, ed., T. Gehrels (Univ. Ariz. Press, Tucson), 132
Bowell, E., Chapman, C.R., Gradie, J.C., et al. 1978, Icarus, 35, 313
Brown, M.E., Barkume, K.M., Ragozzine, D., & Schaller, E.L. 2007, Nature, 446, 294
Brown, W.R. & Luu, J.X. 1998, Icarus, 135, 415
Bus, S.J. & Binzel, R.P. 2002a, Icarus, 158, 106
Bus, S.J. & Binzel, R.P. 2002b, Icarus, 158, 146
Chambers, J.E. 1999, MNRAS, 304, 793
Chapman, C.R., Morrison, D., & Zellner, B. 1975, Icarus, 25, 104
Chiang, E., Lithwick, Y., Murray-Clay, R., et al. 2007, in Protostars & Planets V, ed., B. Reipurth, D. Jewitt, & K. Keil (Univ. Ariz. Press, Tucson), 895
Connors, M., Stacey, G., Brasser, R., & Wiegert, P. 2005, Planet. Space Sci., 53, 617
Cook, J.C., Desch, S.J., Roush, T.L., et al. 2007, ApJ, 663, 1406
Cruikshank, D.P. 1977, Icarus, 30, 224
Cruikshank, D.P., Roush, T.L., Bartholomew, M.J., et al. 1998, Icarus, 135, 389
Cruikshank, D.P., Dalle Ore, C.M., Roush, T.L., et al. 2001, Icarus, 153, 348
Dones, L., Weissman, P.R., Levison, H.F., Duncan, M.J. 2004, in Comets II, ed., M.C. Festou, H.U. Keller, & H.A. Weaver (Univ. Ariz. Press, Tucson), 153
Dotto, E., Fornasier, S., Barucci, M.A., et al. 2006, Icarus, 183, 420
Duncan, M., Levison, H.F., & Budd, S.M. 1995, AJ, 110, 3073
Duncan, M.J., Levison, H.F., & Man Hoi, L. 1998, AJ, 116, 2067
Dumas, C., Owen, T.C., & Barucci, M.A. 1998, Icarus, 132, 80
Emery, J.P. & Brown, R.H. 2003, Icarus, 164, 104
Emery, J.P. & Brown, R.H. 2004, Icarus, 170, 131
Emery, J.P., Cruikshank, D.P., & Van Cleve, J. 2006, Icarus, 182, 496

Emery, J.P., Cruikshank, D.P., Brown, R.H., & Burr, D.M. 2007, 38th LPSC (abs. 1426)
Fanale, F.P. & Salvail, J.R. 1997, Icarus, 125, 397
Fernández, J.A. & Ip, W.-H. 1984, Icarus, 58, 109
Fernández, Y.R., Sheppard, S.S., & Jewitt, D.C. 2003, AJ, 126, 1563
Fernández, Y.R., Jewitt, D.C., Grisetti, R., & Igyarto, C. 2006, 38th DPS (abs. 50.01)
Fornasier, S., Dotto, E., Hainaut, O., et al. 2007, Icarus, 190, 622
Gladman, B. & Duncan, M. 1990, AJ, 100, 1680
Gladman, B.J., Migliorini, F., & Morbidelli, A., et al. 1997, Science, 277, 197
Gladman, B., Michel, P., & Froeschlé, C. 2000, Icarus, 146, 179
Gradie, J. & Veverka, J. 1980, Nature, 283, 840
Gradie, J.C. & Tedesco, E.F. 1982, Science, 216, 1405
Gradie, J.C., Tedesco, E.F., & Chapman, C.R. 1989, in Asteroids II, ed., R. Binzel, T. Gehrels, M. Matthews, (Univ. Ariz. Press, Tucson), 316
Gulbis, A.A.S, Elliot, J.L., & Kane, J.F. 2006, Icarus, 183, 168
Ivezić, Ž, Tabachnik, S., Rafikov, R., et al. 2001, AJ, 122, 2749
Jewitt, D.C. 2002, AJ, 123, 1039
Jewitt, D.C. & Luu, J.X. 1993, Nature, 362, 730
Jewitt, D.C., Trujillo, C.A., & Luu, J.X. 2000, AJ, 120, 1140
Lacerda, P. & Jewitt, D.C. 2007, AJ, 133, 1393
Levison, H.F. & Duncan, M.J. 1997, Icarus, 127, 13
Levison, H., Shoemaker, E.M., & Shoemaker, C.S. 1997, Nature, 385, 42
Levison, H.F., Morbidelli, A., Gomes, R., & Backman, D. 2007, in Protostars & Planets V ed., B. Reipurth, D. Jewitt, & K. Keil (Univ. Ariz. Press, Tucson), 669
Luu, J.X., Jewitt, D.C., & Cloutis, E. 1994, Icarus, 109, 133
Lykawka, P.S. & Mukai, T. 2007, Icarus, 189, 213
Malhotra, R. 1993, Nature, 365, 819
Malhotra, R. 1995, AJ, 110, 420
Marchis, F., Hestroffer, D., Descamps, P., et al. 2006a, Nature, 439, 565
Marchis, F., Berthier, J., Wong, M.H., et al. 2006b, 38th DPS (abs. 65.07)
Marzari, F., Farinella, P., & Vanzani, V. 1995, A&A, 299, 267
Mastrapa, R.M.E. & Brown, R.H. 2006, 183, 207
McCord, T.B., Adams, J.B., & Johnson, T.V. 1970, Science, 168, 1445
Merline, W.J., Weidenschilling, S.J., Durda, D.D., et al. 2002, in Asteroids III, ed. W. Bottke, P. Paolicchi, R. Binzel, & A. Cellina (Univ. Ariz. Press, Tucson), 289
Morbidelli, A., Levison, H.F., Tsiganis, K., & Gomes, R. 2005, Nature, 435, 462
Moroz, L., Baratta, G., Strazzulla, G., et al. 2004, Icarus, 170, 214
Prialnik, D., Brosch, N., & Ianovici, D. 1995, MNRAS, 276, 1148
Richardson, D.C. & Walsh, K.J. 2006, Ann. Rev. Earth & Plan. Sci., 34, 47
Rivkin, A.S., Trilling, D.E., Thomas, C.A., et al. 2007, Icarus, 192, 434
Schaller, E.L. & Brown, M.E. 2007, ApJ, 659, L61
Sheppard, S.S. & Trujillo, C.A. 2006, Sci, 313, 511
Stansberry, J., Grundy, W., Brown, M., et al. 2008, in Kuiper Belt, eds., M. Barucci, H. Boehnhardt, D. Cruikshank, A. Morbidelli (Univ. Ariz. Press, Tucson), in press
Stephens, D.C. & Noll, K.S. 2006, AJ, 131, 1142
Szabó, Gy.M., Ivezić, Ž., Jurić, M., & Lupton, R. 2007, MNRAS, 377, 1393
Tedesco, E.F., Noah, P.V., Noah, M., & Price, S.D. 2002, AJ, 123, 1056
Tsiganis, K., Gomes, R., Morbidelli, A., & Levison, H.F. 2005, Nature, 435, 459
Yang, B. & Jewitt, D. 2007, AJ, 134, 223
Yoshida, F. & Nakamura, T. 2005, AJ, 130, 2900

Josh Emery contributing to the panel discussion.

Expanding and Improving the Search for Habitable Worlds

Avi M. Mandell[1]

NASA Goddard Space Flight Center, Greenbelt, MD, USA

Abstract. This review focuses on recent results in advancing our understanding of the location and distribution of habitable exo-Earth environments. We first review the qualities that define a habitable planet/moon environment. We extend these concepts to potentially habitable environments in our own Solar System and the current and future searches for biomarkers there, focusing on the primary targets for future exploratory missions: Mars, Europa, and Enceladus. We examine our current knowledge on the types of planetary systems amenable to the formation of habitable planets, and review the current state of searches for extra-solar habitable planets as well as expected future improvements in sensitivity and preparations for the remote detection of the signatures of life outside our Solar System.

1. Introduction

We currently have concrete evidence of life on only one planet in the Universe: Earth. Over the last decade we have taken great strides in the quest to expand this tally, both through the investigation of other planets and moons beyond Earth as well as the investigation of exotic environments on Earth as analogs to potentially habitable environments elsewhere. However, the broad interdisciplinary research field known as "astrobiology", roughly defined to include any investigation that expands our understanding of the origin, evolution and distribution of life in the universe, encompasses such a vast range of research topics and core scientific disciplines that it cannot be adequately covered in a single review; in fact, only a brief sampling of research in a few core disciplines will be covered here.

This review will discuss the status of current research in areas of astrobiology related to astronomy and planetary science, focusing on improvements in our understanding of the constraints on the locations of potentially habitable environments as well as current and future efforts to detect life in our own Solar System and beyond. In §2 I introduce current theory on the constraints that define the term "habitable", and in §3 I discuss the potential for habitable planetary environments elsewhere in our own Solar System. We must then place our own planetary system in a wider context; in §4 I present the current paradigms and conflicts in our understanding of planet formation, and specifically in the origin of habitable planets, and I present a review of current and future searches for habitable, and inhabited, environments outside our own Solar System.

---

[1] NASA Post-doctoral Fellow

2. Factors in Assessing Planetary Habitability

The standard definition for a habitable planet has traditionally been one that can sustain life similar to that on Earth on its surface or subsurface for a significant period of time. However, this definition is based on our understanding of the current locales for life on Earth and our current understanding of the environments present on other planets; it is therefore constantly being redefined as we hypothesize or discover new environments in which life can sustain itself. The main requirements for life as we know it are:

- The presence and stability of liquid water over long time periods
- The availability of the basic organic building blocks of life (cellular building blocks and nutrients)
- The availability of energy for assembly of biological structures and metabolic processes

As we evaluate potentially habitable environments we must keep these fundamental requirements in mind. Also, to decrease confusion with respect to the characteristics of rocky bodies in various environments, in this section the word "planet" will be used to signify any large rocky body, either orbiting the central star (a traditional "terrestrial planet") or orbiting another large body in the system (traditionally called a "satellite" or "moon").

2.1. Characteristics Affecting Surface and Subsurface Habitability

The fundamental requirement for liquid water is a clement mean environmental temperature. Temperatures between $0\,C$ ($273\,K$) and $100\,C$ ($373\,K$) are necessary for pure water to form a liquid at standard temperature and pressure, but pressure and solutes can dramatically change these limits; extant life has been detected in water between $-20\,C$ ($253\,K$) and $121\,C$ ($394\,K$) (Rivkina et al. 2000; Kashefi & Lovley 2003). On rocky planets or moons this environment can be either on or below the surface, depending on the incident radiation and the subsurface heat source; the temperature of a planetary environment is a function of the balance between heating and radiation into space. This balance is affected by both surface processes and the bulk characteristics of the planet itself.

Surface Temperature For surface environments, the temperature is modulated primarily by the ratio of radiation absorbed from the central star (or stars) to the radiation radiated to space by the planet's surface and atmosphere. Stellar heat input will be sensitive to the size of the central star and the orbital distance. We can estimate a "Habitable Zone" or range in orbital semi-major axis for a given stellar type where water can exist as a liquid (Hart 1978); for a G-type star such as our Sun, the traditional Habitable Zone lies between $0.95\,AU$ and $1.37\,AU$ (Kasting et al. 1993). The surface temperature can be increased if the incident radiation can be retained by the planet's atmosphere (known as the "greenhouse effect"); the potential for heat retention depends on the composition of the atmosphere, and for thick atmospheres of molecules such as carbon dioxide and methane the surface temperature may be raised significantly. The greenhouse effect may therefore extend the habitable zone out to $2.4\,AU$ for G-type stars (Forget & Pierrehumbert 1997).

Subsurface Temperature In subsurface environments the local temperature is defined by heat transfer through the planet's interior. The heat source is either internal latent heat from accretion (as on Earth) or an external force such as tidal compression (as on Io), while cooling is limited by insulation from surface layers. The decay of radioactive isotopes also makes a small contribution, and may be enough to maintain liquid water in a subsurface layer in the absence of additional heat sources (a possibility for Enceladus; Schubert et al. 2007). Heat can also be generated by tidal forces in moons orbiting a giant planet. As the moon travels closer to and farther from the parent planet, the change in gravity causes it to expand towards the giant planet. The internal structure of the moon continually compresses and expands, creating frictional heating that can be conducted throughout the moon. This heat may also be sufficient to maintain mantle convection and subsurface liquid water (a possibility on Europa and/or Enceladus; see §3.2).

Planet Mass and Structure The internal structure of a planet will have a direct impact on the stability of its climate. On Earth, internal heating mechanisms result in plate tectonics and volcanism, both of which play an integral role in the exchange of materials (especially carbonates) in the atmosphere and oceans (Berner & Raiswell 1983). On Mars, the cessation of plate tectonics may have been critical in the loss of a thick atmosphere and surface water (Kasting & Catling 2003). Additionally, the differentiation of interior layers of a planet can affect energy and material transport. This differentiation is one of the primary requisites for liquid water layers in the moons of Jupiter and Saturn (see §3.2). Finally, the composition and thickness of the outer crustal layers of a planet, such as an outer ice layer or crust, can contribute to the transport of nutrients from the surface to subsurface locales (Greenberg & Geissler 2002).

Biological Feedback Once life forms, its impact on the environment will have profound affects on habitability. It is hypothesized that biological production of methane on early Earth may have been critical for maintaining a high surface temperature and exposed liquid water (Pavlov et al. 2001). Similarly, the rise of oxygen may have triggered global extinctions of anaerobic organisms; however, the presence of oxygen also produced an ozone layer, which provides a barrier against harmful UV radiation. Thus an oxygenic atmosphere opened the way for aerobic and multi-cellular life as well as land-based life, which does not have water to protect it from solar radiation (see Kasting & Catling 2003 for a review of the evolution of the early Earth's atmosphere).

2.2. Characteristics Affecting Long-Term Habitability

Even if a planet's characteristics result in habitable conditions at a specific time, features of a planet's orbital evolution (e.g., inclination, eccentricity, obliquity) and interactions with its neighbors, the evolution of its host star as well as its climactic evolution (e.g., ice ages, carbon sinks) may result in temperature changes in time, with variability ranging from months to 10^6 years or greater. It remains unclear how severe a lapse in habitability must be to make the continued survival of life impossible, but evidence from mass extinction events on Earth suggest that once the distribution of simple life forms reaches a threshold it is difficult to extinguish completely.

Dynamics of the Planetary System To maintain a consistent temperature, an Earth-like planet must continually maintain a nearly circular orbit or it will undergo extreme temperature changes. In addition, the parameters of a planet's orbit can be affected by interactions with other planets in the system on both short and long timescales. Eccentric orbits increase the likelihood of collisions between planets; an eccentric giant planet can strongly inhibit stable orbits of other planets (see §4.1). All the planets in our system have relatively circular orbits; however, many extrasolar planetary systems contain planets with highly eccentric orbits; the current distribution of extra-solar giant planet eccentricities is evenly distributed up to ~ 0.7 (see the Extrasolar Planets Encyclopedia for current results; Schneider 2007). Smaller bodies such as asteroids and comets can also affect the formation and survival of life: the composition of outer Solar System planetesimals makes them ideal for delivering water and organics during planet formation (see §4.1); on the other hand, bombardment of a planet may kill developing biospheres, a process known as "impact frustration".

Host Star Mass and Evolution The radiative effects of the host star change with both stellar mass and the age of the star. More massive stars have shorter lifetimes than less massive stars, which may limit the probability of life arising on a planet around an early-type star. More massive stars also emit a larger fraction of their light at UV and X-ray wavelengths, which would have a detrimental impact on organic processes not protected by a thick atmosphere. Less massive stars than the Sun give off most of their light at longer wavelengths, potentially inhibiting or drastically modifying biological processes such as photosynthesis (Raven 2007), and tidal locking at very small radii could cause atmospheric freeze-out on the dark side of the planet. Additionally, stars change their luminosity over their lifetime; the Sun was 70% of its current luminosity when settled onto the main sequence (Gough 1981). Therefore planets must have a sufficient feedback system and atmospheric volume to compensate for the changes in energy input.

Presence of Satellites Earth is unusual with respect to the other rocky planets in our Solar System in that it has one large moon in a circular orbit. This could play a significant role in stabilizing Earth's rotation; the tilt of Earth towards the Sun's "obliquity" is relatively stable over very long time periods. Without the Moon, Earth's obliquity could potentially vary drastically over million-year timescales, causing major surface temperature variations (Laskar et al. 1993). In addition, the formation of the Moon by a giant impact would have strongly affected the orbit and rotation of Earth, playing a major role in the final characteristics of the temperature and composition of the planet.

3. Assessing Habitability in our Solar System

We can apply our understanding of the qualities that define habitable environments to the known planets and satellites in our own planetary system to gauge the potential for the origin and evolution of life, either ancient or extant. The current surface and subsurface conditions for planets interior to Earth (Mercury and Venus) suggest a very low probability for the origin and survival of life forms due to the very high surface temperatures (~ 700 K) and very low water contents of both planets. The other terrestrial planet in the inner system, Mars, is a much better candidate, and recent results and upcoming investigations will be discussed in §3.1. Beyond the Asteroid Belt, the potential for habitable environments is much less clear. The best candidates appear to be the icy moons of Jupiter and Saturn, where evidence suggests a liquid water environment below an ice crust on both Europa and Enceladus; recent results will be discussed in §3.2.

3.1. Mars

The enigmatic sister to Earth, Mars has long been considered our best candidate for detecting life on other planets; however, its promise as a habitable environment has fluctuated as we learn more about habitability in general and about the properties of Mars itself.

Mars is approximately 1/2 the size of Earth and has 1/10 th the mass; it is also 1.52 times the distance to the Sun. These factors contribute to a thin atmosphere (6/1000 th as dense as Earth) and a mean surface temperature of -63 C (210 K). However, the tilt of Mars' rotational axis (25°) is nearly equal to that of Earth, resulting in similar seasonal variations. The variation in solar insolation results in temperature fluctuations of $\sim$80 K between summer and winter, as well as polar caps that vary in size by a factor of 4. It is clear from both the atmospheric conditions and high-resolution observations that significant amounts of stable liquid water, and therefore any macroscopic life that would rely on it, is not currently present on the surface of Mars. Additionally, more exotic metabolism and cellular structures would be necessary to survive high doses of UV and X-ray radiation (Dartnell et al. 2007; Smith & Scalo 2007) and oxidation reactions (Quinn et al. 2005); microbes on Earth do exist under these conditions (Houtkooper & Schulze-Makuch 2007), but it is unclear if similar organisms could evolve on Mars.

However, this does not preclude the presence of subsurface life. Recent results from thermal imaging of the Martian surface suggest that water ice is stable as close as 20 cm from the surface (Bandfield 2007), and images of outflow gullies taken by the Mars Global Surveyor demonstrate that a large amount of fluid, most likely water, was released from only tens of meters below the surface (Malin & Edgett 2000). Additionally, investigations of surface conditions by the Mars rovers have found mineral sequences and sedimentology indicative of evaporation processes, as well as iron-rich mineral inclusions commonly known as "blueberries" that commonly precipitate out of ground water (Squyres et al. 2004). These results give credence to the idea that microbial life could be present in subsurface environments, and has sparked a resurgence of interest in both remote searches for atmospheric biomarkers as well as new in-situ experiments for local biomarkers below the surface.

Remote Detection of Biomarkers Though it is difficult to probe the subsurface of Mars directly, we can sensitively test the atmospheric composition to search for evidence of byproducts of biological activity through orbiting spacecraft and ground-based observations from Earth. However, discerning the difference between a biologically-produced atmospheric constituent and a gas produced abiotically is not trivial, and searches primarily focus on evidence of disequilibrium in the atmospheric chemistry compared with models.

One of the primary signatures of disequilibrium is the coincidence of both oxidizing species (such as O_2) and reducing species (such as CH_4). The Martian atmosphere is dominated by CO_2 (95%), and photodissociation easily produces trace amounts of other oxidizing species such as O_2 and OH. Therefore the presence of a stable abundance of a reduced species in the Martian atmosphere would suggest ongoing production, either abiotic or biotic. Summers et al. (2002) suggest most biologically relevant gas species would have characteristic lifetimes of less than a year; methane, however, would survive for much longer ($\sim$300 years) and is thought to provide the best chance for detection. Recent spectroscopic searches for methane in the Martian atmosphere have yielded ambiguous results, primarily due to the difficulty of searching for extremely weak features. Formisano et al. (2004) reported a marginal detection of 10$\pm$5 ppb methane using the Planetary Fourier Spectrometer onboard the Mars Express orbiter, but the severe limitations in spectral resolution and sensitivity have cast doubt on these results. Krasnopolsky et al. (2004) published a similar result (10$\pm$3 ppb) using ground-based NIR spectroscopy, but this study was also hampered by instrumental uncertainties and the difficulty of removing the terrestrial methane signature; a more recent search produced only upper limits (Krasnopolsky 2007). Considering the uncertainties in these results a judgement on the presence of methane in the Martian atmosphere would clearly be premature, and we must wait for improved observing resources and/or analysis; several sensitive campaigns are currently being undertaken (i.e. Mumma et al. 2007).

Interpreting these and any future detections or non-detections is further complicated by the possibility of abundance variations due to seasonal or local release. Even if methane were to be securely detected, its provenance would be unclear; there are a variety of abiotic production mechanisms for methane that have been proposed, ranging from cometary delivery (Kress & McKay 2004) to the photolysis of water and CO in the atmosphere (Bar-Nun & Dimitrov 2006) and low-temperature alteration of basalts ("serpentinization"; Oze & Sharma 2005). Future studies must therefore focus on discerning the observable signatures of various production pathways before we can elucidate the true origin of any detected species.

In-situ Detection of Biomarkers Finding evidence of life from the surface of Mars is no less fraught with pitfalls than the remote detection of life. The first experiments to test for Martian life on the surface, conducted on the Viking Lander 1 in 1976, led to more questions than they answered (see Schuerger & Clark 2007 for a recent review). Three different biology experiments, two gas release experiments and a pyrolytic release experiment, yielded what were considered positive results but were later attributed to abiotic processes. The most damning case against biology came from the results of the gas chromatograph-mass spectrometer which failed to find any trace of organic material, the basic building blocks of all life on Earth. The validity of these conclusions is still under debate (i.e. Benner et al. 2000; Houtkooper & Schulze-Makuch 2007), but they continue to inform designs for future searches for evidence of life on or below the surface of Mars.

The first upcoming Mars lander mission to test for evidence of surface habitability will be the NASA Phoenix Lander, currently set to touch down in May of 2008. Phoenix will not conduct any experiments to directly detect extant or extinct life, but it will further characterize the organic material and level of oxidation of the Martian soil with much greater sensitivity than the Viking experiments using a wet soil chemistry experiment (MECA) and a gas-release mass spectrometer experiment (TEGA) (Shotwell 2005). More importantly, Phoenix is equipped with a robotic arm that can dig up to 0.5 m below the surface. Tests of subsurface material will measure the hydration level as well as the degree of oxidation and photolysis of organic material, adding significantly to our understanding of the potential for extant subsurface life.

The first post-Phoenix missions to test for life will be the NASA Mars Science Laboratory (MSL) and the ESA ExoMars mission with both rovers carrying a wide variety of instruments. Both missions will extend Phoenix's investigations of the organic inventory in both the sensitivity and the scope of the experiments performed. The primary life-detection experiment planned for the MSL, expected to launch is 2009, is known as SAM (Sample Analysis at Mars). SAM will be a combination of a gas chromatograph and both a mass spectrometer and a tunable laser spectrometer (Mahaffy 2007). The gas chromatograph will reach higher temperatures than previous experiments ($\sim$1100 C), sufficient to observe both volatile species indicative of extant biology and moderately refractory species indicative of past life, and the coupled GC-MS will be sensitive down to a fraction of a picomole of organic material. ExoMars, expected to launch in 2013, will aim to include a Raman-LIBS (laser-induced breakdown spectrometer) for organic analysis, an oxidant sensor, a "life marker chip" which would use antibody reactions to detect specific molecules with extreme sensitivity, as well as a drill to reach $\sim$2 m below the surface (The Rover Team et al. 2006). These two missions are expected to fully characterize the astrobiological properties of the surface and immediate sub-surface of Mars; future improvements would be left to either a sample return mission or a manned mission.

3.2. Europa & Enceladus

In the outer Solar System, the low solar insolation level results in surface temperatures far below the freezing point of water. Additionally, volatile inventories for most bodies are high compared with the inner Solar System, resulting in low bulk densities and icy surfaces (first confirmed through Voyager imaging; Smith et al. 1979) for many of the satellites of the giant planets. In addition, tidal stresses related to orbital resonances were shown to be a potential heat source for generating a liquid ocean below the ice crust of at least one Galilean moon, Europa (Cassen et al. 1979; Squyres et al. 1983). More recently, direct evidence of liquid water from an icy moon was confirmed through Cassini observations of a plume of water-rich material flowing from the south pole of Enceladus, a moon of Saturn (Porco et al. 2006). Investigations of these potentially habitable environments are still in their infancy, but they offer a tantalizing new option for finding life beyond Earth.

Europa has a radius of 1560 km, close to that of the Earth's moon. In 1998, radio Doppler data from the Galileo spacecraft demonstrated a high probability for a differentiated internal structure and a thick outer ice shell on Europa (Anderson et al. 1998), and the lack of significant cratering and other signatures of resurfacing (Zahnle et al. 1998) suggest an active geologic history possibly aided by a liquid water layer below the ice. This hypothesis was further supported by magnetometer results from Galileo requiring a near-surface global conducting layer, most plausibly in the form of a saline water ocean (Kivelson et al. 2000).

Current analysis of topography data suggests an ice shell thickness of approximately 15 - 25 km (Nimmo et al. 2003), while the magnetometer data suggests an ice shell thickness of less than 15 km (Hand & Chyba 2007). Beyond liquid water, the two primary uncertainties with regard to habitability of a subsurface ocean are the availability of biologically-important compounds and the free energy to assemble and maintain them. Volcanic and tectonic activity as well as hydrothermal vents may contribute to the ocean mineralogy (Reynolds et al. 1983); additionally, cometary impacts can deliver small amounts of biogenic elements (Pierazzo & Chyba 2002). However, without knowledge of the water chemistry or composition and structure of the sub-ocean mantle it is difficult to estimate the contribution of material or heat through such processes.

Enceladus, only 500 km in diameter, was not considered to be a viable astrobiological target due to a lack of internal heating until Cassini revealed an outburst of material emanating from a hot spot near the moon's south pole in 2005. Images of the south pole suggest a recent resurfacing, with "tiger striped" ridges indicative of tectonic activity; temperatures measured for the ridges by the Visual and Infrared Mapping Spectrometer are in the range of 140 K (Brown et al. 2006). During Cassini's passage through the plume, measurements from the Ion and Neutral Mass Spectrometer showed the composition to be mostly water, with traces of methane and N_2 (Waite et al. 2006); the chemical composition shows evidence of production in a hot ($T \sim 500$ K) catalytic environment (Matson et al. 2007). The outgassing may result either from shallow liquid water reservoirs (Porco et al. 2006) or from clathrate disruption (Kieffer et al. 2006); however, it is unclear how the heating necessary to produce the surface features would be generated and transported in the first place. Tidal shearing presents a viable option for heat transport, but would most likely require a global sub-surface ocean (Nimmo et al. 2007); the flattened shape of the south pole also suggests warming due to a local subsurface sea (Collins & Goodman 2007), but alternative models suggest a low-density ice flow (Nimmo & Pappalardo 2006). However, the effects of both past and present tidal heating and radiogenic heating may be insufficient to produce the required heat flux under realistic conditions (Meyer & Wisdom 2007; Schubert et al. 2007); without a clear understanding of the temporal nature of the heating processes it is difficult to assess the validity of specific models. Additional data on the gravitational anomaly from future Cassini fly-bys will help to resolve some of the ambiguities.

The next step in exploring the habitability of both Europa and Enceladus would be a thorough investigation of the surface ice characteristics through instruments for ground-penetrating radar, altimetry, high-resolution imagery and near-infrared spectroscopy to further discern the surface and subsurface chemistry (Chyba & Phillips 2002). For Europa, ESA is considering an orbiter mission known as the Jovian Minisat Explorer, while NASA has considered both a Europa Orbiter as well as the Jupiter Icy Moons Orbiter; however, both concepts are still highly uncertain and will most likely be reconsidered due to the new Enceladus results.

4. Finding Habitable and Inhabited Planets Around Other Stars

For almost 500 years, from the time of Copernicus' discovery of the heliocentric nature of our Solar System until the discovery of the first extra-solar planet in

1995 (Mayor & Queloz 1995), humankind attempted to understand the origin of our living planet and the rest of the celestial bodies through the narrow lens of our single planetary system. We therefore developed complex theories on the formation of planets that naturally result in a high probability of producing a Solar System just like our own. However, it is not altogether surprising that with the discovery of a second planetary system, these theories were abruptly turned upside down. With more than 250 giant planets now known to orbit main-sequence stars (Butler et al. 2006; see Schneider 2007 for recent results), theories on the planet formation and evolution developed for our own planetary system must be re-examined.

4.1. Understanding Habitable Planet Formation

To predict where to look for habitable planets, and what types of characteristics the planets we find will have, we must understand the evolution of the initial gas-and-dust-rich protoplanetary environments in which planets grow and the forces that shape the formation and evolution of these planets into habitable worlds.

The primary factors that define whether habitable planets can form and remain stable over biologically significant timescales are:

- A mass density near the Habitable Zone sufficient to form planets capable of sustaining an atmosphere and geologic activity

- A volatile contribution (most importantly water) sufficient to sustain the origin and evolution of life

- A sufficiently stable planetary system such that any orbital variations do not result in long periods of inhabitability

The first two factors are determined primarily by the initial conditions of the protoplanetary nebula and the subsequent evolution of raw material into a young planetary system as solid bodies accrete and the gaseous disk dissipates. The stability of a planetary system is primarily determined by the post-accretion dynamics of the system when planets begin to interact and scatter each other within (and out of) the system. Our own Solar System clearly passed all three tests: sufficient mass and volatile material was available to create at least one planet capable of harboring life, and our planetary system was stable enough that Earth was able to remain on an almost circular orbit for billions of years. The question remains as to whether our Solar System represents the norm, or just a rare aberration.

Standard Planet Formation Theory Standard theories of the evolution of planetary cores suggest that solid material in a circumstellar disk will proceed through various accretionary stages culminating in the final architecture of a stable planetary configuration. Coagulation of the inceptive dust particles occurs through collisional sticking to produce meter-sized objects on timescales of 10^4 years (Lissauer 1993). Once the largest bodies reach ~ 1 km in size, their gravitational cross-section becomes larger than their geometric cross-section and they begin the phase known as "runaway growth" (Greenberg et al. 1978; Wetherill & Stewart 1989), eventually leading to a binomial distribution of "embryos"

($M \sim 0.1\,\mathrm{M_\oplus}$) and "planetesimals" ($M < 10^{-3}\,\mathrm{M_\oplus}$) after approximately 10^6-10^7 years. The final "chaotic phase" of planet growth proceeds through scattering and collisions between the large protoplanets and final clearing of the remaining planetesimals to produce a stable planetary system after more than 10^8 years (Wetherill 1996; Kenyon & Bromley 2006).

According to traditional theories of planet formation, these processes occur primarily in localized regions, with little mass transport between the inner and outer disk. The initial composition of solids in a circumstellar disk is expected to follow a basic condensation sequence (Grossman 1972); therefore, planets formed beyond the current "snow line" at approximately 2.7 AU, where the enhanced mass density due to volatile freeze-out leads to more massive embryos (Stevenson & Lunine 1988), would be able to reach a critical core mass and initiate run-away gas accretion to form gas giants (Pollack et al. 1996). Rocky planets formed inside the snow line would be relatively water-poor; water-rich material could then be delivered through late-stage cometary impacts (Owen & Bar-Nun 1995) or inward scattering of icy asteroids (Morbidelli et al. 2000). Planets would remain on almost circular orbits due to the relatively quiescent formation process, with the only major rearrangements occurring due to impacts between embryos such as the Moon-forming impact (Hartmann & Davis 1975).

New Results from Extrasolar Planets and Planet Formation Studies The discoveries of extrasolar planetary systems, new results on evolutionary timescales for our own Solar System and other protoplanetary disks, and the advent of sophisticated hydrodynamic and N-body investigations of protoplanetary disks have shed new light on the processes at work in planet formation, while also leading to new conundrums. Though radial velocity (RV) searches are only complete out to approximately 3 AU (Butler et al. 2006), there is already clearly a pile-up of planets at very small semi-major axes: 44% of RV-detected extrasolar planets have orbital radii less than 0.3 AU, and 18% are located within 0.05 AU of the central star. Recent models of gas-rich accretion disks support the theory that giant planets form beyond the snow line and migrate inwards in a process known as "Type II migration" (Papaloizou & Lin 1984; Lin et al. 1996); additional migration processes have also been shown to function on smaller bodies embedded in gaseous disk ("Type I migration" for Earth-sized bodies Ward 1997; Masset et al. 2006 and drag forces for planetesimals Ciesla & Cuzzi 2006). These processes will have a profound affect on mass transport during planet formation, and when combined with models that demonstrate the effects of variable condensation fronts throughout the disk lifetime (Garaud & Lin 2007; Kennedy et al. 2007) these results transform our understanding of the transport and delivery of water and volatiles to the terrestrial planets.

Additionally, both the orbital eccentricities and masses of known extrasolar planets with semi-major axes between 0.1 AU and 3 AU are remarkably evenly distributed up to $e = 0.7$ and $M = 4\,\mathrm{M}_J$ (Butler et al. 2006; Schneider 2007). The upper mass limit is mostly likely a result of initial disk mass; however, there are a number of plausible mechanisms for exciting eccentricities such as planet-disk interactions (Goldreich & Sari 2003; Kley & Dirksen 2006; D'Angelo et al. 2006), perturbations from stellar companions or field stars (Zakamska & Tremaine 2004; Takeda & Rasio 2005) and planet-planet scattering (Marzari & Weidenschilling 2002; Ford & Rasio 2007), and it is unclear what role (if any) each process plays in each system.

These new results present challenges for both explaining our own Solar System's architecture as well as predicting the characteristics of terrestrial planets in other planetary systems. It is still unclear whether our system suffered migration of a giant planet or rocky cores during the earliest stages of formation, though the mass and semi-major axis ratios of Jupiter and Saturn could have acted to stop or reverse Type II migration (Masset & Snellgrove 2001; Morbidelli & Crida 2007). The role of inward migration of water-rich material on the volatile content of the inner system has not been adequately addressed, though modeling improvements have been made (Raymond & Meadows 2007); further work is also necessary to incorporate the effects of fragmentation and heating processes.

Initial investigations of terrestrial planet formation in extrasolar planetary systems are encouraging. Models of planetary systems that experience migration of a giant planet to the inner system have demonstrated that habitable planets can form and survive in these systems (Mandell & Sigurdsson 2003; Fogg & Nelson 2005); the planets formed are usually more massive and water-rich than those in our own Solar System (Raymond et al. 2006; Mandell et al. 2007). Systems which experience the scattering of a giant planet into an eccentric orbit near the Habitable Zone can clear out planetesimals (Veras & Armitage 2005), but terrestrial planet formation in a system with giant planets in stable orbits beyond 2.5 AU will be uninhibited (Raymond 2006). However, many recent advances in our understanding of the evolution of circumstellar disks are not yet fully incorporated into these models, and results must be regarded as preliminary.

4.2. Current and Future Searches for Habitable / Inhabited Planets

Current extrasolar planet detection techniques are not yet sensitive enough to detect Earth-like planets around Sun-like stars in the Habitable Zone; the current detection limit for solar-type stars is approximately 10 $M_\oplus$, and that is only for close-in planets in multi-planet systems (Schneider 2007). However, low-mass M stars present a much better chance for habitable planet detection: the lower stellar mass allows for detection of smaller planets, and the lower luminosity results in a Habitable Zone close to the parent star. The only currently known M-star planetary system which may be habitable is GL581, with a 5 $M_\oplus$ planet at 0.073 AU and an 8 $M_\oplus$ planet at 0.25 AU (Udry et al. 2007). The two planets straddle the traditional Habitable Zone defined solely by the stellar insolation; however, atmospheric circulation models suggest a thick atmosphere on the tidally-locked outer planet could increase the temperature to within the habitable range (von Bloh et al. 2007; Selsis et al. 2007).

Discoveries of habitable planets more like Earth will most likely occur through future space missions such as the Kepler transit search (Basri et al. 2005) with and estimated launch date of 2009 and the SIM PlanetQuest astrometry mission (Catanzarite et al. 2006) with an estimate launch date of 2016. The 1-meter Kepler telescope will monitor ~100,000 main-sequence stars ($m_v = 9-14$) continuously for 4 years, with precision sufficient to detect Earth-mass planets at 1 AU around solar-mass stars with $m_v=12$. Kepler has the potential to find hundreds of Earth-like planets, but the host stars will be beyond the range of most follow-up ground-based techniques. SIM PlanetQuest will most likely be a visible-wavelength interferometer with two 0.3 m telescopes separated by a 9-meter baseline, capable of detecting proper motions with 1 μas precision. Though its detection numbers for Earth-like planets ($\sim$10) would be much smaller than Kepler, it could observe nearby stars with and without detected planets, furthering the characterization of nearby planetary systems currently being observed by ground-based campaigns.

The search for life on these planets will have to wait until the launch of the Terrestrial Planet Finder (NASA) and/or Darwin (ESA) missions, which will seek to simultaneously image nearby Earth-like planets and analyze their atmospheres for biomarkers using low-resolution visible or near-infrared spectroscopy. Both missions are in the early design phases, but considerable work has been done on predicting the spectral signatures of planets with a variety of surface features (see Kaltenegger & Selsis 2007 for a recent review). Primary molecular biomarkers focus on disequilibrium chemistry between oxygen species and reducing organics such as methane, with studies showing variations with cloud cover, surface ice and vegetation fraction. However, the current lack of constraints on potential variations in the characteristics of planetary surfaces and atmospheres, variations in biological evolution under different conditions, and firm plans for instrumentation make constraining potential biosignatures difficult.

Though we cannot currently search for biosignatures, the search for abiotic signs of life continues through various SETI (Search for Extraterrestrial Intelligence) programs. Due to rapid advances in signal processing sophistication and sensitivity, searches now include all regions of the temporal (i.e. pulses versus continuous) and frequency domains (from optical to radio wavelengths). A huge leap in coverage of the GHz regime will be accomplished with the Allen Telescope Array and the future Square Kilometer Array (SKA), with the added benefit of flexible resource allocation based on target availability and priority (Tarter 2004). The ability to target large numbers of stars, combined with improved constraints based on planet detection missions, will vastly improve the SETI search efficiency. It may be a fruitless search, but one confirmed detection would alter our perception of our place in the universe forever.

5. Conclusion

While this review is in no way an exhaustive account of the state of astrobiology-related research, it profiles the most recent results in the search for life beyond the confines of our planet, both within our own planetary system and beyond. The prospects for detecting evidence of life elsewhere in the next few decades, or at least constraining the planetary environments in which life cannot evolve, are bright: within 10 years we will explore the immediate subsurface of Mars, conduct sensitive searches for Earth-like planets around other stars, and drastically improve our understanding of the characteristics of these potentially habitable planetary systems. In the subsequent decade we will hopefully begin further exploration of the icy moons in the outer Solar System, discovery and characterize nearby Earth-like planets, and possibly detect life on these planets through atmospheric biomarkers. Though we currently appear as a single oasis in a vast empty desert, by the year 2025 the Earth may be only one of many examples of locales amenable for the origin and evolution of life.

Acknowledgments. I would like to thank the Astronomy Department at the University of Texas at Austin and the McDonald Observatory for giving me the opportunity to participate in the Bash Symposium, and G. Villanueva for stimulating conversations during the preparation of this manuscript. Support for this work was provided by the NASA through the NASA Post-doctoral Program.

References

Anderson, J. D., Schubert, G., Jacobson, R. A., Lau, E. L., Moore, W. B., & Sjogren, W. L. 1998, Science, 281, 2019
Bandfield, J. L. 2007, Nature, 447, 64
Bar-Nun, A. & Dimitrov, V. 2006, Icarus, 181, 320
Basri, G., Borucki, W. J., & Koch, D. 2005, New Astronomy Review, 49, 478
Benner, S. A., Devine, K. G., Matveeva, L. N., & Powell, D. H. 2000, Proceedings of the National Academy of Science, 97, 2425
Berner, R. A. & Raiswell, R. 1983, Geochim. Cosmochim. Acta, 47, 855
Brown, R. H., Clark, R. N., Buratti, B. J., Cruikshank, D. P., Barnes, J. W., Mastrapa, R. M. E., Bauer, J., Newman, S., Momary, T., Baines, K. H., Bellucci, G., Capaccioni, F., Cerroni, P., Combes, M., Coradini, A., Drossart, P., Formisano, V., Jaumann, R., Langevin, Y., Matson, D. L., McCord, T. B., Nelson, R. M., Nicholson, P. D., Sicardy, B., & Sotin, C. 2006, Science, 311, 1425
Butler, R. P., Wright, J. T., Marcy, G. W., Fischer, D. A., Vogt, S. S., Tinney, C. G., Jones, H. R. A., Carter, B. D., Johnson, J. A., McCarthy, C., & Penny, A. J. 2006, ApJ, 646, 505
Cassen, P., Reynolds, R. T., & Peale, S. J. 1979, Geophys. Res. Lett., 6, 731
Catanzarite, J., Shao, M., Tanner, A., Unwin, S., & Yu, J. 2006, PASP, 118, 1319
Chyba, C. F. & Phillips, C. B. 2002, Origins of Life and Evolution of the Biosphere, 32, 47
Ciesla, F. J. & Cuzzi, J. N. 2006, Icarus, 181, 178
Collins, G. C. & Goodman, J. C. 2007, Icarus, 189, 72
D'Angelo, G., Lubow, S. H., & Bate, M. R. 2006, ApJ, 652, 1698
Dartnell, L. R., Desorgher, L., Ward, J. M., & Coates, A. J. 2007, Biogeosciences, 4, 545
Fogg, M. J. & Nelson, R. P. 2005, A&A, 441, 791
Ford, E. B. & Rasio, F. A. 2007, astro-ph/0703163
Forget, F. & Pierrehumbert, R. T. 1997, Science, 278, 1273
Formisano, V., Atreya, S., Encrenaz, T., Ignatiev, N., & Giuranna, M. 2004, Science, 306, 1758
Garaud, P. & Lin, D. N. C. 2007, ApJ, 654, 606
Goldreich, P. & Sari, R. 2003, ApJ, 585, 1024
Gough, D. O. 1981, Sol. Phys., 74, 21
Greenberg, R. & Geissler, P. 2002, Meteoritics and Planetary Science, 37, 1685
Greenberg, R., Hartmann, W. K., Chapman, C. R., & Wacker, J. F. 1978, Icarus, 35, 1
Grossman, L. 1972, Geochim. Cosmochim. Acta, 36, 597
Hand, K. P. & Chyba, C. F. 2007, Icarus, 189, 424
Hart, M. H. 1978, Icarus, 33, 23
Hartmann, W. K. & Davis, D. R. 1975, Icarus, 24, 504
Houtkooper, J. M. & Schulze-Makuch, D. 2007, International Journal of Astrobiology, 6, 147
Kaltenegger, L. & Selsis, F. 2007, astro-ph/0710.0881
Kashefi, K. & Lovley, D. R. 2003, Science, 301, 934
Kasting, J. F. & Catling, D. 2003, ARA&A, 41, 429
Kasting, J. F., Whitmire, D. P., & Reynolds, R. T. 1993, Icarus, 101, 108
Kennedy, G. M., Kenyon, S. J., & Bromley, B. C. 2007, Ap&SS, 311, 9

Kenyon, S. J. & Bromley, B. C. 2006, AJ, 131, 1837
Kieffer, S. W., Lu, X., Bethke, C. M., Spencer, J. R., Marshak, S., & Navrotsky, A. 2006, Science, 314, 1764
Kivelson, M. G., Khurana, K. K., Russell, C. T., Volwerk, M., Walker, R. J., & Zimmer, C. 2000, Science, 289, 1340
Kley, W. & Dirksen, G. 2006, A&A, 447, 369
Krasnopolsky, V. A. 2007, Icarus, 190, 93
Krasnopolsky, V. A., Maillard, J. P., & Owen, T. C. 2004, Icarus, 172, 537
Kress, M. E. & McKay, C. P. 2004, Icarus, 168, 475
Laskar, J., Joutel, F., & Robutel, P. 1993, Nature, 361, 615
Lin, D. N. C., Bodenheimer, P., & Richardson, D. C. 1996, Nature, 380, 606
Lissauer, J. J. 1993, ARA&A, 31, 129
Mahaffy, P. 2007, Space Science Reviews, 132
Malin, M. C. & Edgett, K. S. 2000, Science, 288, 2330
Mandell, A. M., Raymond, S. N., & Sigurdsson, S. 2007, ApJ, 660, 823
Mandell, A. M. & Sigurdsson, S. 2003, ApJ, 599, L111
Marzari, F. & Weidenschilling, S. J. 2002, Icarus, 156, 570
Masset, F. & Snellgrove, M. 2001, MNRAS, 320, L55
Masset, F. S., D'Angelo, G., & Kley, W. 2006, ApJ, 652, 730
Matson, D. L., Castillo, J. C., Lunine, J., & Johnson, T. V. 2007, Icarus, 187, 569
Mayor, M. & Queloz, D. 1995, Nature, 378, 355
Meyer, J. & Wisdom, J. 2007, Icarus, 188, 535
Morbidelli, A., Chambers, J., Lunine, J. I., Petit, J. M., Robert, F., Valsecchi, G. B., & Cyr, K. E. 2000, Meteoritics and Planetary Science, 35, 1309
Morbidelli, A. & Crida, A. 2007, Icarus, 191, 158
Mumma, M. J., Villanueva, G. L., Novak, R. E., Hewagama, T., Bonev, B. P., DiSanti, M. A., & Smith, M. D. 2007, in AAS/Division for Planetary Sciences Meeting Abstracts, Vol. 39, #31.02
Nimmo, F., Giese, B., & Pappalardo, R. T. 2003, Geophys. Res. Lett., 30, 37
Nimmo, F. & Pappalardo, R. T. 2006, Nature, 441, 614
Nimmo, F., Spencer, J. R., Pappalardo, R. T., & Mullen, M. E. 2007, Nature, 447, 289
Owen, T. & Bar-Nun, A. 1995, Icarus, 116, 215
Oze, C. & Sharma, M. 2005, Geophys. Res. Lett., 32, 10203
Papaloizou, J. & Lin, D. N. C. 1984, ApJ, 285, 818
Pavlov, A. A., Brown, L. L., & Kasting, J. F. 2001, Journal of Geophysical Research (Planets), 106, 23267
Pierazzo, E. & Chyba, C. F. 2002, Icarus, 157, 120
Pollack, J. B., Hubickyj, O., Bodenheimer, P., Lissauer, J. J., Podolak, M., & Greenzweig, Y. 1996, Icarus, 124, 62
Porco, C. C., Helfenstein, P., Thomas, P. C., Ingersoll, A. P., Wisdom, J., West, R., Neukum, G., Denk, T., Wagner, R., Roatsch, T., Kieffer, S., Turtle, E., McEwen, A., Johnson, T. V., Rathbun, J., Veverka, J., Wilson, D., Perry, J., Spitale, J., Brahic, A., Burns, J. A., DelGenio, A. D., Dones, L., Murray, C. D., & Squyres, S. 2006, Science, 311, 1393
Quinn, R. C., Zent, A. P., Grunthaner, F. J., Ehrenfreund, P., Taylor, C. L., & Garry, J. R. C. 2005, Planet. Space Sci., 53, 1376
Raven, J. 2007, Nature, 448, 418
Raymond, S. N.and Scalo, J. & Meadows, V. S. 2007, ApJ, submitted
Raymond, S. N. 2006, ApJ, 643, L131
Raymond, S. N., Mandell, A. M., & Sigurdsson, S. 2006, Science, 313, 1413
Reynolds, R. T., Squyres, S. W., Colburn, D. S., & McKay, C. P. 1983, Icarus, 56, 246
Rivkina, E. M., Friedmann, E. I., McKay, C. P., & Gilichinsky, D. A. 2000, Appl. Environ. Biol., 66, 3230
Schneider, J. 2007, http://exoplanet.eu/catalog.html.
Schubert, G., Anderson, J. D., Travis, B. J., & Palguta, J. 2007, Icarus, 188, 345

Schuerger, A. C. & Clark, B. C. 2007, Space Science Reviews, OnlineFirst
Selsis, F., Kasting, J. F., Levrard, B., Paillet, J., Ribas, I., & Delfosse, X. 2007, A&A, 476, 1373
Shotwell, R. 2005, Acta Astronautica, 57, 121
Smith, B. A., Soderblom, L. A., Beebe, R., Boyce, J., Briggs, G., Carr, M., Collins, S. A., Johnson, T. V., Cook, II, A. F., Danielson, G. E., & Morrison, D. 1979, Science, 206, 927
Smith, D. S. & Scalo, J. 2007, Planet. Space Sci., 55, 517
Squyres, S. W., Arvidson, R. E., Bell, J. F., Brückner, J., Cabrol, N. A., Calvin, W., Carr, M. H., Christensen, P. R., Clark, B. C., Crumpler, L., Des Marais, D. J., d'Uston, C., Economou, T., Farmer, J., Farrand, W., Folkner, W., Golombek, M., Gorevan, S., Grant, J. A., Greeley, R., Grotzinger, J., Haskin, L., Herkenhoff, K. E., Hviid, S., Johnson, J., Klingelhöfer, G., Knoll, A. H., Landis, G., Lemmon, M., Li, R., Madsen, M. B., Malin, M. C., McLennan, S. M., McSween, H. Y., Ming, D. W., Moersch, J., Morris, R. V., Parker, T., Rice, J. W., Richter, L., Rieder, R., Sims, M., Smith, M., Smith, P., Soderblom, L. A., Sullivan, R., Wänke, H., Wdowiak, T., Wolff, M., & Yen, A. 2004, Science, 306, 1698
Squyres, S. W., Reynolds, R. T., & Cassen, P. M. 1983, Nature, 301, 225
Stevenson, D. J. & Lunine, J. I. 1988, Icarus, 75, 146
Summers, M. E., Lieb, B. J., Chapman, E., & Yung, Y. L. 2002, Geophys. Res. Lett., 29, 24
Takeda, G. & Rasio, F. A. 2005, ApJ, 627, 1001
Tarter, J. C. 2004, New Astronomy Review, 48, 1543
The Rover Team, Barnes, D., Battistelli, E., Bertrand, R., Butera, F., Chatila, R., Del Biancio, A., Draper, C., Ellery, A., Gelmi, R., Ingrand, F., Koeck, C., Lacroix, S., Lamon, P., Lee, C., Magnani, P., Patel, N., Pompei, C., Re, E., Richter, L., Rowe, M., Siegwart, R., Slade, R., Smith, M. F., Terrien, G., Wall, R., Ward, R., Waugh, L., & Woods, M. 2006, International Journal of Astrobiology, 5, 221
Udry, S., Bonfils, X., Delfosse, X., Forveille, T., Mayor, M., Perrier, C., Bouchy, F., Lovis, C., Pepe, F., Queloz, D., & Bertaux, J.-L. 2007, A&A, 469, L43
Veras, D. & Armitage, P. J. 2005, ApJ, 620, L111
von Bloh, W., Bounama, C., Cuntz, M., & Franck, S. 2007, A&A, 476, 1365
Waite, J. H., Combi, M. R., Ip, W.-H., Cravens, T. E., McNutt, R. L., Kasprzak, W., Yelle, R., Luhmann, J., Niemann, H., Gell, D., Magee, B., Fletcher, G., Lunine, J., & Tseng, W.-L. 2006, Science, 311, 1419
Ward, W. R. 1997, Icarus, 126, 261
Wetherill, G. W. 1996, Icarus, 119, 219
Wetherill, G. W. & Stewart, G. R. 1989, Icarus, 77, 330
Zahnle, K., Dones, L., & Levison, H. F. 1998, Icarus, 136, 202
Zakamska, N. L. & Tremaine, S. 2004, AJ, 128, 869

Lucas Cieza and Avi Mandell discuss planet formation at the reception.

The Evolution of Primordial Circumstellar Disks

Lucas A. Cieza[1]

Institute for Astronomy, University of Hawaii at Manoa, Honolulu, HI, USA

Abstract. Circumstellar disks are an integral part of the star formation process and the sites where planets are formed. Understanding the physical processes that drive their evolution, as disks evolve from optically thick to optically thin, is crucial for our understanding of planet formation. Disks evolve through various processes including accretion onto the star, dust settling and coagulation, dynamical interactions with forming planets, and photo-evaporation. However, the relative importance and timescales of these processes are still poorly understood. In this review, I summarize current models of the different processes that control the evolution of primordial circumstellar disks around low-mass stars (mass $< 2\,M_\odot$). I also discuss recent observational developments on circumstellar disk evolution with a focus on new *Spitzer* results on transition objects.

1. Introduction

It is currently believed that virtually all stars form surrounded by a circumstellar disk, even if this disk is sometimes very short lived. This conclusion follows from simple conservation of angular momentum arguments and is supported by mounting evidence. This evidence ranges from the excess emission, extending from the near-IR (Strom et al. 1989) to the sub-millimeter (Osterloh & Beckwith 1995), that is observed in most young (age $< 1\,$Myr) pre-main-sequence (PMS) stars to direct *Hubble* images of disks (McCaughrean & O'Dell 1996) seen as silhouettes in front of the Orion Nebula. Also, even though there is not *direct* evidence that planets actually grow from circumstellar material, it has become increasingly clear that they are in fact the birthplaces of planets since their masses, sizes, and compositions are consistent with the theoretical minimum-mass solar nebula (Hayashi 1981). Recently, the discovery of exo-planets orbiting nearby main sequence stars has confirmed that the formation of planets is a common process and not a rare phenomenon exclusive to our Solar System. Thus, any theory of planet formation should be robust enough to account for the high incidence of planets and cannot rely on special conditions or on unlikely processes to convert circumstellar dust and gas into planets.

Star and planet formation are intimately related. Standard low-mass star formation models (e.g. Shu et al. 1987) describe the free fall collapse of a slowly rotating molecular cloud core followed by the development of a hydrostatic protostar surrounded by an envelope and a disk of material supported by its residual angular momentum. This early phase is expected to occur on a timescale of

---

[1]Spitzer Fellow

about 10^5 years (Beckwith 1999) and results in an optically revealed classical T Tauri star (CTTS, low-mass PMS star that shows clear evidence for accretion of circumstellar material). This stage is characterized by intense accretion onto the star, strong winds, and bipolar outflows. As the system evolves, presumably into a weak-lined T Tauri star (WTTS, PMS star mostly coeval with CTTSs but that do not show evidence for accretion), accretion ends, and the dust settles into the mid-plane of the disk where the solid particles are believed to stick together and to grow into planetesimals as they collide. Once the objects reach the kilometer scale, gravity increases the collision cross-section of the most massive planetesimals, and runaway accretion occurs (Lissauer 1993). In the standard core accretion model (e.g., Pollack et al., 1996), massive enough proto-planets still embedded in the disk can accrete the remaining gas and become giant planets. The early stages are the most uncertain, and many people suspect that grains will not grow into planetesimals by collisions alone at the rate necessary to go through all stages of giant planet formation before the gas nebula has dissipated. Also, even assuming that grains can grow into macroscopic bodies at the necessary rate, it is still doubtful whether they can form large planetesimals. Once objects reach the meter-size scale, they are expected to rapidly spiral inward due to the dynamical interactions with the gas in the disk, which rotates at a slightly sub-Keplerian velocity because it is supported against gravity by pressure in addition to the centrifugal force.

However, since current statistics of extra-solar planets indicate that giant planets are common, the difficulties of the standard core accretion model have led some researchers (e.g., Boss 2000) to revisit an alternative planet formation mechanism that had been put aside for several decades, namely, gravitational instability. In the gravitational instability scenario, giant planets form through the direct gravitational collapse of a massive unstable disk over timescales $< 10^4$ yrs. In some hybrid models (e.g., Youdin & Shu 2002), solid particles settle to the mid-plane of the circumstellar disk and form a dense gravitationally unstable sub-disk. In this manner, planetesimals form from the gravitational collapse of the material in the mid-plane. The collision of planetesimals leads to runaway accretion, and the process continues in a way analogous to the core accretion model.

Thus, while the existence of planets around a significant fraction of all the stars is considered verified, the precise mechanisms through which planets form still remain largely unknown. So far, none of the proposed theories have proven satisfactory. On one hand, the standard model of continuous accretion of solid particles relies on doubtful sticking properties of rocks and on unknown processes to prevent the migration of meter-size objects. On the other hand, gravitational instability relies on unproven mechanisms to enhance the surface density of the disk's mid-plane to trigger the process of planet formation. Clearly, more observational constraints are necessary to help the theoretical work on planet formation to proceed forward.

2. Dissipation Timescales

Until the Atacama Large Millimeter Array (ALMA) becomes operational in the next decade, direct detection of forming planets will remain beyond our

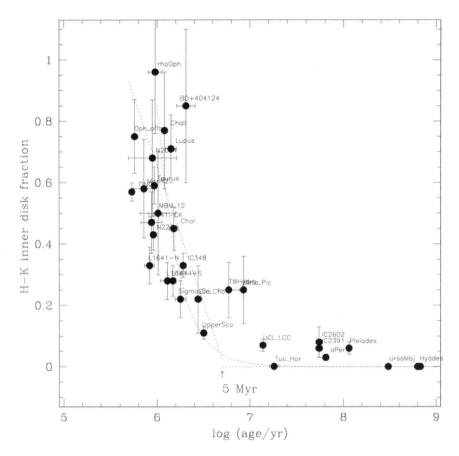

Figure 1. Inner accretion disk fraction vs. stellar age inferred from H-K excess measurements, binned by cluster or association, for ~3500 stars from the literature. (Figure taken from Hillenbrand 2006).

capabilities, but circumstellar disks are easier to detect and study because the surface area of a planetary mass dispersed into small grains is *many* orders of magnitude greater than the surface area of a planet. For this reason, much of the information about the properties of disks such as size, mass, density, and evolution timescales has been obtained by observing the thermal emission of the dust gains. These particles absorb and re-radiate the light mostly in the $1\,\mu m - 1\,mm$ range. Since the temperature of the disk decreases with the distance from the central star, different wavelengths probe different disk radii.

One of the most important quantities on disk evolution is the lifetime of the disk itself, not only because it establishes the relevant timescale of the physical processes controlling disk evolution, but also because it sets a limit for the time available for planet formation. In what follows, I review the constraints on the dissipation timescale of different regions of the disk obtained from different surveys, performed at wavelengths ranging from the near-IR to the sub-millimeter.

2.1. Inner Disk

Since there is an almost 1 to 1 correlation between the presence of near-IR excess ($1-5\,\mu m$) and the occurrence of spectroscopic signatures of accretion (Hartigan et al. 1995), it is possible to investigate the lifetime of inner accretion disks ($r < 0.05 - 0.1\,\mathrm{AU}$) by studying the fraction of stars with near-IR excess as a function of stellar age. Early studies of nearby star-forming regions (e.g., Strom et al. 1989) found that 60-80% of the stars younger than 1 Myr present measurable near-IR excesses, and that just 0-10% of the stars older than 10 Myr do so. It has been argued that individual star-forming regions lack the intrinsic age spread necessary to investigate disk lifetimes from individually derived ages (Hartmann 2001). However, similar disk studies, based on the disk frequency in clusters with different mean ages and extending to the $3.4\,\mu m$ L-band (Haisch et al. 2001, Hillenbrand 2006), have led to results similar to those presented by previous groups. It is now well established that the frequency of inner accretion disks steadily decreases from <1 to 10 Myr. This decrease is illustrated in Figure 1 for a sample of over 3500 PMS stars in nearby clusters and associations. The disk fractions are consistent with *mean* disk lifetimes on the order of 2-3 Myr and a wide dispersion: some objects lose their inner disk at a very early age, even before they become optically revealed and can be placed in the HR diagram, while other objects retain their accretion disks for up to 10 Myr.

2.2. Planet-forming Regions of the Disk

Disk lifetime studies based on near-IR excesses always left room for the possibility that stars without near-IR excess had enough material to form planets at larger radii not probed by near-IR wavelengths. The IRAS and ISO observatories had the appropriate wavelength range to probe the planet-forming regions of the disk ($r \sim 0.05 - 20\,\mathrm{AU}$) but lacked the sensitivity needed to detect all but the strongest mid- and far-IR excesses in low-mass stars at the distances of nearest star-forming regions. *Spitzer* provides, for the first time, the wavelength coverage and the sensitivity needed to detect very small amounts of dust in the planet-forming regions of a statistically significant number of low-mass PMS stars. Recent *Spitzer* results (e.g., Padgett et al. 2006; Silverstone et al. 2006; Cieza el al. 2007) have shown that PMS stars lacking near-IR excess are also very likely to show $24\,\mu m$ fluxes consistent with bare stellar photospheres and that optically thick primordial disks are virtually non-existent beyond an age of 10 Myr. Cieza et al. (2007) also find that over 50% of the WTTS younger than $\sim$1-2 Myr show no evidence for a disk, suggesting that the inner $\sim 10\,\mathrm{AU}$ of a significant fraction of all PMS stars becomes extremely depleted of dust (mass $< 10^{-4}\,M_\oplus$) by that early age.

2.3. Outer Disk

Recent sub-millimeter results extend the conclusions on the survival time of the material in the inner disk ($r < 0.1$ AU) and the planet-forming region of the disk ($r \sim 0.5$–20 AU) to the outer disk ($r \sim$50-100 AU). Andrews & Williams (2005, 2007) study over 170 Young Stellar Objects (YSOs) in the Taurus and Ophiuchus molecular clouds and find that $< 10\%$ of the objects lacking inner disk signatures are detected at sub-mm wavelengths. Given the mass sensitivity of their survey ($M_{DISK} \sim 10^{-4} - 10^{-3}$ $M_\odot$), they conclude that the dust in the inner and the outer disk dissipates nearly simultaneously.

2.4. Implications of Disk Lifetimes

Taken together, the results of the surveys from the near-IR to the sub-millimeter imply that a sizable fraction of all PMS stars lose their disks completely before the star reaches an age of $\sim$1 Myr, while some CTTSs maintain a healthy accretion disk for up to $\sim$10 Myr, which seems to be the upper limit for the survivability of primordial disks. The reason why, within the same molecular cloud or stellar cluster, some primordial disks survive over 10 times longer than others is still unknown. The spread in disk lifetimes is likely to be related to the wide range of initial conditions, the planet formation process, and/or the effect of unseen companions.

Since recent core accretion models (e.g., Alibert et al. 2004) can accommodate planet formation within 10 Myr, the disk survivability limit alone can not distinguish between the competing planet formation mechanisms, core accretion and gravitational instability. On one hand, it is possible that planets form through core accretion around the few CTTS disks that manage to survive for $\sim$5-10 Myr. On the other hand, it is also possible that the youngest WTTSs *without* a disk are objects that have already formed planets through gravitational instability. Thus, establishing the incidence of planets around "young WTTSs" and "old CTTSs" could provide a crucial observational discriminant between the core accretion and the gravitational instability models. This goal is is one of the central objectives of the "Young Stars and Planets" Key Project of the Space Interferometry Mission (Beichman et al. 2002), a mission for which launch has unfortunately been deferred indefinitely at the time of this writing.

3. Transition Timescales and Transition Objects

The fact that very few objects lacking near-IR excess show mid-IR or sub-millimeter excess emission implies that, once accretion stops, the entire disk dissipates very rapidly. Based on the relative numbers of these objects, the transition/dissipation timescale is estimated to be < 0.5 Myr (Skrutskie et al. 1990; Wolk & Walter 1996; Cieza et al. 2007). From the observational point of view, this short transition timescale means that the vast majority of PMS stars in any given population are either accreting CTTSs with excess emission extending all the way from the near-IR to the sub-millimeter or bare stellar photospheres. This also means that any T Tauri star whose SED does not look like a typical CTTS or a bare stellar photosphere can be broadly characterized as a "transition object". It should be noted, however, that precise definitions of what constitutes a transition object found in the disk evolution literature are far from homogeneous.

Since there is a very strong correlation between accretion and the presence of near-IR excess (Hartigan et al. 1995), WTTS very rarely show near-IR excess; therefore, any WTTS that do show IR excesses at longer wavelengths is a transition disk according to the broad definition stated above. Padgett et al. (2006) and Cieza et al. (2007) show that the few WTTS that do have a disk present a wide diversity of SED morphologies and disk to stellar luminosity ratios that

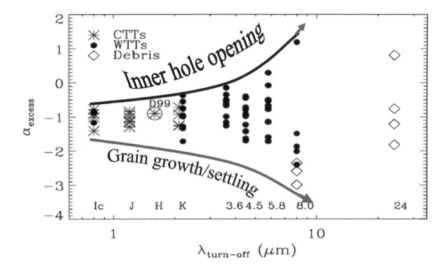

Figure 2. Distribution of excess slopes, α_{excess}, vs. the wavelength at which the infrared excess begins, ($\lambda_{turn-off}$, for the sample of WTTSs (filled circles) from Cieza et al. (2007), a sample of CTTSs in Chamaeleon (asterisk) from Cieza et al. (2005), the median SED of CTTSs in Taurus (marked as D99) from D'Alessio et al. (1999), and a sample of debris disks (diamonds) from Chen et al. (2005). The diagram shows a much larger spread in inner disk properties of WTTSs with respect to those of CTTSs. The spread in disk properties suggests that different physical processes dominate on different objects. Objects at the top of the figure are consistent with disk evolution dominated by the opening of an inner hole, while objects at the bottom of the figure are more consistent with evolution dominated by dust settling. (Figure adapted from Cieza et al. 2007).

bridge the gap observed between the CTTS and the debris disk regime. The ratio of the disk luminosity to the stellar luminosity, L_{DISK}/L_*, is a measurement of the fraction of the star's radiation that is intercepted and re-emitted by the disk plus any accretion luminosity. This quantity is intimately related to the evolutionary status of a circumstellar disk. On the one hand, the primordial, gas rich, disks around CTTSs have typical L_{DISK}/L_* values $> 10-20\%$, mostly because they have optically thick disks that intercept $\sim$10-20% of the stellar radiation. On the other hand, gas poor debris disks have optically thin disks that intercept a much smaller fraction of the star's light and thus have L_{DISK}/L_* values that range from 10^{-3} to 10^{-6} (Beichman et al. 2005).

The fact that WTTSs have L_{DISK}/L_* values intermediate between those of CTTSs and debris disks suggests that they are an evolutionary link between these two well studied stages. These rare WTTS disks are key for our understanding of disk evolution because they seem to trace the dissipation process as it rapidly occurs across the entire disk. The wide range of SED morphologies seen in WTTS disks suggests that the disk dissipation process does not follow the same path for every star. The diversity of SED morphologies of WTTS disks is quantified in Figure 2, which shows the slope of the IR excess, α_{excess}, against the wavelength at which the IR becomes significant, $\lambda_{turn-off}$. Objects in the top right of the figure are WTTS that have lost their short wavelength excess while keeping strong excesses at longer wavelengths. The SEDs of these objects are consistent with the formation of an inner hole in their disks (Calvet et al. 2002, 2005). This possible "inner hole opening" path is illustrated by the SEDs of the objects shown in the top row of Figure 3.

Objects in the bottom right of Figure 2 are WTTS in which the IR excess seems to have decreased simultaneously at every wavelength. These objects are more consistent with an evolution dominated by grain growth and dust settling. As the grains grow and settle into the mid-plane of the disk, the flared disk becomes flatter and hence intercepts a smaller fraction of the stellar radiation. As a result, a smaller excess is expected at every mid-IR wavelength (Dullemond & Dominik 2004). This possible "dust settling" path is illustrated by the SEDs shown in the bottom row of Figure 3.

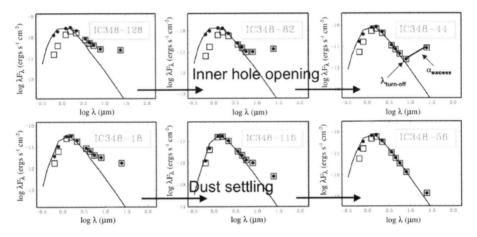

Figure 3. The SEDs of several WTTSs from Figure 2. The open boxes are observed fluxes, the filled circles are extinction corrected values, and the solid lines represent the stellar photospheres. The SEDs illustrate two possible evolutionary sequences, one dominated by the opening of an inner hole (top row) and the other dominated by dust settling (bottom row).

3.1. The Full Diversity of Transition Disks

It is important to note that while most CTTSs occupy a very restricted region of the α_{excess} vs. $\lambda_{turn-off}$ diagram, some CTTSs do show transition SEDs similar to those of WTTS disks. The best known examples of such CTTSs include TW Hydra (Calvet et al. 2002), DM Tau, and GM Aur (Calvet et al. 2005). These systems likely represent yet another path for disk dissipation, a path in which the inner regions of the disk become depleted of dust while gas is still accreting onto the star.

The standard YSO evolutionary class proposed by Lada (1987) and extended by Greene et al. (1994) is based on the slope, α, of the SED between 2 and 25 μm. However, this classification scheme can only capture a single evolutionary path in which α decreases monotonically as young stellar objects evolve from Class I into Class II, and then into Class III objects. The mid-IR data points from *Spitzer*'s camera IRAC (3.6, 4.5, 5.8, and 8.0 μm) sample the SEDs of YSOs at intermediate wavelengths between 2 and 25 μm and reveal that transition objects present a much larger diversity of SED morphologies than what can be described by the α classification. The α_{excess}–$\lambda_{turn-off}$ classification seems more appropriate when abundant mid-IR broad-band photometry is available, but it still does not uncover the full range of SED shapes presented by transition objects.

Spitzer's Infrared Spectrograph (IRS), which allows us to sample the SED of YSOs at hundreds of wavelengths between 5 and 38 μm, has recently revealed a new family of transitions objects whose SEDs show a distinctive "dip" around ∼10-20 μm (Brown et al. 2007). The SEDs of these objects can be modeled as disks having wide gaps with small inner radii (0.2-0.8 AU) and large outer radii (15-50 AU). The presence of small IR excess short-ward of 10 μm in these objects requires the presence of small amounts of dust (M $\sim 10^{-6}\,M_\odot$) in the inner disk, unlike objects such as Coku Tau/4 (D'Alessio et al. 2005) and DM Tau (Calvet et al. 2005), whose inner holes seem to be completely depleted of dust. Since the family of disks with SEDs suggestive of gaps contains both accreting objects (e.g., LkHα 330 and HD 135344) and not accreting objects (SR 21 and T Cha), it significantly expands the diversity of transition disks.

4. Disk Evolution Processes

The processes that drive disk evolution and the different pathways that disks follow as they evolve from optically thick primordial disks to optically thin debris disks are critical for our understanding of planet formation. These processes include accretion onto the star, photo-evaporation, dust settling and coagulation, and dynamical interactions with forming planets. In what follows, I review the main processes believed to control the evolution of circumstellar disks around low-mass stars.

4.1. Viscous Evolution

To first order, the evolution of primordial disks is driven by viscous accretion. Circumstellar material can only be accreted onto the star if it loses angular momentum. Conservation of angular momentum implies that, while most of the mass in the disk moves inward, some material should move outward, increasing the size of the initial disk. The source of the viscosity required for disk accretion and the mechanism by which angular momentum is transported remain a matter of intense debate. As a result, most viscous evolution models describe viscosity, ν, using the α parameterization introduced by Shakura & Sunyaev (1973), according to which $\nu = \alpha H_p C_s$, where H_p is the pressure scale height of the disk and C_s is the isothermal sound speed. The parameter α hides the uncertainties associated with the source of the viscosity and is often estimated to be of the order of 0.01.

Viscous evolution models (Hartmann et al. 1998; Hueso & Guillot 2005) are broadly consistent with the observational constraints for disk masses, disk sizes, and accretion rates as a function of time; however, they also predict a smooth, power-law, evolution of the disk properties. This smooth disk evolution is inconsistent with the very rapid disk dissipation ($\tau < 0.5$ Myr) that usually occurs after a much longer disk lifetime. Pure viscous evolution models also fail to explain the variety of SEDs observed in transition objects discussed above. These important limitations of the viscous evolution models suggest that they are in fact just a first-order approximation of a much more complex process.

4.2. Photo-evaporation

Recent disk evolution models, known as "UV-switch" models, combine viscous evolution with photo-evaporation by the central star (Clarke et al. 2001; Alexander et al. 2006) and are able to account for both the disk lifetimes of several million years and the short disk dissipation timescales ($\tau < 0.5$ Myr).

According to these models, extreme ultraviolet (EUV) photons originating in the accretion shock close to the stellar surface ionize and heat the circumstellar hydrogen to $\sim 10^4$ K. Beyond some critical radius, the thermal energy of the ionized hydrogen exceeds its escape velocity and the material is lost in the form of a wind.

At early stages in the evolution of the disk, the accretion rate dominates over the evaporating rate and the disk undergoes standard viscous evolution: material from the inner disk is accreted onto the star, while the outer disk behaves as a reservoir that resupplies the inner disk, spreading as angular momentum is transported outwards. Later on, as the accretion rate drops to the photo-evaporation rate, the outer disk is no longer able to resupply the inner disk with material. At this point, the inner disk drains on a viscous timescale and an inner hole is formed in the disk. Once this inner hole has formed, the EUV radiation very efficiently photo-evaporates the inner edge of the disk and the disk rapidly dissipates from the inside out. Thus, the UV-switch model naturally accounts for the lifetimes and dissipation timescales of disks as well as for SEDs of some PMS stars suggesting the presence of large inner holes.

Photo-evaporation, however, is not the only mechanism that has been proposed to explain the large opacity holes of some circumstellar disks. Dynamical interactions with planets and even grain growth can deplete the inner disk of small grains and result in SEDs virtually indistinguishable from those expected for photo-evaporating disks.

4.3. Dynamic Interaction with Planets

Since theoretical models of the dynamical interactions of forming planets with the disk (Lin & Papaloizous 1979, Artymowicz & Lubow 1994) predict the formation of inner holes and gaps, planet formation quickly became one of the most exciting explanations proposed for the inner holes of transition disks (Calvet et al. 2002; D'Alessio et al. 2005). Using a combination of hydrodynamical simulations and Monte Carlo radiative transport, Rice et al. (2003) model the broad-band SED of the transition disk around the CTTS GM Aur. They argue that the ~ 4 AU inner hole they infer for the disk could be produced by a ~ 2 $M_{Jupiter}$ mass planet orbiting at 2.5 AU. Modeling of the *Spitzer*-IRS data of GM Aur places the inner wall of the hole significantly farther away from the star, at 24 AU (Calvet et al. 2005); nevertheless, the planetary origin of the gap still remains one of the favored hypotheses (Najita et al. 2007).

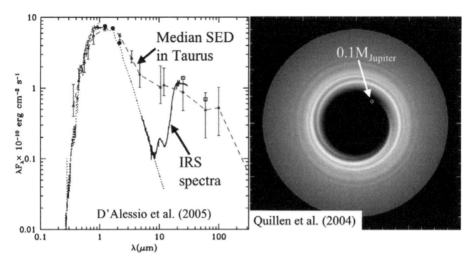

Figure 4. The left panel shows the spectral energy distribution of CoKu Tau/4 (the IRS spectra is from Forrest et al. 2004), compared to the median SED of CTTSs in Taurus (triangles and dashed line). The error bars of the median points represent the upper and lower quartiles of the Taurus CTTS SEDs (D'Alessio et al. 1999). (Figure adapted from D'Alessio et al. 2005). The right panel shows an hydrodynamic simulation of a planet disk system with a planet mass of $0.1 M_{Jupiter}$. This putative planet is suggested to be responsible for the evacuation of the inner disk of Coku Tau/4. (Figure taken from Quillen et al. 2004).

Similarly, hydrodynamic simulations (Quillen et al. 2005) suggest that the 10 AU hole inferred for the disk around the WTTS Coku Tau/4 (Figure 4, left panel) could be produced by the presence of a $\sim 0.1\, M_{Jupiter}$ mass planet orbiting very close to the edge of the hole's wall (Figure 4, right panel). However, the very low accretion rate and the low disk mass render the disk around Coku Tau/4 one of the prime candidates for ongoing photo-evaporation (Najita et al. 2007).

Given the extreme youth of some of the transition objects that have disks with inner holes (age ~ 1 Myr), the confirmation of the planetary origin of these holes would place very strong constraints on the time required for the formation of giant planets. In the not too distant future, ALMA will provide the sensitivity and angular resolution necessary to image the putative planets responsible for the observed holes (Narayanan et al. 2006).

4.4. Grain Growth

Even though the formation of a giant planet can account for the inner holes of some transitional disks, the planet formation process is not required to be far along in order to produce a similar effect in the SED of a circumstellar disk. In fact, once primordial sub-micron dust grains grow into somewhat larger bodies ($r \gg \lambda$), most of the solid mass never interacts with the radiation, and the opacity function, k_ν (cm^2/gr), decreases dramatically. Dullemond & Dominik (2005) model the settling and coagulation of dust in disks and investigate their effect on the resulting SEDs. They find that grain growth is a strong function of radius, it is more efficient in the inner regions where the surface density is higher and the dynamical timescales are shorter, and hence can produce opacity holes. As a result, grain growth, the very first stage of planet formation, can in principle mimic the effect of a fully grown giant planet on the SED of a YSO. Dullemond & Dominik (2005) also find that, in their models, grain growth is *too efficient* to be consistent with the observed persistence of IR excess over a timescale of a few Myrs. They conclude that small grains need to be replenished by fragmentation and that the size distribution of solid particles in a disk is likely to be the result of a complicated interplay between coagulation and fragmentation. As such, the overall importance of the dust coagulation process on the evolution of SEDs is still not well understood.

4.5. Inside-out Evacuation by Magneto-rotational Instability

The formation of an opacity hole, due to grain growth or the presence of a planet, can have not only a dramatic effect on the appearance of the SED of a transition object, but also on disk evolution itself. Chiang & Murray (2007) recently propose that, once an opacity hole of the order of 1-10 AU is formed in a disk, the inner rim of the disk will be rapidly drained from the inside out due to the onset of the magneto rotational instability (MRI). The MRI is one of the mechanisms that has been historically proposed to explain the accretion process in circumstellar disks. Nevertheless, it is believed that the gas in CTTSs is usually too cold and too weakly ionized for the MRI to be efficient (Hartman et al. 2006). Typical CTTS disks are too dusty for the stellar X-rays to penetrate deep enough into the disk to ionize material to the level required by the MRI. Chiang & Murray (2007) argue that an opacity hole in a disk produces the right conditions for the MRI to be activated and sustained. According to their models, the wider the rim, the larger the mass of the MRI-active region of the disk and the higher the accretion rate. Therefore, once the MRI is activated, the entire inner disk is rapidly evacuated, while the outer disk is photo-evaporated by UV radiation.

4.6. Models vs Observations

To date, most disk evolution models have treated each one of the processes discussed above independently of each other. In reality, it is clear that all these processes are likely to operate simultaneously and interact with one another. Hence, the relative importance of these processes for the overall evolution of disks still remains to be established. The diversity of SEDs morphologies discussed in Section 3 strongly suggests that disk evolution can follow different paths, each one of which is likely to be dominated by one or more different physical processes.

In order to investigate the relative incidence of different evolutionary paths, Najita et al. (2007) compare the observed accretion rates and disks masses of transition objects, which they define as objects with weak or no excess shortward of $10\,\mu m$ and strong excess at longer wavelengths, against the values expected for several disk evolution models. They find that transition disks occupy a restricted region of the accretion rate vs. disk mass plane. In particular, they find that transition objects tend to have significantly lower accretion rates, for a given disk mass, than non-transition objects and also tend to have larger median disk masses than regular T Tauri disks.

Najita et al. (2007) argue that most of the objects in their sample have properties consistent with a scenario in which a jovian mass planet has created a gap that isolates the inner disk from a still massive outer disk (Lubow & D'Angelo 2006; Varniere et al. 2006) and suppresses accretion onto the star. They also argue that a minority of their objects have both small accretion rates and small disk masses that make them more consistent with the photo-evaporation model. Finally, they propose that the lack of transition objects with large accretion rates and large disk masses implies that grain growth and the formation of planetesimals in the inner disk is an unlikely explanation for the opacity holes of transition disks. While very provocative, the results of Najita et al. (2007) are based on sample of only 12 transition objects. Fortunately, *Spitzer* is likely to increase the number of known transition disks by an order of magnitude by the end of its mission. Follow-up observations of these objects, required to obtain accretion rates and disk masses, will soon allow to extend the study of Najita et al. (2007) to a much larger sample.

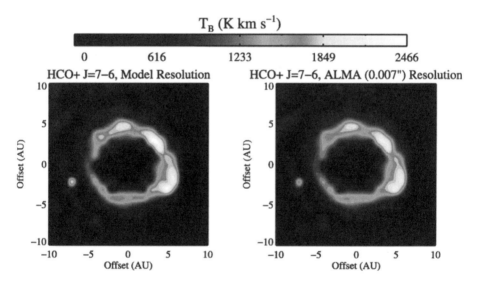

Figure 5. The left panel shows the synthesized image of the HCO+ (J=7-6) transition for a hydrodynamic snapshot of the vicinity of a self gravitating proto-giant planet with a mass of $1.4\,M_{Jupiter}$. The right panel is a simulated ALMA image at $0.007''$ resolution (corresponding to the most extended baseline of ALMA, Bastian 2002) for a source at a distance of 140 pc. The forming giant planet is at 8 o'clock in each panel. The intensity is in units of K km/s and is on a fixed scale for the entire figure. (Figure taken from Narayanan et al. 2006).

5. Conclusions and the Future

The many transition disks discovered by *Spitzer* are likely to be the subjects of many follow-up observations and studies for years to come. These are objects that are undergoing rapid transformations and have properties that place them somewhere in between two much better defined evolutionary states: those of regular CTTSs and debris disks. Understanding the physical processes responsible for the diversity of transition disks and its implications for planet formation will be one of main challenges for the field.

Another key outstanding disk evolution issue is the question of when the transition from the primordial to the debris disk stage actually occurs. As mentioned in Section 3, some of the WTTS disks have very small fractional luminosities and thus seem to be optically thin. These objects *could* be younger analogs of the β Pic and AU Mic debris disks, and thus some of the youngest debris disks ever observed. However, this interpretation depends on the assumption that these young WTTS disks are gas poor. It is also possible that some of the WTTS disks have very low fractional disk luminosities as a result of most of their grains growing to sizes $\gg$10-20 μm, in which case they would still be primordial disks. Since a real debris disks requires the presence of second generation of dust produced by the collision of much larger objects, the ages of the youngest debris disks can constrain the time it takes for a disk to form planetesimals.

It has even been proposed that detectable second generation dust is not produced until Pluto-sized objects form and trigger a collision cascade (Dominik & Decin , 2003). In that case, the presence of debris disks around $\sim$1-3 Myrs old stars would have even stronger implications for planet formation theories. The fact that, as discussed in Section 3, the incidence of 24 μm excesses around low-mass PMS stars with ages $\sim$10 Myr is $\sim$0 %, is consistent with the idea that a quiescent period exists between the dissipation/coagulation of the primordial dust and the onset of the debris phenomenon. However, the confirmation of such a scenario will require far-IR observations sensitive enough to detect, at the distance of the nearest star-forming regions (150-200 pc), debris disks as faint as those observed in the solar neighborhood (10-30 pc). In the near future, *Herschel* will provide such sensitivity. It will also provide the information on the gas content and grain size distribution of optically thin disks around WTTS required to establish whether they represent the end of the primordial disk phase or the beginning of the debris disk stage.

Even though much still remains to be learned about planet formation from the study of circumstellar disks, ultimately, we would like to be able to directly observe planets as they form. ALMA could achieve this milestone in the next decade. With its most extended baseline, ALMA will have a resolution of 0.007″ at 625 GHz. This corresponds to an angular resolution of less than 1 AU at the distance of nearby star forming regions like Ophiuchus and Taurus. Narayanan et al. (2006) model the molecular emission line from a gravitationally unstable protoplanetary disk (Figure 5, left panel) and conclude that forming giant planets could be directly image-able using dense gas tracers such as HCO+ (Figure 5, right panel). Such observations could provide the ultimate tests needed to distinguish among the competing theories of planet formation.

Acknowledgments. Support for this work was provided by NASA through the *Spitzer* Fellowship Program Under an award from Caltech. I would like to thank Trent Dupuy, Jonathan Swift, and Michael Liu for their helpful suggestions.

References

Alexander, R. D., Clarke, C. J., & Pringle, J. E. 2006, MNRAS, 369, 229
Alibert, Y., Mordasini, C., & Benz, W. 2004, A&A, 417, L25
Andrews, S. M., & Williams, J. P. 2005, ApJ, 631, 1134
Andrews, S. M., & Williams, J. P. 2007, Ap&SS, 363
Artymowicz, P., & Lubow, S. H. 1994, Circumstellar Dust Disks and Planet Formation, 339
Bastian, T. S. 2002, Astronomische Nachrichten, 323, 271
Beckwith, S. V. W. 1999, NATO ASIC Proc. 540: The Origin of Stars and Planetary Systems, 579
Beichman, C. A., et al. 2002, Science with the Space Interferometry Mission, 1
Beichman, C. A., et al. 2005, ApJ, 622, 1160
Boss, A. P. 2000, ApJ, 536, L101
Brown, J. M., et al. 2007, ApJ, 664, L107
Calvet, N., D'Alessio, P., Hartmann, L., Wilner, D., Walsh, A., & Sitko, M. 2002, ApJ, 568, 1008
Calvet, N., et al. 2005, ApJ, 630, L185
Chiang, E., & Murray-Clay, R. 2007, Nature Physics, 3, 604
Cieza, L. A., Kessler-Silacci, J. E., Jaffe, D. T., Harvey, P. M., & Evans, N. J., II 2005, ApJ, 635, 422
Cieza, L., et al. 2007, ApJ, 667, 308
Clarke, C. J., Gendrin, A., & Sotomayor, M. 2001, MNRAS, 328, 485
Clarke, C. J., & Pringle, J. E. 2006, MNRAS, 370, L10
D'Alessio, P., et al. 2005, ApJ, 621, 461
Dominik, C., & Decin, G. 2003, ApJ, 598, 626
Dullemond, C. P., & Dominik, C. 2004, Extrasolar Planets: Today and Tomorrow, 321, 361
Dullemond, C. P., & Dominik, C. 2005, A&A, 434, 971
Forrest, W. J., et al. 2004, ApJS, 154, 443
Greene, T. P., Wilking, B. A., Andre, P., Young, E. T., & Lada, C. J. 1994, ApJ, 434, 614
Haisch, K. E., Jr., Lada, E. A., & Lada, C. J. 2001, ApJ, 553, L153
Hartigan, P., Edwards, S., & Ghandour, L. 1995, ApJ, 452, 736
Hartmann, L., Calvet, N., Gullbring, E., & D'Alessio, P. 1998, ApJ, 495, 385
Hayashi, C. 1981, Progress of Theoretical Physics Supplement, 70, 35
Hillenbrand, L. A., 2006, A Decade of Discovery: Planets Around Other Stars (eds. M. Livio, STScI Symposium Series #19
Hueso, R., & Guillot, T. 2005, A&A, 442, 703
Lada, C. J. 1987, Star Forming Regions, IAU Symposium 115, 1
Lin, D. N. C., & Papaloizou, J. 1979, MNRAS, 188, 191
Lissauer, J. J. 1993, ARA&A, 31, 129
Lubow, S. H., & D'Angelo, G. 2006, ApJ, 641, 526
McCaughrean, M. J., & O'dell, C. R. 1996, AJ, 111, 1977
Najita, J. R., Strom, S. E., & Muzerolle, J. 2007, MNRAS, 378, 369
Narayanan, D. T., et al. 2006, Bulletin of the American Astronomical Society, 38, 1209
Osterloh, M., & Beckwith, S. V. W. 1995, ApJ, 439, 288
Padgett, D. L., et al. 2006, ApJ, 645, 1283
Pollack, J. B., Hubickyj, O., Bodenheimer, P., Lissauer, J. J., Podolak, M., & Greenzweig, Y. 1996, Icarus, 124, 62
Quillen, A. C., Varnière, P., Minchev, I., & Frank, A. 2005, AJ, 129, 2481

Rice, W. K. M., Wood, K., Armitage, P. J., Whitney, B. A., & Bjorkman, J. E. 2003, MNRAS, 342, 7
Shakura, N. I., & Sunyaev, R. A. 1973, X- and Gamma-Ray Astronomy, 55, 155
Shu, F. H., Adams, F. C., & Lizano, S. 1987, ARA&A, 25, 23
Silverstone, M. D., et al. 2006, ApJ, 639, 1138
Skrutskie, M. F., Dutkevitch, D., Strom, S. E., Edwards, S., Strom, K. M., & Shure, M. A. 1990, AJ, 99, 1187
Strom, K. M., Strom, S. E., Edwards, S., Cabrit, S., & Skrutskie, M. F. 1989, AJ, 97, 1451
Varnière, P., Bjorkman, J. E., Frank, A., Quillen, A. C., Carciofi, A. C., Whitney, B. A., & Wood, K. 2006, ApJ, 637, L125
Youdin, A. N., & Shu, F. H. 2002, ApJ, 580, 494
Wolk, S. J., & Walter, F. M. 1996, AJ, 111, 2066

Kelle Cruz explains the fine line between brown dwarfs and planets.

New Horizons in Astronomy: Frank N. Bash Symposium 2007
ASP Conference Series, Vol. 393, © 2008
A. Frebel, J. R. Maund, J. Shen, and M. H. Siegel, eds.

Connecting the Dots: Low-Mass Stars, Brown Dwarfs, and Planets

Kelle Cruz

Department of Astronomy, California Institute of Technology, Pasadena, CA, USA

Abstract. The lowest mass object that Mother Nature makes through the process of "star formation" is currently unknown. While numerous very low-mass stars, brown dwarfs, and planets have been found, their relation to each other remains unclear. Here I describe how the study of brown dwarfs has the potential to help us understand both star and planet formation mechanisms. I describe the physical traits attributed to stars, brown dwarfs, and planets; compare the mass functions of brown dwarfs and planets; and discuss how studies of brown dwarfs in both young clusters and in the field can be used to challenge and constrain star and planet formation theories.

1. What are they and what do they look like?—Brown Dwarf Introduction

Over the past decade, there have been great advances in the exploration of the low-mass end of the main sequence including the discovery of brown dwarfs—star-like objects that are not massive enough to maintain hydrogen burning in their core. Two new spectral classes cooler than M (2200–4000 K) have been defined and characterized: the L (1400–2100 K) and T dwarfs (700–1300 K) (Kirkpatrick 2005, and references therein).

An artist rendition of three brown dwarfs compared to the Sun and Jupiter is shown in Figure 1. Due to the competing effects of coloumb repulsion ($R \propto M^{1/3}$) and electron degeneracy ($R \propto M^{-1/3}$) all very low-mass stars and brown dwarfs have a radius of ~ 1 $R_{Jupiter}$. Also similar to Jupiter, L dwarf photospheres are dominated by condensate clouds. There is still a large temperature gap between the coolest observed T dwarf and Jupiter. Since brown dwarfs cool with time, there is good reason to expect a class of objects cooler than the T dwarfs likely comprised of the least massive and oldest brown dwarfs (see Figure 2. This class has been tentatively dubbed "Y" but no candidates have yet been found.

Unlike stars, brown dwarfs gradually cool and evolve down the spectral sequence as shown in Figure 2. A 0.075 $M_\odot$ object, just below the hydrogen-burning limit, will start as a mid-M dwarf at 3100 K, but after 10 Gyr, it will be a late-L dwarf at 1300 K. Similarly, a 7 $M_{Jupiter}$ (0.007 $M_\odot$) object that is first visible as an early-L dwarf at 2200 K is only 400 K after 10 Gyr—significantly cooler than the latest T dwarf. Thus, there is no mass-luminosity relation for brown dwarfs as there is for stars. In addition, not all M and L dwarfs are brown dwarfs. Current theories suggest that all objects cooler than about spectral type

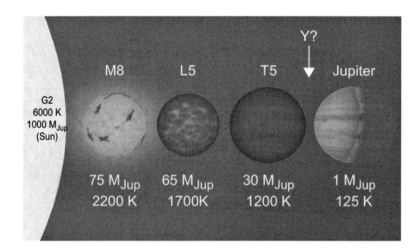

Figure 1. Artist rendition of the Sun, a late-M dwarf, an L dwarf, a T dwarf, and Jupiter. The M, L, and T dwarfs are shown at the age of 1 Gyr and all of the objects are shown on the same physical scale. Illustration by Dr. Robert Hurt of the Infrared Processing and Analysis Center.

L4 are brown dwarfs while mid-M to early-L dwarfs are a mix of stars and young brown dwarfs (Burrows et al. 2001).

While brown dwarfs were first hypothesized to exist in the 1960s (Kumar 1963a,b; Hayashi & Nakano 1963), it was not until the 1980s and the advent if infrared-sensitive CCDs that late-M dwarfs began to be discovered in significant numbers (Reid & Gilmore 1981; Probst & Liebert 1983; Hawkins & Bessell 1988; Bessell 1991). At the time, these objects were thought to be very low-mass stars, not brown dwarfs. It turns out the first L dwarf, GD 165B, was discovered in 1988, but nobody was sure what it was (Becklin & Zuckerman 1988; Kirkpatrick et al. 1993).

Finally in 1995, two objects were identified that were generally accepted to be the first brown dwarfs. Gl 299B was identified and with strong CH_4 absorption bands (the signature feature of the T dwarf class), was confirmed as the first brown dwarf found in the field and the first T dwarf (Nakajima et al. 1995; Oppenheimer et al. 1995). Tiede 1 is a late-type M dwarf with lithium absorption discovered in the Pleiades. Based on the measured abundance of lithium and the age of the Pleiades, Tiede 1 was the first young brown dwarf found (Rebolo et al. 1995, 1996). The spectral sequence was formally extended to include types L and T in 1999 by Kirkpatrick et al.. As of October 2007 there were 500 L dwarfs and 120 T dwarfs listed on the brown dwarf online repository.[1]

Since the discovery of brown dwarfs, there has been substantial debate surrounding the formation of brown dwarfs and several contentious assertions have been made:

- *Brown dwarfs must form differently than stars because they are so much lighter than the Jeans Mass.* Star formation theory cannot easily produce low mass objects and as a result, other theories have been put forward to explain the existence of low-mass stars and brown dwarfs.

- *Objects below 13 $M_{Jupiter}$ could not possibly have formed as stars and therefore must be planets.* The Shu et al. (1987) model for low-mass star formation relies on deuterium burning to start convection which then produces the stellar winds necessary to halt the infall of the collapsing proto-star. As shown in Figure 2, objects more massive than 13 $M_{Jupiter}$ (solid lines) burn deuterium causing their temperatures at young ages to plateau; objects with masses below 13 $M_{Jupiter}$ (dashed lines) do not burn deuterium and cool rapidly during their first 50 Myr. As a result, it has been proposed that a mass of 13 $M_{Jupiter}$ be the delineation between planets and brown dwarfs.

- *Brown dwarfs are proto-star cores whose accretion was halted due to being ejected from the nebula.* This theory, proposed by Reipurth & Clarke (2001), provides a different scenario for halting the accretion onto the proto-star and predicts truncated and/or missing disks around brown dwarfs.

One discovery in particular has challenged brown dwarf formation theories and shed light on properties that might be useful for distinguishing planets from brown dwarfs. Gizis (2002) identified 2MASS J12073346−3932539 (hereafter 2M1207−39), a M8 brown dwarf in the ∼10 Myr-old TW Hydrae Association (TWA) with an estimated mass of ∼25 $M_{Jupiter}$ (Mohanty & Basri 2003). Chauvin et al. (2004) detected a ∼5–10 $M_{Jupiter}$ mid-to-late L dwarf companion (hyped as the first direct detection of an exo-planet) with a ∼70 AU separation from the primary. Also, the primary has a disk, is actively accreting, and does not support the ejection model for brown dwarf formation (Riaz et al. 2006; Mohanty et al. 2007). The planetary-mass secondary could not have formed in the disk of the primary due to the wide separation and the (relatively) large mass of the secondary (compared to the mass expected to be in the primary's disk). In most respects, this system appears to have formed in the same way as a binary star system even though the secondary is well below the 13 $M_{Jupiter}$ deuterium burning limit. The 2M1207−39 system has demonstrated: 1) the mass and temperature regimes of planets and brown dwarfs indeed overlap; 2) observables such as separation and mass ratio point toward either a planet or star

---

[1]http://www.dwarfarchives.org

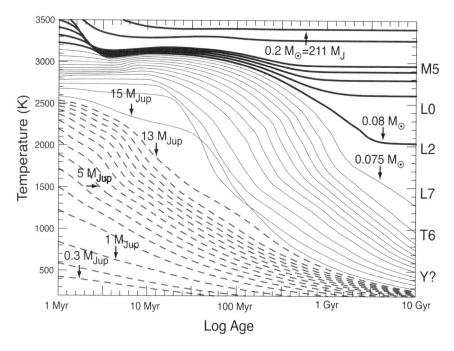

Figure 2. Effective temperature evolution of low-mass stars (*thick lines*), brown dwarfs (*thin lines*) and planetary-mass brown dwarfs (*dashed lines*). Adapted with permission from Burrows et al. (2001). Copyright by the American Physical Society (DOI: 10.1103/RevModPhys.73.719).

formation scenario; and 3) the disks and companions of nearby, intermediate-age brown dwarfs might play an important role in determining the range of properties spanned by planetary-mass objects regardless of formation method.

2. How many are there?—Brown Dwarf and Planet Luminosity and Mass Functions

The field luminosity function of the Solar Neighborhood (8 and 20 pc) is shown in the left panel of Figure 3. M dwarfs are by far the most numerous stellar constituents of the Solar Neighborhood. The L dwarf luminosity function (shaded) is composed of both stars and brown dwarfs and has only recently been measured (Cruz et al. 2003, 2007). These new data show that the luminosity function continues to decline sharply beyond $M_J = 10$ and reaches a minimum at $M_J \sim 13$. Formally, these results indicate that the space densities remain constant at fainter magnitudes, however, since the measurements are lower limits (as indicated by arrows) the luminosity function likely increases for $M_J > 14$.

Simulated luminosity functions with different underlying mass functions for brown dwarfs and the lowest-mass stars from Allen et al. (2005) are shown in

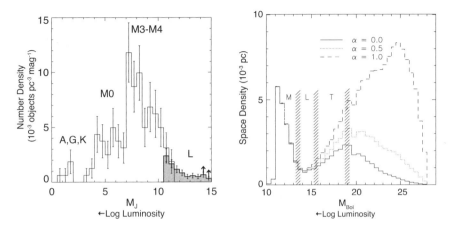

Figure 3. *Left*: J-band luminosity function of the Solar Neighborhood based on an 8-pc sample (Reid et al. (2003), *unshaded*) and a 20-pc sample for the coolest objects (Cruz et al. (2007), *shaded*). The last two magnitude bins are incomplete and the lower limits on the space densities are indicated with arrows. *Right*: Simulated luminosity functions of low-mass stars and brown dwarfs derived assuming underlying mass functions with different α parameters. Recent T dwarf space densities indicate $\alpha \sim 0.5$. Adapted from Allen et al. (2005) and reproduced by permission of the American Astronomical Society (DOI: 10.1086/429548).

the right panel of Figure 3. Unfortunately, the L dwarf luminosity function does not constrain the underlying mass function and T dwarf space densities are required. Estimates of T dwarf space densities indicate $\alpha \sim 0.5$ (Burgasser et al., in preparation) and imply that Y dwarfs are not very numerous. Reliable discrimination between different models for the underlying mass function must await observational surveys that probe brown dwarfs at temperatures below 600 K, likely near the boundary between T and Y dwarfs.

The overall morphology of the J-band luminosity function for $M_J < 10$ (unshaded histogram in the left panel of Figure 3) reflects the convolution of the underlying mass function and the M_J-mass relation. Qualitatively, the luminosity function increases for $0 < M_J < 7$ since the mass function increases with decreasing mass ($\alpha = 2.35$ at high masses and 1 at lower masses). At fainter magnitudes, the J-band luminosity function turns over not because the mass function changes drastically, but because the slope of the M_J-mass relation changes; while $\delta \text{mass}/\delta M_J \sim 0.4 M_\odot$ mag^{-1} for $M_J < 7$, $\delta \text{mass}/\delta M_J \sim 0.07 M_\odot$ mag^{-1} for $7 < M_J < 10$ (Delfosse et al. 2000).

The morphology of the luminosity function for $M_J > 10$ (right panel of Figure 3 and the shaded histogram in the left panel) is due to the mix of stars and brown dwarfs in the ultracool regime and is qualitatively in accord with theoretical expectations. The drop in number density with increasing magnitude reflects a further contraction in $\delta \text{mass}/\delta M_J$. The population of the very lowest mass stars that appear as L dwarfs span an extremely small range in mass

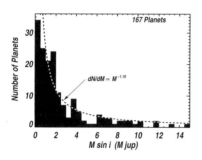 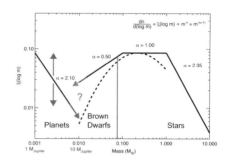

Figure 4. *Left:* Minimum mass distribution of the 167 known radial velocity planets. Originally in Butler et al. (2006) and reproduced by permission of the American Astronomical Society (DOI: 10.1086/504701). *Right:* Mass functions of stars, brown dwarfs, and planets represented as segmented power-laws (*solid black lines*) and the lognormal system mass function found by Chabrier (2003) (*dashed line*). The lowest mass brown dwarf, the highest mass planet, and the absolute scale for the planet mass function are unknown.

($0.075 < M < 0.085 M_\odot$) and, as a result, are rare. The brown dwarfs in this effective temperature regime are relatively young and are at the high-mass extreme, near the hydrogen-burning limit. Brown dwarfs dominate the counts beyond $M_J > 13.5$, and the upturn in number densities reflects the slowdown in cooling rates at lower temperatures. For example, a 0.07 $M_\odot$ brown dwarf takes 2.7 Gyr to evolve down the L dwarf sequence, but remains a (cooling) T dwarf (T_{eff} 1400–600 K) for 30 Gyr, or more than two Hubble times; a low-mass, 0.025 $M_\odot$ brown dwarf spends only 120 Myr as an L dwarf, but 1.5 Gyr as a T dwarf (Burrows et al. 2001).

The mass distribution of radial velocity planets is shown in the left panel of Figure 4. Despite the numerous biases and selection effects, the mass function of planets is strongly increasing toward smaller masses. The best estimates for the mass functions of stars, brown dwarfs, and planets is shown in the right panel of Figure 4 (solid lines). The absolute normalization of the planet mass function is not yet known: Are there more or less planets than brown dwarfs? Also unknown are the masses of the highest-mass planet and the lowest-mass brown dwarf, however it is likely that the mass regimes overlap. The very different mass distributions of the radial velocity planets and brown dwarfs (increasing versus decreasing toward lower mass) indicate two independent populations with different formation mechanisms and probably other observable differences.

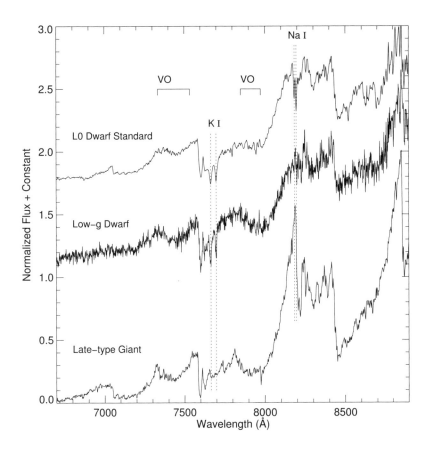

Figure 5. Optical spectra of a normal L0 dwarf, a low-gravity dwarf, and a late-type giant (*top-to-bottom*). The VO, K I, and Na I absorption strengths of the low-gravity dwarf are between the normal dwarf and the giant.

3. What can they tell us?—A New Population of Juvenile Brown Dwarfs

Until recently, studies of brown dwarfs (BD) have largely focused on two stages of evolution: the very young ($\sim$1 Myr, e.g., Taurus) and the mature ($\gtrsim$1 Gyr). Brown dwarfs in young clusters are studied because they are still fairly luminous (typically M type) and the age of the cluster can be adopted for the brown dwarf. A major drawback of these clusters is their rather large distance from the Sun ($\sim$100–500 pc) and reliably identifying the lowest-mass components of these clusters has proven to be a significant observational challenge.

We believe that we have uncovered a nearby ($\sim$30–60 pc), juvenile (5–50 Myr) population of brown dwarfs that mediate the current bi-polar situation. The youth of our targets is inferred from the presence of conspicuous low-gravity features in their optical and/or near-infrared spectra not seen in hundreds of other M and L dwarfs; one example is shown in Figure 5. Compared to old field dwarfs of similar spectral type (equivalent temperature), low gravity is indicative of both a lower mass and larger radius—hallmarks of young brown dwarfs still undergoing gravitational contraction. The spectral features present in our targets are similar to those seen in members of very young clusters (e.g., Taurus) but since none of our targets are near any tightly bound groups, they are most certainly older than 1 Myr. The upper limit on our age estimate is based on the stronger low-gravity spectral features exhibited in our targets than those seen in members of 100 Myr old clusters (e.g., Pleiades).

Figure 6 shows the location of our candidate young brown dwarfs (five-pointed stars) with confirmed members of the 8–50 Myr nearby associations AB Dor, β Pic, Tuc/Hor, and TWA as identified by Zuckerman & Song (2004). The spatial distributions of the two populations, widely distributed and clumped in the south, are suggestively similar. This is not too surprising since the age and distance estimates of our young brown dwarfs are consistent with those of the moving groups.

Our new-found population of brown dwarfs with older ages has the potential to lend insight on disk evolution and planet formation. It is now known that it is not unusual for young brown dwarfs to harbor disks and there is evidence that brown dwarf disks are longer lived than those of more massive stars (Carpenter et al. 2006; Scholz et al. 2007). Any disks found around juvenile objects are particularly interesting because their age is coincident with the epoch of planet formation (10–30 Myr). Our candidates are currently being targeted with Spitzer IRAC and 24 μm imaging to investigate the frequency and properties of brown dwarf disks at juvenile ages.

The new population is also ideal for searching for planetary-mass companions — counterparts to the 2M1207−39 system. Tight systems more easily resolved due to their relative proximity (within 60 pc). Additionally, since the objects are still fairly young, they have not cooled too much and are still relatively bright compared to their older counterparts of the same mass.

To confirm that our candidates are both young and members of the southerly associations, significant follow-up observations are being undertaken. High signal-to-noise spectra covering 0.8–2.5 μm is being compiled for all of the young candidates in order to fully study the low-gravity spectral features. Proper motions, radial velocities and trigonometric parallaxes are being obtained in order to derive accurate space motions and determine cluster membership.

Acknowledgments. The results discussed are a result of efforts undertaken with my collaborators Adam Burgasser, Jackie Faherty, Davy Kirkpatrick, Dagny Looper, Eric Mamajek, Subhanjoy Mohanty, Lisa Prato, and Neill Reid. Support for this work was provided by NASA through the Spitzer Space Telescope Fellowship Program, through a contract issued by the Jet Propulsion Laboratory, California Institute of Technology under a contract with NASA. I would particularly like to thank the conference organizers and the other invited speakers for their hard work and for making the conference a success. Many thanks to the McDonald Observatory and Astronomy Department Board of Visitors for their generous support of this unique and symposium—I am very grateful to have been a part of it.

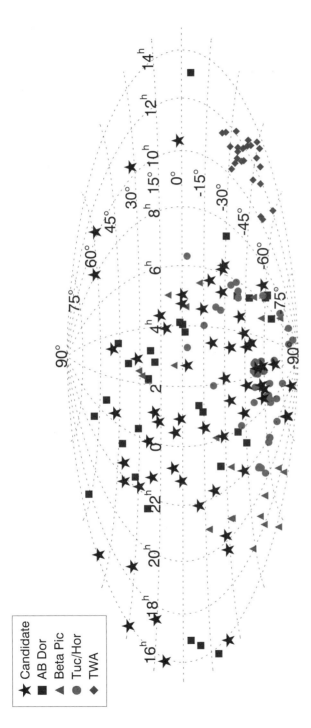

Figure 6. Celestial distribution of young brown dwarf candidates (*five-pointed stars*) on the sky. Also shown are known members of the AB Dor (*squares*), β Pic (*triangles*), Tuc/Hor (*circles*), and TWA (*diamonds*) young stellar associations (Zuckerman & Song 2004).

References

Allen, P. R., Koerner, D. W., Reid, I. N., & Trilling, D. E. 2005, ApJ, 625, 385
Becklin, E. E., & Zuckerman, B. 1988, Nature, 336, 656
Bessell, M. S. 1991, AJ, 101, 662
Burrows, A., Hubbard, W. B., Lunine, J. I., & Liebert, J. 2001, Reviews of Modern Physics, 73, 719
Butler, R. P., Wright, J. T., Marcy, G. W., Fischer, D. A., Vogt, S. S., Tinney, C. G., Jones, H. R. A., Carter, B. D., Johnson, J. A., McCarthy, C., & Penny, A. J. 2006, ApJ, 646, 505
Carpenter, J. M., Mamajek, E. E., Hillenbrand, L. A., & Meyer, M. R. 2006, ApJ, 651, L49
Chabrier, G. 2003, ApJ, 586, L133
Chauvin, G., Lagrange, A.-M., Dumas, C., Zuckerman, B., Mouillet, D., Song, I., Beuzit, J.-L., & Lowrance, P. 2004, A&A, 425, L29
Cruz, K. L., Reid, I. N., Kirkpatrick, J. D., Burgasser, A. J., Liebert, J., Solomon, A. R., Schmidt, S. J., Allen, P. R., Hawley, S. L., & Covey, K. R. 2007, AJ, 133, 439
Cruz, K. L., Reid, I. N., Liebert, J., Kirkpatrick, J. D., & Lowrance, P. J. 2003, AJ, 126, 2421
Delfosse, X., Forveille, T., Ségransan, D., Beuzit, J.-L., Udry, S., Perrier, C., & Mayor, M. 2000, A&A, 364, 217
Gizis, J. E. 2002, ApJ, 575, 484
Hawkins, M. R. S., & Bessell, M. S. 1988, MNRAS, 234, 177
Hayashi, C., & Nakano, T. 1963, Progress of Theoretical Physics, 30, 460
Kirkpatrick, J. D. 2005, ARA&A, 43, 195
Kirkpatrick, J. D., Henry, T. J., & Liebert, J. 1993, ApJ, 406, 701
Kirkpatrick, J. D., Reid, I. N., Liebert, J., Cutri, R. M., Nelson, B., Beichman, C. A., Dahn, C. C., Monet, D. G., Gizis, J. E., & Skrutskie, M. F. 1999, ApJ, 519, 802
Kumar, S. S. 1963a, ApJ, 137, 1126
—. 1963b, ApJ, 137, 1121
Mohanty, S., & Basri, G. 2003, ApJ, 583, 451
Mohanty, S., Jayawardhana, R., Huélamo, N., & Mamajek, E. 2007, ApJ, 657, 1064
Nakajima, T., Oppenheimer, B. R., Kulkarni, S. R., Golimowski, D. A., Matthews, K., & Durrance, S. T. 1995, Nature, 378, 463
Oppenheimer, B. R., Kulkarni, S. R., Matthews, K., & Nakajima, T. 1995, Science, 270, 1478
Probst, R. G., & Liebert, J. 1983, ApJ, 274, 245
Rebolo, R., Martin, E. L., Basri, G., Marcy, G. W., & Zapatero-Osorio, M. R. 1996, ApJ, 469, L53+
Rebolo, R., Zapatero-Osorio, M. R., & Martin, E. L. 1995, Nature, 377, 129
Reid, I. N., Cruz, K. L., Laurie, S. P., Liebert, J., Dahn, C. C., Harris, H. C., Guetter, H. H., Stone, R. C., Canzian, B., Luginbuhl, C. B., Levine, S. E., Monet, A. K. B., & Monet, D. G. 2003, AJ, 125, 354
Reid, L. N., & Gilmore, G. 1981, MNRAS, 196, 15P
Reipurth, B., & Clarke, C. 2001, AJ, 122, 432
Riaz, B., Gizis, J. E., & Hmiel, A. 2006, ApJ, 639, L79
Scholz, A., Jayawardhana, R., Wood, K., Meeus, G., Stelzer, B., Walker, C., & O'Sullivan, M. 2007, ApJ, 660, 1517
Shu, F. H., Adams, F. C., & Lizano, S. 1987, ARA&A, 25, 23
Zuckerman, B., & Song, I. 2004, ARA&A, 42, 685

Anna Frebel advocates for the oldest stars.

Metal-poor Stars

Anna Frebel[1]

McDonald Observatory, University of Texas, Austin, TX, USA

Abstract. The abundance patterns of metal-poor stars provide a wealth of chemical information about various stages of the chemical evolution of the Galaxy. In particular, these stars allow us to study the formation and evolution of the elements and the involved nucleosynthetic processes. This knowledge is invaluable for our understanding of cosmic chemical evolution and the onset of star- and galaxy formation. Metal-poor stars are the local equivalent of the high-redshift Universe, and offer crucial observational constraints on the nature of the first stars. This review presents the history of the first discoveries of metal-poor stars that laid the foundation to this field. Observed abundance trends at the lowest metallicities are described, as well as particular classes of metal-poor stars such as r-process and C-rich stars. Scenarios on the origins of the abundances of metal-poor stars and the application of large samples of metal-poor stars to cosmological questions are discussed.

1. Introduction

The first stars that formed from the pristine gas left after the Big Bang were very massive (Bromm et al. 2002). After a very short life time these co-called Population III stars exploded as supernovae, which then provided the first metals to the interstellar medium. All subsequent generations of stars formed from chemically enriched material. Metal-poor stars that are observable today are Population II objects and belong to the stellar generations that formed from non-zero metallicity gas. In their atmospheres these objects preserve information about the chemical composition of their birth cloud. They thus provide archaeological evidence of the earliest times of the Universe. In particular, the chemical abundance patterns provide detailed information about the formation and evolution of the elements and the involved nucleosynthesis processes. This knowledge is invaluable for our understanding of the cosmic chemical evolution and the onset of star- and galaxy formation. Metal-poor stars are the local equivalent of the high-redshift Universe. Hence, they also provide us with observational constraints on the nature of the first stars and supernovae. Such knowledge is invaluable for various theoretical works on the early Universe.

Focusing on long-lived low-mass ($\sim 0.8\ M_\odot$) main-sequence and giant metal-poor stars, we are observing stellar chemical abundances that reflect the composition of the interstellar medium during their star formation processes. Main-sequence stars only have a shallow convection zone that preserves the stars' birth composition over billions of years. Stars on the red giant branch have

---

[1] W. J. McDonald Fellow

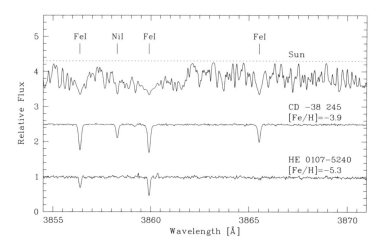

Figure 1. Spectral comparison of the Sun with the metal-poor stars CD −38° 245 and HE 0107−5240. Figure taken from Christlieb et al. (2004).

deeper convection zones that lead to a successive mixing of the surface with nuclear burning products from the stellar interior. In the lesser evolved giants the surface composition has not yet been significantly altered by any such mixing processes. The main indicator used to determine stellar metallicity is the iron abundance, [Fe/H], which is defined as [A/B]= $\log_{10}(N_A/N_B)_\star - \log_{10}(N_A/N_B)_\odot$ for the number N of atoms of elements A and B, and $\odot$ refers to the Sun. With few exceptions, [Fe/H] traces the overall metallicity of the objects fairly well.

To illustrate the difference between younger metal-rich and older metal-poor stars, Figure 1 shows spectra of the Sun and two of the most metal-poor stars, CD −38° 245 (Bessell & Norris 1984) and HE 0107−5240 (Christlieb et al. 2002). The number of atomic absorption lines detected decreases with declining metallicity. In HE 0107−5240, only the intrinsically strongest metal lines are left to observe. Compared with the metal-poor stars presented in the figure, a spectrum of a similarly unevolved Population III object would have no metal features since it contains no elements other than H, He and Li at its surface.

Large numbers of metal-poor Galactic stars found in objective-prism surveys in both hemispheres have provided enormous insight into the formation and evolution of our Galaxy (e.g., Beers & Christlieb 2005). However, there is only a very small number of stars known with [Fe/H] < -3.5. The number of known metal-poor stars decreases significantly with decreasing metallicity as illustrated in Figure 2. Objects at the lowest metallicities are extremely rare but of utmost importance for a full understanding of the early Universe. Abundance trends are poorly defined in this metallicity range, and the details of the metallicity distribution function (MDF) at the lowest metallicity tail remains unclear. A major recent achievement in the field was the push down to a new, significantly lower limit of [Fe/H] measured in a stellar object: From a longstanding [Fe/H] $= -4.0$ (CD −38° 245; Bessell & Norris 1984) down to [Fe/H] $= -5.3$ (HE 0107−5240; Christlieb et al. 2002), and very recently, down

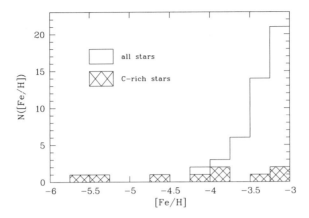

Figure 2. Low-metallicity tail of the observed halo metallicity distribution function. Only few objects are known, and many of the most iron-poor stars appear to be enriched with unusually large amounts of carbon. Figure adapted from Norris et al. (2007).

to [Fe/H] = −5.4 (HE 1327−2326; Frebel et al. 2005). Overall, only three stars are known with iron abundances of [Fe/H] < −4.0. The recently discovered star HE 0557−4840 (Norris et al. 2007) with [Fe/H] < −4.8 bridges the gap between [Fe/H] = −4.0 and the aforementioned two objects with [Fe/H] < −5.0. These extremely iron-deficient objects have opened a new, unique observational window of the time very shortly after the Big Bang. They provide key insight into the very beginning of the Galactic chemical evolution.

This review discusses the history of the first discoveries of metal-poor stars that laid the foundation to this field in § 2, and observed abundance trends at the lowest metallicities in § 3. § 4, § 5, and § 6 describe particular classes of metal-poor stars. The origins of metal-poor stars are reviewed in § 7 and the application of these stars to cosmological questions in § 8.

2. First Discoveries of Metal-Poor Stars

It was long believed that all stars would have a similar chemical composition to the Sun. In the late 1940s, however, some metal lines observed in stars appeared to be unusually weak compared with the Sun. It was first suspected that these stars could be hydrogen-deficient or might have peculiar atmospheres, but Chamberlain & Aller (1951) concluded that "one possibly undesirable factor" in their interpretation would be the "prediction of abnormally small amounts of Ca and Fe" in these stars. They found about 1/20th of the solar Ca and Fe values. Subsequent works on such metal-weak stars during the following few decades confirmed that stars do indeed have different metallicities that reflect different stages of the chemical evolution undergone by the Galaxy. HD140283 was one object observed by Chamberlain & Aller (1951). Repeated studies in the past half century have shown that this subgiant has [Fe/H] ∼ −2.5 (e.g., Norris et al.

2001). HD140282 is what one could call a "classical" metal-poor halo star that now often serves as reference star in chemical abundance analyses.

In 1981, Bond (1981) asked the question "Where is Population III?" At the time, Population III stars were suggested to be stars with [Fe/H] < −3.0 because no stars were known with metallicities lower than 1/1000 of the solar iron abundance. Even at [Fe/H] ∼ −3.0, only extremely few objects were known although more were predicted to exist by a simple model of chemical evolution in the Galactic halo. Bond (1981) reported on unsuccessful searches for these Population III stars and concluded that long-lived low-mass star could not easily form from zero-metallicity gas, and hence were extremely rare, if not altogether absent. Today, we know that star formation in zero-metallicity gas indeed does not favor the creation of low-mass stars due to insufficient cooling processes (Bromm et al. 2002). It has also become apparent that the number of stars at the tail of the MDF is extremely sparsely populated (Figure 2). All those stars are fainter, and hence further away, than those among which Bond was searching ($B = 10.5$ to 11.5). As will be shown in this review, the current lowest metallicity has well surpassed [Fe/H] ∼ −3.0 although it is still unclear what the lowest observable metallicity of halo stars may be.

A few years later, a star was serendipitously discovered, CD −38° 245, with a record low Fe abundance of [Fe/H] = −4.5 Bessell & Norris (1984). It was even speculated that CD −38° 245 might be a true Population III star, although Bessell & Norris (1984) concluded an extreme Population II nature for the star. The red giant was later reanalyzed and found to have [Fe/H] = −4.0 (Norris et al. 2001). CD −38° 245 stayed the lowest metallicity star for almost 20 years.

Large objective-prism surveys were carried out beginning in the early 1980s to systematically identify metal-poor stars. For a review of surveys and techniques, the reader is referred to Christlieb (2006) and Beers & Christlieb (2005).

3. Abundance Trends at Low Metallicity

The quest to find more of the most metal-poor stars to investigate the formation of the Galaxy lead to the first larger samples of stars with metallicities down to [Fe/H] ∼ −4.0. Abundance ratios [X/Fe] as a function of [Fe/H] were extended to low metallicities for the lighter elements ($Z < 30$) and neutron-capture elements ($Z > 56$). McWilliam et al. (1995) studied 33 stars with −4.0 < [Fe/H] < −2.0. The α-elements Mg, Si, Ca, and Ti are enhanced by ∼ 0.4 dex with respect to Fe. α-elements are produced through α-capture during various burning stages of late stellar evolution, before and during supernova explosions. This enhancement is found down to the lowest metallicities, although with few exceptions. Recently, some stars were discovered that are α-poor (Ivans et al. 2003), whereas others are strongly overabundant in Mg and Si (Aoki et al. 2002a). The iron-peak elements are produced during supernova type II explosions and their yields depend on the explosion energy. The abundance ratios of Cr/Fe and Mn/Fe become more underabundant with decreasing [Fe/H] and below [Fe/H]=-2.5, up to ∼ −0.7 at [Fe/H] ∼ −3.5 for Cr and ∼ −1.0 for Mn. Co becomes overabundant, up to ∼ 0.8. Sc/Fe and Ni/Fe have a roughly solar ratio, even at [Fe/H] ∼ −4.0. Compared with these elements, the neutron-capture elements appear to behave differently. Sr has an extremely large scatter (∼ 2 dex),

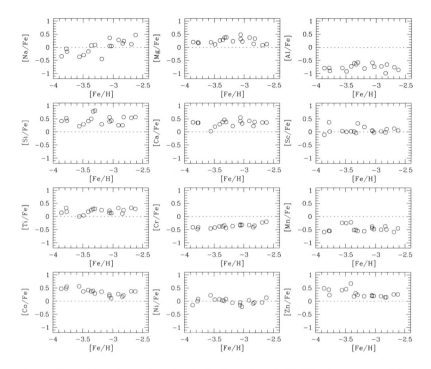

Figure 3. Abundance trends at the lowest metallicities by means of the Cayrel et al. (2004) stars.

indicating that different nucleosynthetic process may contribute to its Galactic inventory. Ba has less but still significant scatter at various metallicities. These abundance trends were confirmed by Ryan et al. (1996) who used 22 stars in the same metallicity regime. These authors furthermore proposed that the chemical enrichment of the early interstellar medium was primarily due to the explosion energies of the first supernovae that determine their yield distribution.

Several stars at the lowest metallicities showed significant deviations from the general trends. This level of abundance scatter at the lowest metallicities led to the conclusion that the interstellar medium was still inhomogeneous and not well mixed at [Fe/H] ~ -3.0 (e.g., Argast et al. 2000).

The claim for an inhomogeneous interstellar medium was diminished when Cayrel et al. (2004) presented a sample of 35 extremely metal-poor stars. Figure 3 shows the abundance trends of those stars for 12 different elements. These authors found very little scatter in the abundance trends of elements with $Z < 30$ among their stars, down to [Fe/H] ~ -4.0. This result suggests that the interstellar medium was already well-mixed at early times, leading to the formation of stars with almost identical abundance patterns.

4. r-Process Stars

All elements except H and He are created in stars during stellar evolution and supernova explosions. The so-called r-process stars formed from material previously enriched with heavy neutron-capture elements. About 5% of stars with [Fe/H] < −2.5 show strong enhancement of neutron-capture elements associated with the rapid (r-) nucleosynthesis process(Beers & Christlieb 2005) that is responsible for the production of the heaviest elements in the Universe. In these stars, we can observe the majority (i.e., ∼ 70 of 94) of elements in the periodic table: the light, α, iron-peak, and light and heavy neutron-capture elements. So far, the nucleosynthesis site of the r-process has not been unambiguously identified, but supernova explosions are the most promising location. In 1996, the first r-process star was discovered, CS 22892-052 (Sneden et al. 1996). The heavy neutron-capture elements observed in this star follow the scaled solar r-process pattern. The abundance patterns of the heaviest elements with $56 < Z < 90$ of the few known r-process stars indeed *all* follow the scaled *solar* r-process pattern (Figure 4). This behavior suggests that the r-process is universal – an important empirical finding that could not be obtained from any laboratory on earth. However, there are deviations among the lighter neutron-capture elements. Since it is not clear if the stellar abundance patterns are produced by only a single r-process, an additional new process might be needed to explain all neutron-capture abundances (e.g., Aoki et al. 2005).

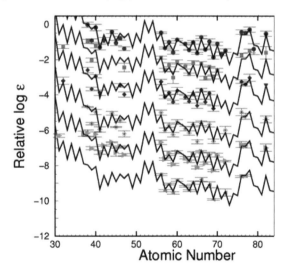

Figure 4. r-process abundance patterns of the most strongly-enhanced r-process metal-poor stars (various symbols). Overplotted are scaled solar r-process patterns. There is excellent agreement between the stellar data and the solar r-process pattern. All patterns are arbitrarily offset to allow a visual comparison. Figure kindly provided by J. J. Cowan.

Among the heaviest elements are the long-lived radioactive isotopes ^{232}Th (half-life 14 Gyr) or ^{238}U (4.5 Gyr). While Th is often detectable in r-process

Table 1. Stellar ages derived from different abundance ratios

Star	Age (Gyr)	Abundance ratio	Ref.
HD15444	15.6 ± 4	Th/Eu	Cowan et al. (1999)
CS31082-001	14.0 ± 2	U/Th	Hill et al. (2002)
BD +17° 3248	13.8 ± 4	average of several	Cowan et al. (2002)
CS22892-052	14.2 ± 3	average of several	Sneden et al. (2003)
HE 1523−0901	13.2 ± 2	average of several	Frebel et al. (2007a)
HD22170	11.7 ± 3	Th/Eu	Ivans et al. (2006)

stars, U poses a real challenge because *only one*, extremely weak line is available in the optical spectrum. By comparing the abundances of the radioactive Th and/or U with those of stable r-process nuclei, such as Eu, stellar ages can be derived. Through individual age measurements, r-process objects become vital probes for observational "near-field" cosmology by providing an independent lower limit for the age of the Universe. This fortuity also provides the opportunity of bringing together astrophysics and nuclear physics because these objects act as "cosmic lab" for both fields of study.

Most suitable for such age measurements are cool metal-poor giants that exhibit strong overabundances of r-process elements[2]. Since CS 22892-052 is very C-rich, the U line is blended and not detectable. Only the Th/Eu ratio could be employed, and an age of 14 Gyr was derived (Sneden et al. 2003). The U/Th chronometer was first measured in the giant CS 31082-001 (Cayrel et al. 2001), yielding an age of 14 Gyr. The recently discovered giant HE 1523−0901 ([Fe/H] = −3.0; Frebel et al. 2007a) has the largest measured overabundance of r-process elements. It is only the *third* star with a U detection at all, and is has the most reliable measurement of all three. HE 1523−0901 is the first star that has been dated with seven different "cosmic clocks", i.e. abundance ratios such as U/Th, Th/Eu, U/Os. The average age obtained is 13 Gyr, which is consistent with the WMAP (Spergel et al. 2007) age of the Universe of 13.7 Gyr. These age measurements also confirm the old age of similarly metal-poor stars.

Since Eu and Th are much easier to detect than U, the Th/Eu chronometer has been used several times to derive stellar ages of metal-poor stars. Table 1 lists the ages derived from the Th/Eu and other abundance ratios measured in the stars that are shown in Figure 4. Compared to Th/Eu, the Th/U ratio is much more robust to uncertainties in the theoretically derived production ratio due to the similar atomic masses of Th and U (Wanajo et al. 2002). Hence, stars displaying Th *and* U are the most desired stellar chronometers.

5. Carbon-Rich Stars & s-Rich Stars

It was first noted by Rossi et al. (1999) that a large fraction of metal-poor stars have an overabundance of carbon with respect to iron (C/Fe > 1.0). This

---

[2]Stars with [r/Fe] > 1.0; r represents the average abundance of elements associated with the r-process.

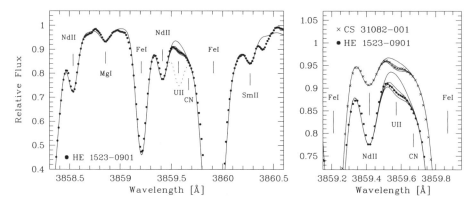

Figure 5. Spectral region around the U RomanII line in HE 1523−0901 (*filled dots*) and CS 31082-001 (*crosses*; right panel only). Overplotted are synthetic spectra with different U abundances. The dotted line in the left panel corresponds to a scaled solar r-process U abundance present in the star if no U were decayed. Figure taken from Frebel et al. (2007a).

suggestion has been widely confirmed, although different samples lead to a range of fractions (10% to 25%; e.g., (Cohen et al. 2005; Lucatello et al. 2006)). At the lowest metallicities, this fraction is even higher, and all three stars with [Fe/H] < -4.0 are extremely C-rich. The fraction of C-rich stars may also increase with increasing distance from the Galactic plane (Frebel et al. 2006b).

Many C-rich stars also show an enhancement in neutron-capture elements. These elements are produced in the interiors of intermediate-mass asymptotic giant branch (AGB) stars through the slow (s-) neutron-capture process. Such material is later dredged up to the star's surface. Contrary to the r-process, the s-process is not universal because two different sites seem to host s-process nucleosynthesis. The so-called "weak" component occurs in the burning cores of the more massive stars, and preferentially produces elements around $Z \sim 40$. The "main" component of the s-process occurs in the helium shells of thermally pulsing lower mass AGB stars and is believed to account for elements with $Z \geq 40$ (e.g., Arlandini et al. 1999). The s-process leads to a different characteristic abundance pattern than the r-process. Indeed, the s-process signature is observed in some metal-poor stars, and their neutron-capture abundances follow the scaled solar s-process pattern. The metal-poor objects observed today received the s-process enriched material during a mass transfer event across a binary system from their more massive companion that went through the AGB phase (e.g., Norris et al. 1997; Aoki et al. 2001). The binary nature of many s-process stars has been confirmed through monitoring their radial velocity variations (Lucatello et al. 2005). Some of the s-process metal-poor stars contain huge amounts of lead (Pb) – in fact they are more enhanced in lead than in any other element heavier than iron (Van Eck et al. 2001). Pb is the end product of neutron-capture nucleosynthesis and AGB models predict an extended Pb, and also Bi, production (Gallino et al. 1998). More generally, the observed patterns

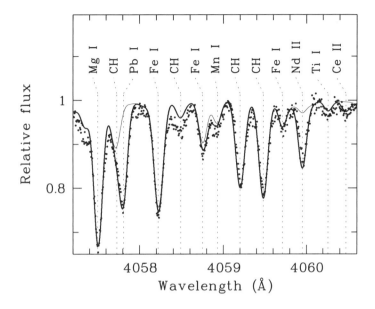

Figure 6. Spectral region around the Pb RomanI line at ~4058 Å in the metal-poor star HD196944 (dots). Overplotted is the best-fit synthetic spectrum. Figure reprinted by permission from Macmillan Publishers Ltd.: Nature, Van Eck et al. (2001).

of many s-process metal-poor stars can be reproduced with these AGB models, which is evidence for a solid theoretical understanding of AGB nucleosynthesis.

Unlike the r-process, it is not clear whether the s-process already operated in the early Universe, since it depends on the presence of seed nuclei (i.e., from Fe-peak elements created by previous generations of stars) in the host AGB star. However, recent evidence from some metal-poor stars (Simmerer et al. 2004) may challenge the late occurrence of the s-process at higher metallicities.

As usual, there are a exceptions. Some C-rich stars do not show any s-process enrichment (Aoki et al. 2002b; Frebel et al. 2007c), and the origin of these chemical signatures remains unclear. Other C-rich stars show r-process enhancement (Sneden et al. 1996), and the two stars with the largest carbon overabundances with respect to iron are the two with [Fe/H] < −5.0. The frequent finding of C-rich stars in combination with a variety of abundance patterns points toward the importance of C in the early Universe. The exact role of C is still unclear but it likely played a crucial role in star formation processes (Bromm & Loeb 2003).

6. [Fe/H] < −5.0 Stars

In 2002, the first star with a new, record-low iron abundance was found. The faint ($V = 15.2$) red giant HE 0107−5240 has [Fe/H] = −5.3 (Christlieb et al.

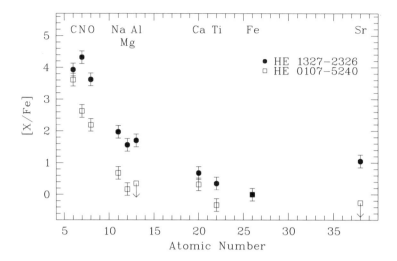

Figure 7. Abundance patterns of HE 0107−5240 and HE 1327−2326. Abundances taken from Christlieb et al. (2004), Bessell et al. (2004), Frebel et al. (2005), and Frebel et al. (2006a), where the LTE Fe abundance of HE 0107−5240 was corrected for non-LTE effects in the same way as for HE 1327−2326.

2002)[3]. In 2004, the bright ($V = 13.5$) subgiant HE 1327−2326 was discovered (Frebel et al. 2005; Aoki et al. 2006). Both objects were discovered in the Hamburg/ESO survey. HE 1327−2327 had an even lower iron abundance of [Fe/H] = −5.4. This value corresponds to ∼ 1/250, 000 of the solar iron abundance. With its extremely large abundances of CNO elements, HE 1327−2326 has a very similar abundance pattern compared with HE 0107−5240 (see Figure 7). No neutron-capture elements are found in HE 0107−5240, whereas, unexpectedly, Sr is observed in HE 1327−2326. The Sr may originate from the neutrino-induced νp-process operating in supernova explosions (Fröhlich et al. 2006). Furthermore, in the relatively unevolved subgiant HE 1327−2326 Li could not be detected ($\log \epsilon$(Li) < 1.6, where $\log \epsilon(A) = \log_{10}(N_A/N_H) + 12$). This is surprising, given that the primordial Li abundance is often inferred from similarly unevolved metal-poor stars (Ryan et al. 1999). The upper limit found from HE 1327−2326, however, strongly contradicts the WMAP value ($\log \epsilon$(Li) = 2.6) from the baryon-to-photon ratio (Spergel et al. 2007). This indicates that the star must have formed from extremely Li-poor material. Figure 8 shows the C and Mg abundance for the two stars in comparison with other, more metal-rich stars. Whereas both [Fe/H] < −5.0 objects have similarly high C abundances that are much higher than C/Fe ratios observed in other stars, there exists a significant difference in the Mg abundances. HE 0107−5240 has an almost solar

---

[3]Applying the same non-LTE correction of +0.2 dex for Fe I abundances for HE 0107−5240 and HE 1327−2326 leads to an abundance of [Fe/H] = −5.2 for HE 0107−5240.

Mg/Fe ratio similar to the other stars. An elevated Mg level is, however, observed not only in HE 1327−2326, but also in a few stars at higher metallicities.

If HE 1327−2327 would have been a few hundred degrees hotter or a few tens of a dex more metal-poor, no Fe lines would have been detectable in the spectrum. Even now, the strongest Fe lines are barely detectable and very high S/N ratio (> 100) was required for the successful Fe measurement. The hunt for the most metal-poor stars with the lowest iron abundances is thus reaching a its technical limit with regard to turn-off stars. Fortunately, in suitable giants, Fe lines should be detectable at even lower Fe abundances than [Fe/H] < -5.5. It remains to be seen, though, what the lowest iron-abundance, or lowest overall metallicity for that matter, in a stellar object might be. Considering a simple accretion model, (Iben 1983) suggested such the observable lower limit to be [Fe/H] < -5.7. Future observations may be able to reveal such a lower limit.

HE 0107−5240 and HE 1327−2326 immediately became benchmark objects to constrain various theoretical studies of the early Universe, such as the formation of the first stars (e.g., Yoshida et al. 2006), the chemical evolution of the early interstellar medium (e.g., Karlsson & Gustafsson 2005) or supernovae yields studies. More such stars with similarly low metallicities are urgently needed. Very recently, a star was found at [Fe/H] < -4.75 (Norris et al. 2007). It sits right in the previously claimed "metallicity gap" between the stars at [Fe/H] ~ -4.0 and the two with [Fe/H] < -5.0. These three stars provide crucial information for the shape of the tail of the metallicity distribution function.

7. What is the Chemical Origin of the Most Metal-Poor Stars?

The two known hyper-metal-poor stars (with [Fe/H] < -5.0; Christlieb et al. 2002; Frebel et al. 2005) provide a new observational window to study the onset of the chemical evolution of the Galaxy. The *highly individual* abundance patterns of these and other metal-poor stars have been successfully reproduced by several supernovae scenarios. HE 0107−5240 and HE 1327−2326 both appear to be early, extreme Population II stars possibly displaying the "fingerprint" of only one Population III supernova. Umeda & Nomoto (2003) first reproduced the observed abundance pattern of HE 0107−5240 by suggesting the star formed from material enriched by a faint 25 $M_\odot$ supernova that underwent a mixing and fallback process. To achieve a simultaneous enrichment of a lot of C and only little Fe, large parts of the Fe-rich yield fall back onto the newly created black hole. Using yields from a supernova with similar explosion energy and mass cut, Iwamoto et al. (2005) then explained the origin of HE 1327−2326.

Meynet et al. (2006) explored the influence of stellar rotation on elemental yields of very low-metallicity supernovae. The stellar mass loss yields of fast rotating massive Pop III stars quantitatively reproduce the CNO abundances observed in HE 1327−2326 and other metal-poor stars. Limongi et al. (2003) were able to reproduce the abundances of HE 0107−5240 through pollution of the birth cloud by at least two supernovae. Suda et al. (2004) proposed that the abundances of HE 0107−5240 would originate from a mass transfer of CNO elements from a potential companion, and from accretion of heavy elements from the interstellar medium. However, neither HE 0107−5240 nor HE 1327−2326 show the radial velocity variations that would indicate binarity. Self-enrichment

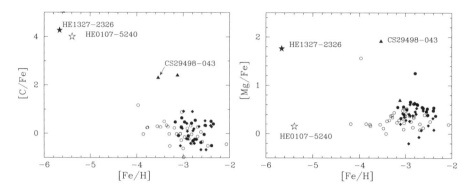

Figure 8. C/Fe and Mg/Fe abundance ratios as function of [Fe/H]. Figure taken from Aoki et al. (2006).

with CNO elements has also been ruled out for HE 0107−524 (Picardi et al. 2004), while this is not a concern for the less evolved subgiant HE 1327−2326.

Tominaga et al. (2007) model the averaged abundance pattern of four non-carbon-enriched stars with $-4.2 <$ [Fe/H] < -3.5 (Cayrel et al. 2004) with the elemental yields of a massive, energetic ($\sim 30 - 50\,M_\odot$) Population III hypernova. Abundance patterns of stars with [Fe/H] ~ -2.5 can be reproduced with integrated (over a Salpeter IMF) yields of normal Population III supernovae.

8. Stellar Archaeology

Numerical simulations show that the first stars in the Universe must have been very massive ($\sim 100\,M_\odot$). However, the observed metal-poor stars all have *low masses* ($< 1\,M_\odot$). To constrain the first low-mass star formation, Bromm & Loeb (2003) put theorized that fine-structure line cooling by C and O nuclei, as provided by the Population III objects, may be responsible for the necessary cooling of the interstellar medium to allow low-mass stars to form. On the other hand, cooling by dust may play a major role in the transition from Population III to Population II star formation. Dust grains created during the first supernova explosions may induce cooling and subsequent fragmentation processes that lead to the formation of subsolar-mass stars (e.g., Schneider et al. 2006).

Such cooling theories can be tested with large numbers of *carbon and oxygen-poor* metal-poor stars. In order for stars to form, a critical metallicity of the interstellar medium is required. Metal-poor stars with the lowest metallicities should thus have values equal or higher than the critical metallicity. Suitable metallicity indicators for the fine-structure line theory are C and O abundances in the most metal-poor stars. Frebel et al. (2007b) collected observational literature data of these two abundances in metal-poor stars. A critical metallicity limit of $D_{\rm trans} = -3.5$ was developed, where $D_{\rm trans}$ parameterizes the observed C and/or O abundances. So far, no star has abundances below the critical value, as can be seen in Figure 9. It is predicted that any future stars with [Fe/H] < -4 to be discovered should have high C and/or O abundances.

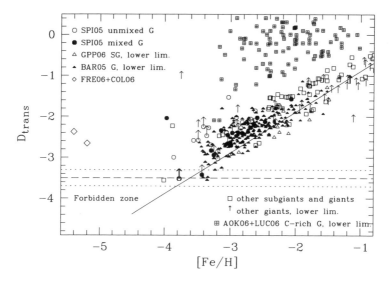

Figure 9. Comparison of metal-poor data with the theory that fine-structure lines of C and O dominate the transition from Population III to Population II in the early Universe. C and O-poor extremely metal-poor stars are invaluable to test this formation scenario. Figure taken from Frebel et al. (2007b).

This agrees with the empirically frequent C-richness among the most metal-poor stars. Since the critical metallicity of the dust cooling theory is a few orders of magnitude below the D_{trans} the current data is also consistent with that theory, although more firm constraints can not be derived at this point. Future observations of large numbers of metal-poor stars will provide the necessary constraints that may lead to identify the dominant cooling process responsible for low-mass star formation in the early Universe.

9. Outlook

More and more metal-poor stars are being discovered, thanks to recent and ongoing large-scale surveys. The hunt for the most metal-poor stars has just begun with the discoveries of the two stars with [Fe/H] < −5.0. Exciting times are ahead that will hopefully lead to observations of new metal-poor stars from which we can learn about the first stars, the onset of the Galactic chemical enrichment and the involved nucleosynthesis processes. More stars at the lowest metallicities are clearly needed for this quest and the next-generations telescopes will be of great interest to this field. There are ongoing targeted searches for the most metal-poor stars with existing 8-10 m optical facilities. Large high-resolution spectroscopy programs have been started with the Hobby-Eberly-Telescope, *Chemical Abundance of Stars in the Halo* (CASH), and also with Subaru, Keck and Magellan. More faint stars (down to $V \sim 17$) are currently being discovered with surveys such as SDSS and SDSS-II (SEGUE). However, many stars are too faint to be followed-up with high-resolution, high S/N spectroscopy with the current 8-10 m telescopes. Future 25 m class telescopes, such as the Giant Magellan Telescope, will be required for discovering the most metal-poor stars in the outskirts of the Galactic halo and nearby dwarf galaxies.

Acknowledgments. I would like to thank the McDonald Observatory Board of Visitors and the meeting organizers for making possible such a unique and inspiring meeting of young astronomers. I feel honored to have been a part of it. Financial support through the W. J. McDonald Fellowship of the McDonald Observatory is gratefully acknowledged.

References

Aoki, W. et al. 2006, ApJ, 639, 897
Aoki, W., Honda, S., Beers, T. C., Kajino, T., Ando, H., Norris, J. E., Ryan, S. G., Izumiura, H., Sadakane, K., & Takada-Hidai, M. 2005, ApJ, 632, 611
Aoki, W., Norris, J. E., Ryan, S. G., Beers, T. C., & Ando, H. 2002a, ApJ, 576, L141
—. 2002b, ApJ, 567, 1166
Aoki, W., Ryan, S. G., Norris, J. E., Beers, T. C., Ando, H., Iwamoto, N., Kajino, T., Mathews, G. J., & Fujimoto, M. Y. 2001, ApJ, 561, 346
Argast, D., Samland, M., Gerhard, O. E., & Thielemann, F.-K. 2000, A&A, 356, 873
Arlandini, C., Käppeler, F., Wisshak, K., Gallino, R., Lugaro, M., Busso, M., & Straniero, O. 1999, ApJ, 525, 886
Beers, T. C. & Christlieb, N. 2005, ARAA, 43, 531
Bessell, M. S., Christlieb, N., & Gustafsson, B. 2004, ApJ, 612, L61
Bessell, M. S. & Norris, J. 1984, ApJ, 285, 622
Bond, H. 1981, ApJ, 248, 606
Bromm, V., Coppi, P. S., & Larson, R. B. 2002, ApJ, 564, 23
Bromm, V. & Loeb, A. 2003, Nature, 425, 812
Cayrel, R. et al. 2004, A&A, 416, 1117
Cayrel, R. et al. 2001, Nature, 409, 691
Chamberlain, J. W. & Aller, L. H. 1951, ApJ, 114, 52
Christlieb, N. 2006, in Astronomical Society of the Pacific Conference Series, Vol. 353, Stellar Evolution at Low Metallicity: Mass Loss, Explosions, Cosmology, ed. H. J. G. L. M. Lamers, N. Langer, T. Nugis, & K. Annuk, 271
Christlieb, N., Bessell, M. S., Beers, T. C., Gustafsson, B., Korn, A., Barklem, P. S., Karlsson, T., Mizuno-Wiedner, M., & Rossi, S. 2002, Nature, 419, 904
Christlieb, N., Gustafsson, B., Korn, A. J., Barklem, P. S., Beers, T. C., Bessell, M. S., Karlsson, T., & Mizuno-Wiedner, M. 2004, ApJ, 603, 708
Cohen, J. G., Shectman, S., Thompson, I., McWilliam, A., Christlieb, N., Melendez, J., Zickgraf, F.-J., Ramírez, S., & Swenson, A. 2005, ApJ, 633, L109
Cowan, J. J., Pfeiffer, B., Kratz, K.-L., Thielemann, F.-K., Sneden, C., Burles, S., Tytler, D., & Beers, T. C. 1999, ApJ, 521, 194
Cowan, J. J., Sneden, C., Burles, S., Ivans, I. I., Beers, T. C., Truran, J. W., Lawler, J. E., Primas, F., Fuller, G. M., Pfeiffer, B., & Kratz, K.-L. 2002, ApJ, 572, 861
Frebel, A. et al. 2005, Nature, 434, 871
Frebel, A., Christlieb, N., Norris, J. E., Aoki, W., & Asplund, M. 2006a, ApJ, 638, L17
Frebel, A., Christlieb, N., Norris, J. E., Beers, T. C., Bessell, M. S., Rhee, J., Fechner, C., Marsteller, B., Rossi, S., Thom, C., Wisotzki, L., & Reimers, D. 2006b, ApJ, 652, 1585
Frebel, A., Christlieb, N., Norris, J. E., Thom, C., Beers, T. C., & Rhee, J. 2007a, ApJ, 660, L117
Frebel, A., Johnson, J. L., & Bromm, V. 2007b, MNRAS, 380, L40

Frebel, A., Norris, J. E., Aoki, W., Honda, S., Bessell, M. S., Takada-Hidai, M., Beers, T. C., & Christlieb, N. 2007c, ApJ, 658, 534
Fröhlich, C., Martínez-Pinedo, G., Liebendörfer, M., Thielemann, F.-K., Bravo, E., Hix, W. R., Langanke, K., & Zinner, N. T. 2006, Physical Review Letters, 96, 142502
Gallino, R., Arlandini, C., Busso, M., Lugaro, M., Travaglio, C., Straniero, O., Chieffi, A., & Limongi, M. 1998, ApJ, 497, 388
Hill, V., Plez, B., Cayrel, R., Nordström, T. B. B., Andersen, J., Spite, M., Spite, F., Barbuy, B., Bonifacio, P., Depagne, E., François, P., & Primas, F. 2002, A&A, 387, 560
Iben, I. 1983, Memorie della Societa Astronomica Italiana, 54, 321
Ivans, I. I., Simmerer, J., Sneden, C., Lawler, J. E., Cowan, J. J., Gallino, R., & Bisterzo, S. 2006, ApJ, 645, 613
Ivans, I. I., Sneden, C., James, C. R., Preston, G. W., Fulbright, J. P., Höflich, P. A., Carney, B. W., & Wheeler, J. C. 2003, ApJ, 592, 906
Iwamoto, N., Umeda, H., Tominaga, N., Nomoto, K., & Maeda, K. 2005, Science, 309, 451
Karlsson, T. & Gustafsson, B. 2005, A&A, 436, 879
Limongi, M., Chieffi, A., & Bonifacio, P. 2003, ApJ, 594, L123
Lucatello, S., Beers, T. C., Christlieb, N., Barklem, P. S., Rossi, S., Marsteller, B., Sivarani, T., & Lee, Y. S. 2006, ApJ, 652, L37
Lucatello, S., Tsangarides, S., Beers, T. C., Carretta, E., Gratton, R. G., & Ryan, S. G. 2005, ApJ, 625, 825
McWilliam, A., Preston, G. W., Sneden, C., & Searle, L. 1995, AJ, 109, 2757
Meynet, G., Ekström, S., & Maeder, A. 2006, A&A, 447, 623
Norris, J. E., Christlieb, N., Korn, A. J., Eriksson, K., Bessell, M. S., Beers, T. C., Wisotzki, L., & Reimers, D. 2007, ApJ, 670, 774
Norris, J. E., Ryan, S. G., & Beers, T. C. 1997, ApJ, 488, 350
Norris, J. E., Ryan, S. G., & Beers, T. C. 2001, ApJ, 561, 1034
Picardi, I., Chieffi, A., Limongi, M., Pisanti, O., Miele, G., Mangano, G., & Imbriani, G. 2004, ApJ, 609, 1035
Rossi, S., Beers, T. C., & Sneden, C. 1999, in ASP Conf. Ser. 165: The Third Stromlo Symposium: The Galactic Halo, 264
Ryan, S. G., Norris, J. E., & Beers, T. C. 1996, ApJ, 471, 254
—. 1999, ApJ, 523, 654
Schneider, R., Omukai, K., Inoue, A. K., & Ferrara, A. 2006, MNRAS, 369, 1437
Simmerer, J., Sneden, C., Cowan, J. J., Collier, J., Woolf, V. M., & Lawler, J. E. 2004, ApJ, 617, 1091
Sneden, C., Cowan, J. J., Lawler, J. E., Ivans, I. I., Burles, S., Beers, T. C., Primas, F., Hill, V., Truran, J. W., Fuller, G. M., Pfeiffer, B., & Kratz, K.-L. 2003, ApJ, 591, 936
Sneden, C., McWilliam, A., Preston, G. W., Cowan, J. J., Burris, D. L., & Amorsky, B. J. 1996, ApJ, 467, 819
Spergel, D. N. et al. 2007, ApJS, 170, 377
Suda, T., Aikawa, M., Machida, M. N., Fujimoto, M. Y., & Iben, I. J. 2004, ApJ, 611, 476
Tominaga, N., Umeda, H., & Nomoto, K. 2007, ApJ, 660, 516
Umeda, H. & Nomoto, K. 2003, Nature, 422, 871
Van Eck, S., Goriely, S., Jorissen, A., & Plez, B. 2001, Nature, 412, 793
Wanajo, S., Itoh, N., Ishimaru, Y., Nozawa, S., & Beers, T. C. 2002, ApJ, 577, 853
Yoshida, N., Omukai, K., Hernquist, L., & Abel, T. 2006, ApJ, 652, 6

Stuart Barnes summons everyone to listen to his review of future instrumentation.

Instrumentation in the ELT Era

Stuart Barnes

McDonald Observatory, University of Texas, Austin, TX, USA

Abstract. At present there are several efforts worldwide which aim to build the next generation of large ground-based telescopes. These extremely large telescopes (ELTs) will ensure the boundaries of astronomical exploration continue to expand. While these future extremely large telescopes are in themselves technically challenging, the greatest challenges may arise from the complexity demanded by those instruments inspired by the prospect of large apertures and high angular resolutions. In this proceedings an overview of the proposed optical to near-infrared future ELTs will be given along with a discussion of some of the proposed instrumentation. Some of the difficulties faced by ELT instrumentation builders and possible solutions will be explored.

1. Introduction

There are currently several groups world wide who are attempting to build the next generation of large telescopes. At the heart of these proposals are a desire to answer the "big questions" in astronomy. These are questions such as the following[1]:

- What is the nature and composition of the Universe?
- When did the first galaxies form and how did they evolve?
- What is the relationship between black holes and galaxies?
- How do stars and planets form?
- What is the nature of extrasolar planets?
- Is there life elsewhere in the Universe?

In the coming decades a number of tools will become available which may help answer these and other questions. These tools include:

- JWST (James Webb Space Telescope): To be launched 2013, 5–10 year lifetime
- ALMA (Atacama Large Millimeter Array): Completion expected 2010–2012
- SKA (Square Kilometer Array): Completed 2015–2020

---

[1]Source: TMT construction proposal, 2007, http://www.tmt.org/news/TMT-Construction Proposal-Public.pdf

In addition to facilities such as the above, there is also a strong desire to add ground-based extremely large telescopes.

The motivation for large optical to near-infrared telescopes lies in the sensitivity gains that result from increased aperture. These gains are shown in Table 1. Depending on the mode in which an ELT is used, the gain in sensitivity (measured either by increased S/N or speed) for a telescope which has a relative diameter D can be anywhere from linear to proportional to the fourth power of D. That is, a 30 m telescope, if equipped with an adaptive optics system which provides the highest Strehl (possible in the foreseeable future only in the NIR), might be expected to be up to 81 times faster than a 10 m telescope (see Figure 1). High order adaptive optics will also allow unprecedented angular resolution (see Figure 2).

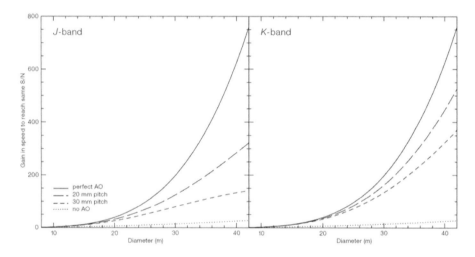

Figure 1. The gain in speed of an ELT relative to an existing 8 m class telescope. The performance gain is dependent on the adaptive optics methods. (From Gilmozzi & Spyromilio (2007))

2. The ELT Projects

At present, there are three credible ELT projects at similar levels of development. These projects are the Giant Magellan Telescope (GMT), the Thirty Meter Telescope (TMT), and the European Extremely Large Telescope (E-ELT).

Table 1. The dependency of the diameter D of a telescope to the sensitivity. Table based on D'Odorico (2006)

	Point source, sky noise limited, natural seeing	Point source, sky noise limited, diffraction limit	Photon noise limited	Detector noise limited
Gain in S/N:[a]	D	D^2	D	D^2
Gain in speed:[b]	D^2	D^4	D^2	D^4

[a] For a given magnitude m_λ and integration time t.
[b] Time t to reach a given S/N for a given magnitude m_λ.

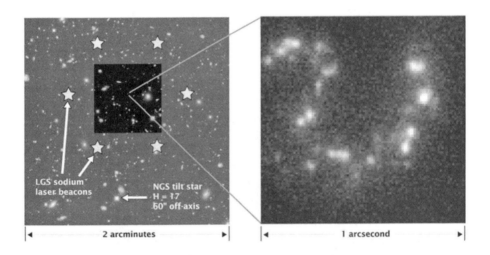

Figure 2. (Right) A simulated one hour GMT image of a starburst galaxy at $z = 1.4$, demonstrating the GMT's 0.016″ H band resolution with adaptive optics, ten times the resolution of the Hubble Space Telescope. This simulation is based on an Hα image of NGC 3089 at a detector plate scale of 0.010″ per pixel. This narrow band image is centered at the redshifted Hα wavelength of 1.58 μm. (Left) The full field of the GMT AO imager is 40″ across, with an area 1600 times larger than the detail to the right. The positions of the constellation of laser guide star beacons and the natural guide star used to recover tip-tilt information are indicated. (Credit: Giant Magellan Telescope - Carnegie Observatories.)

2.1. The Giant Magellan Telescope

The GMT[2], which is the smallest and perhaps most conservative of all the ELT projects, is based on the concept of mosaicking seven 8.4 m mirrors on a single mount. This results in a full aperture diameter of 24.5 m at $f/0.7$ and an effective collecting area of an (unobstructed) 22 m telescope (see Figure 3). The optics follow a Gregorian layout where each primary mirror segment is conjugated to an actively controlled secondary. The $f/8$ focal plane, which allows a field of view 10–20′, lies behind the telescope primary mirror, and all instruments will be mounted on a quasi-cassegrain platform (or possibly in a separate coudé room beside the telescope). The GMT is being developed by a consortium comprising the Carnegie Institution of Washington, Harvard University, the Massachusetts Institute of Technology, Smithsonian Astrophysical Observatory, the University of Arizona, the University of Texas at Austin, Texas A&M University, and the Australian National University. An intriguing possible upgrade for the GMT is to add a second identical telescope, thereby creating the possibility for an interferometer with a ∼150 m baseline (see Figure 4).

Figure 3. An aerial view of the Giant Magellan Telescope. (Credit: Giant Magellan Telescope - Carnegie Observatories.)

---

[2]http://www.gmto.org/

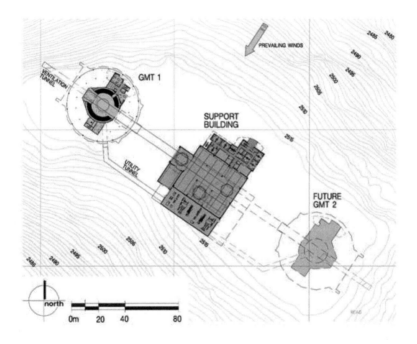

Figure 4. The GMT twin telescope concept. (Credit: Giant Magellan Telescope - Carnegie Observatories.)

2.2. The Thirty Meter Telescope

The TMT[3] is being led by a U.S.-Canadian partnership, that includes the California Institute of Technology, the University of California, and the Association of Canadian Universities for Research in Astronomy (ACURA). This project represents the coming together of several previous ELT projects; namely, the Californian Extremely Large Telescope (CELT) and the Canadian ACURA Very Large Optical Telescope (VLOT). The design of TMT also draws upon elements of the concept by the Association of Universities for Research in Astronomy (AURA) for the Giant Segmented Mirror Telescope (GSMT). TMT is, as the name suggests, a 30 m diameter telescope, and has a Ritchey-Chrétien optical design (see Figure 5). The primary mirror will be made from 492 individual hexagonal segments, and the science instruments will be located on large Nasymth platforms.

Figure 5. The Thirty Meter Telescope concept. (Credit: By kind permission of the Thirty Meter Telescope.)

---

[3]http://www.tmt.org/

2.3. The European Extremely Large Telescope

The E-ELT[4] is primarily a European Southern Observatory (ESO) project which is being developed together with the wider European astronomical community. This project derives in part from an earlier ESO ELT project called the Overwhelmingly Large Telescope (OWL) which was intended to be a 100 m diameter optical telescope. The EURO-50 project (led by the Lund University with assistance from astronomers from Finland, Ireland, Spain, Sweden and the United Kingdom) is also part of the E-ELT heritage. To mitigate against technical and financial risks these projects were combined and scaled down so that the final E-ELT proposal is for a 42 m telescope which has an $f/1$ primary mirror formed from a mosaic of 906 individual mirrors. The "baseline reference design" is a five-mirror combination with a convex secondary mirror although a (folded) three-mirror Gregorian design is considered an alternative. Access to the 10′ field is via a pair of Nasymth platforms with an alternative coudé feed also being possible.

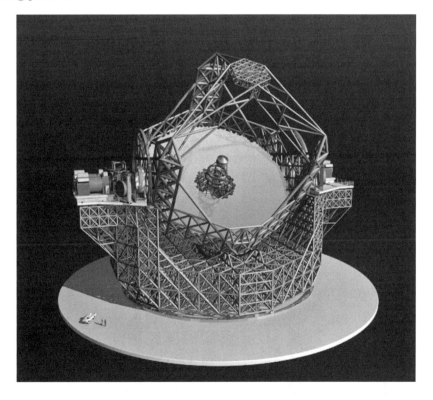

Figure 6. The baseline reference design for the European Extremely Large Telescope structure. (Credit: ESO)

---

[4]http://www.eso.org/projects/e-elt/

Table 2.   The GMT instrument concepts.

Instrument	Wavelength range (μm)	Resolving power (λ/δλ)	Field of view	Description
Concept design review instruments[a]:				
1. GMTNIRS	1–5	50–120K	2″	NIR high resolution spectrograph
2. MIISE	3–28	5–2000	2′ × 2′	Mid-infrared imaging spectrograph
3. HRCAM	1–2.5	5–2000	1′ × 1′	NIR AO imager and spectrograph
4. NIRMOS	0.9–2.5	1500–3500	5′ × 5′	NIR multi-object spectrograph and imager
5. GMACS	0.4–1.0	3500–5000	9′ × 18′	Wide field optical spectrograph
Additional instrument concepts:				
6. Q-Spec	0.3–1.0	30–150K	20″	Optical high resolution spectrograph
7. PVRS	0.4–0.7	150K	10″	Precision radial velocity spectrograph

[a]"Giant Magellan Telescope conceptual design review, 2006 February" http://www.gmto.org/CoDRpublic.

Table 3.   The TMT instrument[a] concepts. Parameters in parentheses are goals.

Instrument	Wavelength range (μm)	Resolving power (λ/δλ)	Field of view	Description
Early light instruments:				
1. WFOS	0.34–1.0	150–7500	2 × 4.5′ × 4.5′	Wide field optical spectrograph
2. IRIS	0.8–2.5	2–4000	2–10″	Infrared integral field spectrometer and imager
3. IRMS	0.9–2.5	< 5000	2′	Infrared multi-slit spectrometer
First decade instruments:				
4. IRMOS	0.8–2.5	2–50/2000–10000	5′	Infrared multi-object spectrograph
5. MIRES	8–18 (5–28)	2–50/5000–100K	10″	Mid-IR échelle spectrometer
6. PFI	1–2.5(4)	50–1000	1.4–4″	Planet formation instrument
7. NIRES	1–2.4	20–100K	2″	Near-IR echelle spectrometer
8. HROS	0.31–1(1.3)	50–90K	10″	High resolution optical spectrograph
9. WIRC	0.8–5 (0.6–5)	5–100	30″	Wide-field infrared camera

[a]For details see "TMT Instrumentation and Performance: A handbook for the July 2007 TMT science workshop" http://www.physics.uci.edu/TMT-Workshop/TMT-Handbook.pdf.

Table 4. The E-ELT "comprehensive instrument suite"[a]. Parameters in parentheses are goals.

	Description	Wavelength range (μm)	Resolution ($\lambda/\delta\lambda$)	Field of view	Notes
1.	High resolution visual spectrograph	0.4–0.7 (0.35–0.9)	150K	5″ (10″)	c.f. CODEX on OWL (Pasquini et al. 2005)
2.	Visual imager	0.35–1.05	N/A	4′ × 4′	Assuming a 42 m, $f/12$ telescope
3.	Multi-object visual spectrograph	0.32–1	700–7,500 (500–20K)	6′ (10′)	
4.	High time resolution instrument	0.4–0.7 (0.4–1)	N/A	1′	c.f. QuantEYE for OWL (Naletto et al. 2006), 10^{-3}–10^{-4} s time resolution
5.	Polarimeter	0.4–2.4/2.4–28	N/A	> 10″	Concept only
6.	Multi-object NIR spectrograph	0.8–2.4	4,000–20,000	2′ × 2′ (5–10′)	c.f. MOMSI for OWL (Evans et al. 2006)
7.	Planet imager and spectrograph	0.6–1.8 (0.6–2.3)	15, ~1000, ~50K	> 4″ (10″)	
8.	Wide-field NIR imager	0.6–2.4	N/A	5–10′	
9.	High resolution NIR spectrograph	1–5	~100K	1–5″	
10.	Mid-IR imager and spectrograph	7–20 (3.5–27)	200, 3000, 50K	2′	
11.	Sub-mm imager	200–850	N/A	5′	

[a] http://www.eso.org/projects/e-elt/Publications/ELT_INSWG_FINAL_REPORT.pdf.

2.4. Other ELTs

In addition to these projects there is some work in Japan which is concentrating on developing technology for zero-expansion pore-free ceramic mirrors[5]. Finally, there are ambitions for a "Future Chinese ELT" (Cui 2006) with a diameter of 30 m to be sited in either the West of China or Antarctica.

3. The ELT Instruments

The instrument concepts for each of the three ELT projects described above are listed in Tables 2, 3, and 4. It should be noted that the E-ELT instrumentation list is an evolution of previous concepts that were proposed for either the EURO-50 or OWL concepts and allows for a suite of instruments that will address the primary science goals for an E-ELT.

4. Challenges and Solutions

Several authors have given reviews of the challenges involved in ELT instrumentation. These include Cunningham & Crampton (2006) and Spanò et al. (2006). The various European groups have taken a consolidated approach to developing advanced technology for ELTs through the "OPTICON Key Technologies Network"[6]. The TMT project has passed an initial phase of funded competitive instrument design studies, while GMT is about to formally enter a similar phase.

4.1. Instrument Size and Complexity

One of the first and most obvious challenges for ELT instrumentation are the inevitable increases in both the size and complexity of the instruments. These are necessary in order to maintain the aperture advantage of the future giant telescopes. That is, in the case of spectroscopic instrumentation, the collimated beam size presented to the diffraction element correlates linearly with the final resolving power. Therefore, to a first approximation, an instrument designed for a 10 m class telescope will be at least 3 times larger on a 30 m telescope, or nearly 10 times the volume. An example of this scaling is shown in Figure 7.

In order to mitigate against the increased costs and risks associated with such large instruments, many of the ELT projects will be built by consortia whose members are spread over a number of institutions. This will allow the design to draw upon the experience of each institute. In some cases it is likely that smaller prototype instruments which demonstrate key technologies will be built and tested on smaller 4–10 m class telescopes.

4.2. Size of Optics

A particular challenge will be the size of the individual optical elements required for an ELT instrument. Some elements reaching between 1.2 m to 2 m in size may be needed. These optics will require high homogeneity and otherwise excellent

---

[5]See http://jelt.mtk.nao.ac.jp/index.html and Iye (2006)

[6]http://www.astro-opticon.org/networking/key_tech.html

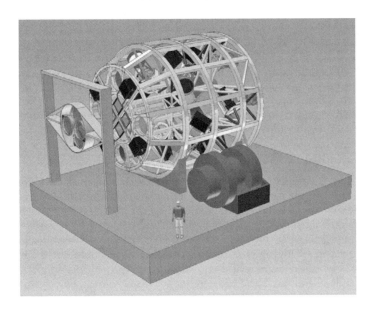

Figure 7. The proposed Wide-Field Optical Spectrograph (WFOS) for TMT is an example of the scale and complexity of the instrumentation being planned for future ELTs. WFOS is eight meters in diameter and nine meters long. The DEIMOS spectrograph at the Keck Observatory is shown in the foreground (solid black form) for comparison. (Credit: by kind permission of the Thirty Meter Telescope.)

optical properties. The choice of materials in such large dimensions will be limited as will the quantities available.

Gratings As an example, the CODEX instrument (originally proposed for OWL but rescaled for the E-ELT) (Pasquini et al. 2005) requires an échelle grating mosaic 200 × 1700 mm in size. This is nearly twice as large as the largest mosaics built to date. The high resolution spectrograph proposed for the TMT calls for an échelle grating mosaic considerably larger than this. Such challenges may in fact force the designers of this class of instrument to seek novel and innovative ways to minimize the growth of their instruments. An example of this is the HROS proposal for TMT, developed by Froning et al. (2006), which calls for an integral field unit fiber-feed (thereby decoupling the size of the diffraction elements from the telescope diameter) followed by a series of dichroics and first order gratings which each disperse only a small fraction of the entire wavelength coverage. The CODEX proposal offers a potentially more elegant solution comprising a combination of anamorphic pre-slit optics, pupil-slicing, and post-dispersive beam-demagnification, which keep the overall optics size under control.

Filters Of particular concern for some of the imaging cameras are the requirements for large filters. This is already an issue prior to any ELT direct imager. For instance, the LSST concepts call for filters up to 740 mm in diameter

(Olivier et al. 2006). The PanSTARRS project (Kaiser et al. 2002) requires up to 514 × 514 mm filters, while the Skymapper telescope (Granlund et al. 2006) has sets of filters 309 × 309 mm in size. According to Döhring et al. (2006), Schott currently produces filters in standard sizes of 50 × 50 mm. Formats up to 165 × 165 mm are available, however the filter glasses GG, OG, and RG may be produced in sizes up to 900 × 900 mm once per year. While other suppliers may be possible, this last limitation hints that while some large filters will be available, careful scheduling of orders will be necessary.

4.3. Demanding Performance Requirements

With the large increase in collecting area comes the opportunity to push the limits of astronomical instrumentation. For instance, the CODEX spectrograph aims to measure directly the expansion of the universe. In order to achieve this, the instruments must obtain a radial velocity precision which is several orders of magnitude greater in precision than is currently possible. While the possibility of extremely high signal-to-noise observations contributes towards this goal, it is likely that new methods of instrument calibration will be necessary. One promising technique is the laser frequency comb described by Murphy et al. (2007). These 'frequency combs' provide a series of calibration lines whose absolute frequencies are known a priori to a precision better than 10^{-12}. Simulations suggest that the photon-limited wavelength calibration will allow radial velocity precisions of $\sim 1\,\mathrm{cm\,s^{-1}}$ given 400 nm of wavelength coverage. Several groups are currently attempting to demonstrate this technique on existing échelle spectrographs.

Another instrument with extremely demanding performance requirements is the QuantEYE concept for the E-ELT (Naletto et al. 2006). The aim of QuantEYE is to explore astrophysical variability on microsecond and nanosecond scales. The instrument concept calls not only for advanced single-photon counting avalanche diodes (SPADs), but will also demand extremely fast real time data processing which can cope with petabytes of data, while also dealing with various instrumental physical effects which become relevant at these time scales.

4.4. Large Focal Planes

All ELTs will naturally have very large focal planes. At the $f/8$ focal plane of GMT the plate scale is approximately $1''/\mathrm{mm}$. Therefore a $10'$–$20'$ field of view results in a focal plane that is between 600 mm and 1.2 m in size. Both TMT and the E-ELT concepts have similarly sized focal planes. This presents a challenge to instruments which rely on having a large field of view for their scientific productivity.

One solution is to design the instruments so that they have access to a subset of the full field. The focal plane is then divided into two or more sections which are accessed by copies of an identical instrument. This is the solution proposed by the GMACS and WFOS instruments for GMT and TMT, respectively.

Another approach is to deploy pick-off mirrors (and/or fibers) across the focal plane. Again, these may be fed into one or more identical instruments. This method has been proposed for the MOMSI instrument for OWL (Evans et al. 2006) and research into possible configurations has resulted in the "Planetary Positioning System" (Hastings et al. 2006). Other "Smart Focal Plane" projects[7] include MOEMS (micro-opto-electromechanical systems) and concepts such as the "Starbug" or "Crowd Surfer". For an overview of these concepts see Cunningham et al. (2005).

4.5. Detectors

As an example of the scale, consider the proposed E-ELT imager. In order to cover a 600 × 600 mm focal plane 100 4k × 4k CCDs would be required. While a considerable number, this does not seem outside the realms of possibility, and detector mosaics of comparable size may be a reality prior to the advent of an ELT (for instance, on Pan-STARRS). Several other instruments require detectors which will push current technology. In cases where detector mosaics are undesirable the largest single CCDs will be sought. The largest commercially available CCD detectors are presently between 4k × 2k and 4k × 4k, however, it is foreseeable that detectors up to 8k × 6k or 8k × 8k will be available in the near future (this limit is set by the maximum diameter of silicon wafers). In fact, the multi-object double spectrograph for the Large Binocular Telescope LBT MODS (Pogge et al. 2006) is likely to have 8k × 3k CCDs installed within a year of commissioning[8].

Infrared instrumentation also desires detectors as large as possible. Due to technical and financial limitations, such detectors are unlikely to be available larger than 4k × 4k for quite some time.

4.6. Adaptive Optics

A summary of the technical challenges of adaptive optics is outside the scope of this proceedings. For an overview of the issues involved see Hubin et al. (2006).

5. Summary

This paper has presented a sample of the challenges facing the designers and builders of the next generation of instrumentation for extremely large telescopes. There are undoubtedly many more challenges than those that have been discussed here. However, given the multitude of innovations that the investigations to date have yielded, it's certain that a range of excellent and competitive instruments can be built for any future giant telescope.

Acknowledgments. The author wishes to thank the Scientific Organizing Committee of the Frank N. Bash Symposium 2007 ("New Horizons in Astronomy") for the invitation to give a talk on this subject.

References

Cui, X. 2006, in IAU Symposium, Vol. 232, The Scientific Requirements for Extremely Large Telescopes, ed. P. Whitelock, M. Dennefeld, & B. Leibundgut, 391

---

[7]http://www.astro-opticon.org/joint_research_activities/focal_planes.html

[8]See http://www.astronomy.ohio-state.edu/MODS/Detectors/

Cunningham, C., Atad, E., Bailey, J., Bortoletto, F., Garzon, F., Hastings, P., Haynes, R., Norrie, C., Parry, I., Prieto, E., Howat, S. R., Schmoll, J., Zago, L., & Zamkotsian, F. 2005, in Proceedings of the SPIE, Vol. 5904, , Cryogenic Optical Systems and Instruments XI, ed. J. B. Heaney & L. G. Burriesci (Bellingham: SPIE), 281
Cunningham, C. & Crampton, D. 2006, in IAU Symposium, Vol. 232, The Scientific Requirements for Extremely Large Telescopes, ed. P. Whitelock, M. Dennefeld, & B. Leibundgut, 443
D'Odorico, S. 2006, in Proceedings of the ESO/Lisbon/Aveiro Conference, Precision Spectroscopy in Astrophysics, ed. N. C. Santos, L. Pasquini, A. C. M. Correia, & M. Romaniello, (Berlin/Heidelberg: Springer), 221
Döhring, T., Loosen, K.-D., & Hartmann, P. 2006, in Proceedings of the SPIE, Vol. 6273, Optomechanical Technologies for Astronomy, ed. E. Atad-Ettedgui, J. Antebi, L. Dietrich (Bellingham: SPIE), 62370U
Evans, C., Cunningham, C., Atad-Ettedgui, E., Allington-Smith, J., Assémat, F., Dalton, G., Hastings, P., Hawarden, T., Hook, I., Ivison, R., Morris, S., Ramsay Howat, S., Strachan, M., & Todd, S. 2006, in Proceedings of the SPIE, Vol. 6269, Ground-based and Airborne Instrumentation for Astronomy, ed. I. S. McLean, M. Iye (Bellingham: SPIE), 62692V
Froning, C., Osterman, S., Beasley, M., Green, J., & Beland, S. 2006, in Proceedings of the SPIE, Vol. 6269, Ground-based and Airborne Instrumentation for Astronomy, ed. I. S. McLean, M. Iye (Bellingham: SPIE), 62691V
Gilmozzi, R. & Spyromilio, J. 2007, The Messenger, 127, 11
Granlund, A., Conroy, P. G., Keller, S. C., Oates, A. P., Schmidt, B., Waterson, M. F., Kowald, E., & Dawson, M. I. 2006, in Proceedings of the SPIE, Vol. 6269, Ground-based and Airborne Instrumentation for Astronomy, ed. I. S. McLean, M. Iye (Bellingham: SPIE), 626927
Hastings, P., Ramsay Howat, S., Spanoudakis, P., van den Brink, R., Norrie, C., Clarke, D., Laidlaw, K., McLay, S., Pragt, J., Schnetler, H., & Zago, L. 2006, in Proceedings of the SPIE, Vol. 6273, Optomechanical Technologies for Astronomy, ed. E. Atad-Ettedgui, J. Antebi, L. Dietrich (Bellingham: SPIE), 62732X
Hubin, N., Ellerbroek, B. L., Arsenault, R., Clare, R. M., Dekany, R., Gilles, L., Kasper, M., Herriot, G., Le Louarn, M., Marchetti, E., Oberti, S., Stoesz, J., Veran, J. P., & Vérinaud, C. 2006, in IAU Symposium, Vol. 232, The Scientific Requirements for Extremely Large Telescopes, ed. P. Whitelock, M. Dennefeld, & B. Leibundgut, 60
Iye, M. 2006, in IAU Symposium, Vol. 232, The Scientific Requirements for Extremely Large Telescopes, ed. P. Whitelock, M. Dennefeld, & B. Leibundgut, 381
Kaiser, N., Aussel, H., Burke, B. E., Boesgaard, H., Chambers, K., Chun, M. R., Heasley, J. N., Hodapp, K.-W., Hunt, B., Jedicke, R., Jewitt, D., Kudritzki, R., Luppino, G. A., Maberry, M., Magnier, E., Monet, D. G., Onaka, P. M., Pickles, A. J., Rhoads, P. H. H., Simon, T., Szalay, A., Szapudi, I., Tholen, D. J., Tonry, J. L., Waterson, M., & Wick, J. 2002, in Proceedings of the SPIE, Vol. 4836, ed. J. A. Tyson & S. Wolff (Bellingham: SPIE), 154
Murphy, M. T., Udem, T., Holzwarth, R., Sizmann, A., Pasquini, L., Araujo-Hauck, C., Dekker, H., D'Odorico, S., Fischer, M., Hänsch, T. W., & Manescau, A. 2007, MNRAS, 380, 839
Naletto, G., Barbieri, C., Dravins, D., Occhipinti, T., Tamburini, F., Da Deppo, V., Fornasier, S., D'Onofrio, M., Fosbury, R. A. E., Nilsson, R., Uthas, H., & Zampieri, L. 2006, in Proceedings of the SPIE, Vol. 6269, Ground-based and Airborne Instrumentation for Astronomy, ed. I. S. McLean, M. Iye (Bellingham: SPIE), 62691W
Olivier, S. S., Seppala, L., Gilmore, K., Hale, L., & Whistler, W. 2006, in Proceedings of the SPIE, Vol. 6273, Optomechanical Technologies for Astronomy, ed. E. Atad-Ettedgui, J. Antebi, L. Dietrich (Bellingham: SPIE), 62730Y

Pasquini, L., Cristiani, S., Dekker, H., Haehnelt, M., Molaro, P., Pepe, F., Avila, G., Delabre, B., D'Odorico, S., Liske, J., Shaver, P., Bonifacio, P., Borgani, S., D'Odorico, V., Vanzella, E., Bouchy, F., Dessauges-Lavadsky, M., Lovis, C., Mayor, M., Queloz, D., Udry, S., Murphy, M., Viel, M., Grazian, A., Levshakov, S., Moscardini, L., Wiklind, T., & Zucker, S. 2005, The Messenger, 122, 10

Pogge, R. W., Atwood, B., Belville, S. R., Brewer, D. F., Byard, P. L., DePoy, D. L., Derwent, M. A., Eastwood, J., Gonzalez, R., Krygier, A., Marshall, J. R., Martini, P., Mason, J. A., O'Brien, T. P., Osmer, P. S., Pappalardo, D. P., Steinbrecher, D. P., Teiga, E. J., & Weinberg, D. H. 2006, in Proceedings of the SPIE, Vol. 6269, Ground-based and Airborne Instrumentation for Astronomy, ed. I. S. McLean, M. Iye (Bellingham: SPIE), 62690I

Spanò, P., Zerbi, F. M., Norrie, C. J., Cunningham, C. R., Strassmeier, K. G., Bianco, A., Blanche, P. A., Bougoin, M., Ghigo, M., Hartmann, P., Zago, L., Atad-Ettedgui, E., Delabre, B., Dekker, H., Melozzi, M., Snÿders, B., & Takke, R. 2006, Astronomische Nachrichten, 88, 789

Dougal Mackey is all about globular clusters

Globular Clusters in the Local Universe

Dougal Mackey

Institute for Astronomy, University of Edinburgh, Royal Observatory, Blackford Hill, Edinburgh, UK

Abstract. In this contribution I review several of the most interesting recent developments in the field of globular cluster research. The sensitivity and resolution of modern observational facilities in combination with recent advances in computational modelling are rapidly revolutionizing our understanding of clusters in the local Universe, with many important implications. On the observational front, large surveys are uncovering more and more new objects, while deep high-precision measurements are revealing previously unsuspected complexities in Galactic globular clusters. Computationally, the advent of special-purpose hardware and ever-more sophisticated codes has heralded the rise of realistic globular cluster N-body simulations. We are now, for the first time, in a position to confront superb observational data with the results of direct and highly detailed numerical modelling.

1. Introduction

From an astrophysical perspective, globular clusters (i.e., star clusters with masses $\sim 10^4 < M < 10^6 M_\odot$) constitute a class of relatively simple objects. The vast majority of those which have been observed in detail as resolved systems are well described as single stellar populations. Each given cluster is composed of a dense, roughly spherically symmetric group of stars which were formed over a very short duration of time, in a very small volume of space, and out of chemically homogeneous material. The central regions of globular clusters are among the densest stellar concentrations observed in the Universe – these are collisional systems where interactions between stars are relatively common.

Globular clusters are observed in almost all galaxies above a certain mass limit, irrespective of morphological type – from dwarf spheroidals (such as the nearby Fornax dSph, which contains five globulars) to giant ellipticals (such as M87 in the Virgo cluster, which contains in excess of ten thousand globulars). As compact, luminous objects, they can be detected to large distances, and serve as excellent tracers of the star formation history, mass distribution, and chemical and structural evolution of a given parent galaxy. Globular clusters are also, arguably, the smallest stellar aggregations which do not appear to be associated with significant amounts of dark matter.

Combined, the above factors lend globular clusters a considerable utility, and they are important to a surprisingly large number of astrophysical fields of research, including star formation and stellar evolution; the study of a wide variety of binary and variable stars, and stellar exotica; the dynamics of self-gravitating systems; and galaxy formation and evolution, with implications for cosmology. In our Galaxy, the globular clusters are ancient objects, many of which possess very low metal abundances. Most were probably involved with the earliest stages of formation of the Milky Way. In this respect, they represent one of the very few direct nearby links to the earliest epochs of the Universe which are still available to observe today.

Unlike many of the review topics in these proceedings (such as exoplanets, astrobiology, or γ-ray bursts) the study of globular clusters, especially those in the Milky Way, is a very old field of astronomy. Several of the Galactic globulars bear Messier numbers, and it was known as far back as the late 1700s (e.g., by Herschel) that these objects could be resolved into stars using telescopic observations. Basic mathematical models for the structures and internal dynamics of globular clusters were in existence by the late 19th and early 20th centuries. Despite this long history, globular clusters still hold a large number of secrets, and their study remains a vigorous field of astrophysics. In recent years, globular cluster astronomy has received significant new impetus due to a number of developments. The advent of high sensitivity instruments with exquisite resolution, particularly the Hubble Space Telescope (HST) and terrestrial 8-10 m class facilities, has driven a number of spectacular observational discoveries. Simultaneously, ever-increasing computational power and sophistication, and particularly the construction of special-purpose hardware for N-body simulations, has heralded a new era of realistic globular cluster modelling.

In this contribution I aim to provide a brief review of selected recent observational developments, as well as the new and exciting possibilities provided by realistic N-body models and their combination with high-quality data. Due to time and space constraints, this discussion will be limited to globular clusters in the local Universe – primarily the Milky Way and Magellanic Clouds – where clusters may be fully resolved observationally, and examined on a star-by-star basis. The study of extra-Galactic globular clusters, while not covered here, is also an exciting and rapidly developing field, again driven by modern telescopic capabilities. The interested reader is referred to a recent comprehensive review of this topic by Brodie & Strader (2006).

2. New Members of an Old Family

Despite the fact that a large fraction of the Galactic family of globular clusters have been known for well over a century, the total population of such objects remains unclear. The standard catalog of Galactic globular cluster properties – that of Harris (1996), last updated in 2003[1] – lists 150 members. One exciting development in the past couple of years has been the significant number of new additions to this family. Primarily these are due to deep wide-area surveys which are able to probe previously unexplored regions of parameter space, whether this be the very faint end of the globular cluster luminosity function (e.g., SDSS) or the Galactic regions heavily obscured by dust and gas (e.g., 2MASS). In all, roughly equal numbers of clusters have been found in the bulge/disk region of the inner Galaxy and in the outer Galactic halo since 2005.

Examples of new inner-Galaxy globular cluster candidates are:

---

[1]See http://www.physics.mcmaster.ca/Globular.html

- GLIMPSE-C01, lying ~ 4 kpc distant, discovered in the Spitzer GLIMPSE survey (Kobulnicky et al. 2005).

- AL 3, an ancient metal-poor object ~ 6 kpc distant (Ortolani et al. 2006).

- Four objects from the 2MASS survey, all identified in 2007 – FSR 1735 (Froebrich et al. 2007b), FSR 1767 (Bonatto et al. 2007), FSR 0584 (Bica et al. 2007), FSR 0190 (Froebrich et al. 2007a).

Two of the 2MASS objects lie surprisingly close to the sun: FSR 1767 and 0584, at 1.5 and 1.4 kpc respectively, placing them as two of the nearest globular clusters. They remained unknown until very recently due to the large amount of absorption in the line-of-sight: $A_V > 9$ mag in the case of FSR 0584.

Examples of new globular cluster candidates in the outer halo are:

- Whiting 1, a rather young globular cluster (~ 6.5 Gyr) located 29 kpc distant, which may well be a former member of the infalling Sagittarius dwarf galaxy (Carraro 2005; Carraro et al. 2007).

- Willman 1, a very diffuse SDSS object lying at 40 kpc, which may in fact be a very faint dwarf galaxy (Willman et al. 2005).

- SDSS J1257+3419, at ~ 150 kpc (Sakamoto & Hasegawa 2006; Belokurov et al. 2007); again this object may possibly be a small dwarf galaxy.

- Segue 1, at 23 kpc (Belokurov et al. 2007).

- Two additional, extremely faint SDSS objects – Koposov 1 and 2 at ~ 50 and ~ 40 kpc respectively (Koposov et al. 2007).

Several of these objects are among the very faintest Galactic globular clusters, with integrated luminosities $M_V > -3$. The recent discovery of so many new globular cluster candidates from SDSS and 2MASS suggests that upcoming wide-field optical and IR survey facilities such as Pan-STARRS, SkyMapper, and VISTA should prove fruitful in this regard.

New globular cluster discoveries have not been limited to the Milky Way. Most notably, several new examples of unusually extended clusters have been discovered in recent years in external galaxies. These include the "faint fuzzies", discovered in lenticular galaxies by Brodie & Larsen (2002); a number of diffuse objects in the Virgo cluster of galaxies (Peng et al. 2006); and, closer to home, a family of luminous extended clusters in M31 (Huxor et al. 2005). None of these objects appear to have counterparts in the Milky Way. Recent HST/ACS imaging of four of the M31 clusters (Mackey et al. 2006) has demonstrated that they are ancient metal-poor objects indistinguishable from regular globular clusters apart from their structures, which are very extended for their observed luminosities (see Fig 1). Explaining the unusual structures of these newly discovered objects, why they apparently form in some galaxies but not others, and their relationship to classical "compact" globular clusters remains an open challenge.

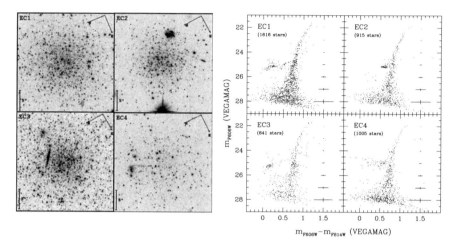

Figure 1. *Left:* HST/ACS images of four of the unusual M31 clusters discovered by Huxor et al. (2005). These objects are very extended for their observed luminosities, with characteristic (half-light) radii $r_h \sim 30$ pc. *Right:* color-magnitude diagrams for the four clusters, which exhibit old metal-poor stellar populations strongly resembling those found in classical globular clusters (figures reproduced by permission from Mackey et al. (2006)).

3. What's in a Globular Cluster?

Modern telescopes and instrumentation, and in particular the superb spatial resolution of HST, have recently been providing dramatic new insights into the stellar populations present in Galactic globular clusters. It is possible to completely resolve even the densest nearby clusters into their constituent stars, and, with long exposure times such clusters are being probed to new depths and with unprecedented photometric precision.

One crucial technique facilitated by HST has been the proper-motion cleaning of cluster color-magnitude diagrams (CMDs). This procedure relies on multi-epoch imaging of the same field, with epochs usually separated by several years. These observations are then used to measure the proper motions of stars relative to some frame of reference. Because the internal velocities of stars in a given cluster are much smaller than the global motion of the cluster about the Galaxy, cluster members form a well-defined co-moving group in proper-motion space. These objects are hence relatively easy to separate from any contaminating population of unresolved background galaxies or non-cluster field stars, resulting in amazingly clean CMDs. An excellent example of this procedure may be seen in Richer et al. (2007).

One exciting development is the possibility of using this technique to go one step further and measure the residual proper motions of cluster members due to their velocities *within the cluster*. This offers new insight into the dynamical state of a given cluster, and, in combination with accurate radial velocity measurements for member stars, opens up the prospect of obtaining dynamical distances to globular clusters completely independently of calibrations due to standard candles such as RR Lyrae stars. Such distances have been obtained for a number of clusters, for example 47 Tuc (McLaughlin et al. 2006).

One major HST project in recent years has been to obtain ultra-deep imaging of several of the closest Galactic globular clusters. Most recently this has been achieved for NGC 6397, a metal-poor globular cluster ([Fe/H] ≈ -2.0) lying 2.5 kpc from the sun. Richer et al. (2007) imaged a single field with HST/ACS for 126 orbits. Their resulting proper-motion-cleaned CMD is the deepest ever obtained for a globular cluster, and exhibits several previously unseen features. The main sequence is extremely narrow, and clearly terminates well above their faint completeness limit. Comparison with theoretical models suggests the drop-off in the main sequence is most likely due to the observations reaching the level of the hydrogen-burning limit in the cluster.

In addition, there is an extended white-dwarf cooling sequence visible in the CMD, which hooks to the blue at its faint end as predicted theoretically due to atmospheric collision-induced absorption. This is the first time this hook has been observed in a star cluster (Richer et al. 2007). Hansen et al. (2007) examined the white dwarf cooling sequence in detail, demonstrating that a cutoff at its faint end is due to the truncation of the cluster white dwarf luminosity function. Via a statistical comparison of the observed data with a number of models, Hansen et al. (2007) were able to estimate an absolute age for NGC 6397: $\tau = 11.47 \pm 0.47$ Gyr (95% confidence limits). Taken at face value, this measurement places the formation of NGC 6397 at a redshift $z \sim 3$, near the peak of the cosmological star formation rate. NGC 6397 apparently formed well after the epoch of reionization, which is generally taken to be at $z \gtrsim 6$. If NGC 6397 is a representative globular cluster, this result has important implications for various models which have associated globular cluster formation with the reionization of the Universe (see e.g., Brodie & Strader 2006). NGC 6397 seems to have formed well before the Galactic disk (Hansen et al. 2007).

Davis et al. (2007) have examined the radial distribution of young white dwarfs in the HST/ACS deep field, and found that these objects have a significantly more extended radial distribution than do the most massive main sequence stars. This is contrary to expectations, because the progenitors of young white dwarfs were more massive than any present main sequence stars, while the white dwarfs themselves were formed less than a dynamical relaxation time ago. These objects should therefore still possess a more centrally concentrated radial distribution than that of upper main sequence stars. The simplest interpretation of this measurement is that white dwarfs are born with a small natal kick of order a few $km\,s^{-1}$, perhaps due to asymmetric mass loss.

Moving on, new high-precision photometry of a number of massive Galactic globular clusters is challenging our notions of how some globular clusters are formed. The most massive Galactic globular cluster, ω Centauri, has long been known as a very unusual object which harbors a peculiar mix of stellar populations. This is contrary to the generally held picture of globular clusters being composed of single stellar populations. The more closely ω Centauri is examined, the more pathological it appears to become. Recent work by Villanova et al. (2007) summarizes most of what is presently known. The CMD for this object exhibits at least four distinct sub-giant branches, a very widely spread red-giant branch, and an extremely extended blue horizontal branch. Spectroscopy of stars on the sub-giant branch has revealed a number of stellar populations: (i) an old metal-poor population; (ii) two $\sim 3-4$ Gyr younger populations, one metal-poor and one of intermediate metallicity; (iii) an old metal-rich population; and (iv) a small residual population about which little is yet known. Clearly these groups cannot have formed in a single progression of star-formation and chemical enrichment. The main sequence of ω Centauri offers further mysteries – it is clearly split in two, and, contrary to all expectations, the blue branch is more metal-rich than the red branch. The only viable explanation for this feature is that the blue metal-rich population is super-enriched in helium (with a mass-fraction of up to $Y \sim 0.4$); this would also help explain the extreme blue horizontal branch. Overall, the formation of ω Centauri must have involved the formation and merger of several "fragments" over an extended period of time. The specifics of this process, and particularly the production of a super He-enriched population, remain an open challenge.

Very recent HST/ACS photometry of other massive Galactic globulars is further complicating the picture. Piotto et al. (2007) have identified a *triple* main sequence in the cluster NGC 2808. This cluster also possesses an extremely extended blue horizontal branch. Again, multiple populations of He-enriched stars appear to be the only viable explanation; even so, stars in NGC 2808 show little evidence for a spread in either age or iron abundance. Milone et al. (2007) have found that the massive globular cluster NGC 1851 also possesses an unusual CMD – in this case in the form of a cleanly split sub-giant branch. However, the red-giant branch and main sequence appear normal in this object, implying that there is likely an age spread among the stars in NGC 1851, but no spread in iron or helium abundance. The most massive globular clusters in our Galaxy are not only far more complex than previously suspected, there does not appear to be any coherent pattern to these complexities from cluster to cluster.

One further recent development worth mentioning is the mounting evidence for intermediate mass black holes (IMBHs; masses $M \gtrsim 1000\,\mathrm{M}_\odot$) in a few globular clusters. This possibility is of considerable interest because it is hypothesized that IMBHs may be the seeds of the supermassive black holes observed at the centers of galaxies; one possible IMBH formation channel is in extremely dense young massive star clusters. The object with the most secure IMBH detection is the very massive globular cluster G1 in M31. This object exhibits a brightness profile and kinematics consistent with the presence of an IMBH of mass $M \sim 2 \times 10^4\,\mathrm{M}_\odot$ (Gebhardt et al. 2005). X-ray and radio emission from G1 have also been detected; these are consistent with the existence of an accreting IMBH of the correct mass (Pooley & Rappaport 2006; Ulvestad et al. 2007); however other possibilities cannot yet be completely ruled out. On a closely related topic, X-ray measurements have very recently provided the first evidence for a much less massive black hole in a globular cluster, albeit in a distant extra-Galactic object (Maccarone et al. 2007; Zepf et al. 2007).

4. New Horizons in Cluster N-body Modelling

In addition to the numerous observational developments in globular cluster astronomy in recent years, there have also been significant advances in the numerical modelling of these systems. While there are a wide variety of computational schemes which are used to model globular cluster evolution (such as Fokker-

Plank and Monte Carlo methods, and conducting gas models – see e.g., Meylan & Heggie (1997) for further details), the method which has undergone perhaps the most significant recent advances is that of N-body modelling.

In this formulation, the equations of motion for each of the N particles in a self-gravitating system are integrated directly. For the i-th particle, of mass m_i and position $\mathbf{r}_i(t)$, these take the form:

$$\ddot{\mathbf{r}}_i = -G \sum_{j=1, j \neq i}^{N} \frac{m_j (\mathbf{r}_i - \mathbf{r}_j)}{|\mathbf{r}_i - \mathbf{r}_j|^3} \ . \tag{1}$$

Unlike in cosmological N-body models, high accuracy integration is required, close interactions between particles are common and important. Furthermore, the size of a typical time-step is short compared to the lifetime of the system. From Eq. 1, the computational requirement increases at least as N^2; in fact the cost scales more like N^3 – the extra factor of N is contributed by the increased time-scale for heat conduction (see e.g., Heggie & Hut 2003). N-body modelling has historically, therefore, been restricted to systems with N of only several thousand particles, or, with massive effort, several tens of thousands of particles. Even so, N-body simulations have the advantage that all relevant cluster physics may be accounted for with a minimum of simplifying assumptions.

There have been two recent developments in the field of cluster N-body modelling which are bringing it into a new era. The first of these is the advent of special-purpose hardware, in particular the GRAPE (GRAvity PipE)[2] series of machines – most recently GRAPE-6 (Makino et al. 2003; Fukushige et al. 2005) – to calculate the N^2 forces necessary for each integration step of Eq. 1. Such a machine acts only as an accelerator for the force summations; the main body of the simulation itself is still run on the host computer. The availability of GRAPE-6 is now routinely allowing cluster models with $N \sim 10^5$ particles (i.e., objects at the lower end of the globular cluster mass function) to be calculated over more than a Hubble time of evolution.

The second major advance has been in the sophistication of the N-body codes themselves (for full details see Aarseth 2003; Heggie & Hut 2003). Various *regularization* algorithms have been implemented to alleviate the computational bottleneck induced by the close encounter of two or more stars (e.g., Mikkola & Aarseth 1993, 1998). This treatment allows stable binary and multiple systems to be included in cluster models as well as the possibility of random strong interactions between stars. In addition, metallicity-dependent stellar evolution prescriptions (e.g., Hurley, Pols, & Tout 2000) have been incorporated into N-body codes such that each star in a model cluster loses mass and changes type in step with the dynamical evolution of the cluster (e.g., main sequence star $\rightarrow$ red giant $\rightarrow$ helium burning star $\rightarrow$ white dwarf for a solar-type member). It is possible to assign an arbitrary initial mass function to any given model; doing so can greatly alter the evolutionary path a cluster follows, since the masses m_i in Eq. 1 become time-dependent when stellar evolution is accounted for. An additional important consequence of the inclusion of stellar evolution

---

[2] See http://astrogrape.org and http://www.manybody.org for further details.

routines is that each star possesses a finite time-dependent radius, which is vital when close encounters between stars are considered. Binary star evolution has also been incorporated into modern N-body codes (e.g., Hurley, Tout, & Pols 2002), including such processes as the tidal circularization of orbits, mass transfer, common envelope evolution, and mergers. Finally, it is possible to place a model cluster on an arbitrary orbit about a host galaxy, resulting in external tidal forces which can again result in a wide variety of evolutionary paths. Two examples of state-of-the-art, freely-distributed N-body codes are the NBODYx family (Aarseth 2003) and the STARLAB software environment[3].

Combined, the hardware and software developments described above mean that we are now, for the first time, in a position to directly confront high-precision star cluster observations with realistic N-body modelling. In particular, because simulations with large N are now feasible, and the stellar evolution of each member in a model cluster can be followed, it is possible to perform simulated observations of model clusters as they evolve. This process may, for example, be used to compare the results of a simulation to the observed properties of a real system, to understand the biases introduced due to the imperfections inherent in observations of clusters with telescopes, or to predict the observational signature of a given process which is being modelled.

5. A Case Study: Cluster Evolution in the Magellanic Clouds

The best way to illustrate the new-found synergy between high quality observations and realistic N-body modelling is by means of a brief case study. This also serves to demonstrate an aspect of globular cluster evolution which has not been previously widely appreciated, but which has been explored by us recently in a series of large N-body simulations (Mackey et al. 2007a,b).

The work presented here concerns massive star clusters in the Large and Small Magellanic Clouds (LMC/SMC), two nearby companion galaxies to the Milky Way. Both the LMC and SMC possess extensive systems of star clusters with masses comparable to the Galactic globulars, but crucially *of all ages*: $10^6 \lesssim \tau \lesssim 10^{10}$ yr. While we can accurately assess the end-points of globular cluster evolution from the exclusively ancient ($\tau \gtrsim 10^{10}$ yr) Galactic population, we must infer the complete long-term development which has brought these objects to their observed states. In contrast, by observing members of the LMC and SMC, we can obtain snapshots of all phases of massive star cluster development.

As with many aspects of globular cluster astronomy, HST has proved central to this endeavor. Early ground-based observations by Elson et al. (1989) suggested that massive LMC star clusters exhibit an unexpected relationship between core radius[4] (r_c) and age – specifically that the spread in core radius increases with increasing age. Building on this result, Mackey & Gilmore

---

[3] See http://www.ids.ias.edu/~starlab/ .

[4] The core radius is a convenient measure of how centrally compact a star cluster is. In this contribution I refer to the observationally defined quantity – the radius at which the surface density (or brightness) of a cluster has decreased to half its central value. The interested reader is referred to Wilkinson et al. (2003) for a more detailed discussion.

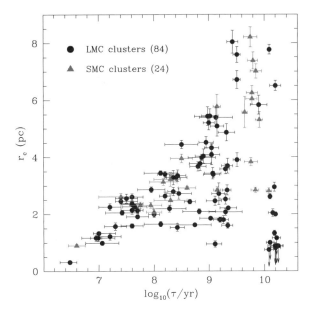

Figure 2. Core radius versus age for massive star clusters in the Large and Small Magellanic Clouds. This Figure includes all clusters from the HST/WFPC2 measurements of Mackey & Gilmore (2003a,b) as well as subsequent HST/ACS measurements. Reproduced by permission from Mackey et al. 2008, MNRAS, *in press*.

(2003a,b) used HST/WFPC2 imaging of 63 massive Magellanic Cloud clusters to demonstrate that a striking trend of this form does indeed exist, in both the LMC and the SMC systems. Subsequent HST/ACS imaging of 45 additional clusters has significantly improved sampling of the radius-age plane. Structural measurements for all 108 LMC and SMC clusters may be seen in Fig 2.

The observed radius-age relationship provides strong evidence that our understanding of massive cluster evolution is incomplete, since standard quasi-equilibrium models do not predict long-term large-scale core expansion (see e.g., Meylan & Heggie 1997). Discerning the origin of the radius-age trend is therefore of considerable importance. Large N-body simulations are an ideal tool to investigate the possibility that the radius-age trend is the result of cluster dynamical evolution, because the flexibility of these models allows various physical processes to be isolated and studied in a straightforward manner. Furthermore, with presently available hardware and software, it is possible to construct direct realistic models of present-day young Magellanic Cloud clusters (this requires $N \sim 10^5$ particles per model). Since the radius-age trend is defined observationally, direct N-body modelling also has the advantage that the observational techniques used to produce Fig. 2 may be reproduced on the simulated clusters, thus implicitly accounting for any inherent biases in the measurement process.

We used the NBODY4 code (Aarseth 2003) in combination with a 32-chip GRAPE-6 special-purpose computer at the Institute of Astronomy in Cambridge, to investigate the effect of stellar-mass black holes on massive star cluster evolution. We generated initial conditions with properties (masses, structures, densities, etc) as similar as possible to those observed for the youngest Magellanic Cloud clusters. We selected the IMF of Kroupa (2001), with a stellar mass range $0.1 - 100 \, M_\odot$, so that with $N \sim 10^5$ particles our cluster models possessed total masses $\log M_{\rm tot} \sim 4.75$. Full details are provided in Mackey et al. (2007b).

Under the assumption that black holes are produced in the supernova explosions of stars with masses initially above $20 \, M_\odot$, we found that ~ 198 black holes formed in each model cluster. We assigned the black holes masses in the range $8 \leq m_{\rm BH} \leq 12 \, M_\odot$. At formation each black hole was given a velocity kick – the magnitude of which determines whether a black hole is retained in a star cluster (if, roughly speaking, the size of the kick is less than the cluster escape velocity) or whether it is ejected. Hence the distribution of velocity kicks determines the retention fraction of black holes in a cluster. In the present contribution I consider only the extreme cases of zero retention and complete retention of the black hole population.

Many very young LMC and SMC clusters exhibit some degree of mass segregation – where the most massive stars in a cluster are preferentially found in the central regions of that cluster. In order to include any effects of this in our models, we developed a method to generate clusters with primordial mass segregation in a self-consistent fashion; again, full details are in Mackey et al. (2007b). Here I again consider the extreme cases – models either with no primordial mass segregation, or a strong degree of primordial mass segregation chosen to match that observed in young Magellanic Cloud objects such as R136, NGC 330, 1805, and 1818.

Four models therefore define the extremities of the parameter space we investigated: the combination of clusters with either zero or complete black hole retention, and with either no primordial mass segregation or strong primordial mass segregation. These simulations were expected to cover the limits of cluster behavior; this was verified with further models, which are not presented here (see Mackey et al. 2007b).

First consider Runs 1 and 2, which have no primordial mass segregation, and zero and complete black hole retention respectively. Their evolution is visible in Fig. 3. Run 1 behaves exactly as expected for a classical massive star cluster. There is an early mass-loss phase ($\tau \leq 100$ Myr) due to the evolution of the most massive cluster stars. The black holes are formed in supernova explosions between 3.5-10 Myr; however, all receive large velocity kicks and escape the cluster. The early mass-loss is not reflected in the evolution of r_c, presumably because it is evenly distributed throughout the cluster. Hence, the general progression is a slow contraction as two-body relaxation proceeds and mass segregation sets in. The cluster core collapses near the end of the simulation.

Run 2 evolves similarly to a point, but, in striking contrast to Run 1, subsequently undergoes dramatic and long-term core expansion. The only difference between this model and Run 1 is that all the black holes are retained in Run 2. Once early stellar evolution is complete, these black holes are more massive than all other cluster members (of mean mass $m_* \approx 0.5 \, M_\odot$) and hence sink to the core on a time-scale of $\sim (m_*/m_{\rm BH}) \, t_{rh} \approx 100$ Myr. By 200 Myr, the

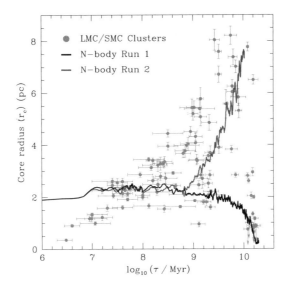

Figure 3. Core radius evolution of N-body Runs 1 and 2 (lower and upper lines, respectively). Neither model has any primordial mass segregation; Run 1 has zero black hole retention, while Run 2 has complete retention. Reproduced by permission from Mackey et al. 2008, MNRAS, *in press*.

mass density of black holes at the cluster center is similar to that of the stars; by 400 Myr it is about three times larger. Soon after, the central black hole subsystem becomes unstable against further contraction and decouples from the stellar core in a runaway collapse. At 490 Myr, the central density of the black hole subsystem is ~ 80 times that of the stars. This is sufficient for the creation of stable black hole binaries in three-body interactions – the first is formed at ~ 510 Myr, and by 800 Myr there are four.

Until this phase the evolution of Run 2 is observationally identical to that of Run 1. Neither the initial retention of the black holes in Run 2, nor the subsequent formation of a central black hole subsystem in this model leads to differential evolution of r_c. Once formed, the binary black holes undergo superelastic collisions with other black holes in the core. The binaries become more tightly bound, and the released binding energy is carried off by the interacting black holes. This leads to black holes being *scattered* outside r_c, often into the cluster halo, as well as to black holes being *ejected* from the cluster. Eventually a black hole binary is sufficiently tightly bound that the recoil velocity imparted to it during a collision is larger than the cluster escape velocity, and the binary is ejected. A black hole that is scattered into the halo gradually sinks back into the center via dynamical friction, thus transferring its newly-gained energy to the stellar component of the cluster. Most is deposited within r_c, where the stellar density is greatest. Ejected black holes also transfer energy to the cluster, since a mass m escaping from a cluster potential well of depth $|\Phi|$ does work $m|\Phi|$ on the cluster. This mechanism is particularly effective in heating the stellar core, since black holes are ejected from the very center of the cluster, and the energy

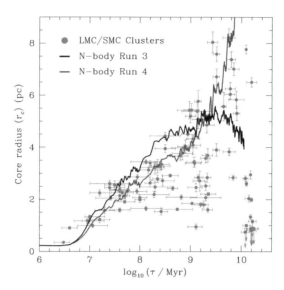

Figure 4. Core radius evolution of N-body Runs 3 and 4. Both models have strong primordial mass segregation; Run 3 has zero black hole retention, while Run 4 has complete retention. Reproduced by permission from Mackey et al. 2008, MNRAS, *in press*.

contributed to each part of the cluster is proportional to the contribution which that part makes to the central potential. Together these two processes result in significant prolonged core expansion, starting at $\tau \approx 650$ Myr. Merritt et al. (2004) observed expansion due to similar processes in their simplified models.

Next, consider the evolution of Runs 3 and 4, which are primordially mass segregated versions of Runs 1 and 2, respectively (Fig. 4). In these models, the most massive stars preferentially reside in the cluster cores. The early mass-loss due to stellar evolution is therefore highly centrally concentrated, and the amount of heating per unit mass lost is maximized. This leads to dramatic early core expansion. Run 3 traces the observed upper envelope of clusters until an age of several hundred Myr. Run 4 retains all its black holes and hence loses less mass than Run 3 during this period – this is reflected in its smaller core radius. After the early mass-loss phase is complete, core expansion stalls in both runs. Two-body relaxation gradually takes over in Run 3, leading to a slow contraction of its core radius. At $\tau = 1$ Gyr, the median relaxation time $t_{rh} \approx 4$ Gyr; hence this cluster is not near core collapse by the end of the simulation.

In Run 4, the black hole population evolves similarly to that in Run 2. One might naively expect the earlier development of a compact black hole subsystem in Run 4, because the black holes are already located in the core due to the primordial mass segregation. However, the centrally concentrated mass-loss acts against the accumulation of a dense black hole core, and the first binary black hole does not form until 570 Myr, a similar time to the equivalent non mass segregated model (Run 2). As in Run 2, the evolution of the black hole subsystem leads to expansion of the core radius. This begins at $\tau \approx 800$ Myr and continues for the remainder of the simulation. By $\tau \approx 10$ Gyr, Run 4 has $r_c \sim 11$ pc, comparable to that observed for the most extended old Magellanic Cloud clusters (e.g., the Reticulum cluster).

The four N-body simulations described above cover the observed cluster distribution in radius-age space, thereby defining a possible dynamical origin for the radius-age trend. At ages less than a few hundred Myr, cluster cores can expand due to centrally concentrated mass-loss from stellar evolution. At later times, expansion can be induced via heating due to a population of stellar-mass black holes. Although we have assumed complete retention for two of the four models presented above, this is not necessary for cluster expansion. Black hole kicks of order $10 \le v_{\text{kick}} \le 20$ km s^{-1} result in roughly fifty per cent retention in our model clusters; evolution in such systems is demonstrably intermediate between that of Runs 1 and 2, or 3 and 4 (Mackey et al. 2007b). Given this, for the scenario described above to be a viable explanation for the radius-age trend, variations in black hole population size between otherwise similar clusters are required. Discussion in this regard is presented by (Mackey et al. 2007b).

Various arguments suggest the dynamical core expansion processes we have investigated will scale efficiently to more massive clusters than we have been able to directly model with N-body simulations, such as most of the Galactic globular cluster population (Mackey et al. 2007a,b). Core expansion due to early mass-loss and prolonged black hole heating has strong implications for the observed properties of globular clusters as well as their survivability. For example, extended clusters are significantly more susceptible to tidal disruption, so it is important to account for expansion effects in studies of, say, the evolution of the globular cluster mass function, or the destruction of globular clusters in the Galactic system. Core expansion due to black holes may also offer a viable explanation for the origin of the various newly-discovered classes of extended globular clusters discussed in Section 2.

It is not straightforward to observationally test the prediction that some massive clusters should harbor retained populations of stellar-mass black holes. The N-body modelling described above has demonstrated that long-lived black hole-star binaries are extremely rare, mainly because clusters with retained black holes tend to be very diffuse objects for most of their lives. It is unsurprising that only one stellar-mass black hole X-ray source is known in a cluster (Maccarone et al. 2007; Zepf et al. 2007). More promising is the possibility of measuring the dynamical effect a black hole population should have on the stellar component of the cluster. Unlike for an IMBH, the N-body models show that no density and velocity cusp will form; however, the velocity dispersion of the cluster should be larger than is to be expected solely from the observed luminous mass (Mackey et al. 2007b). Observations to test this will be difficult, primarily because the target clusters are extended diffuse objects with low velocity dispersions and few bright stars. It may, none the less, be possible to make sufficiently precise observations with multiplexing spectrographs on presently available 8-10 m-class telescopes, or on planned larger facilities.

6. Future Prospects

As described in this review, many recent developments in our understanding of globular clusters have been driven primarily by advances in both observational and computational capabilities. Among the most perplexing open questions at present concern the absolute ages of Galactic globular clusters; the presence (or not) of black holes of various masses within globular clusters; the origin of the multiple stellar populations and enhanced He abundances observed in massive Galactic globular clusters; and the cosmological context of globular cluster formation. It is quite likely that these questions are inextricably linked.

Technology continues to advance apace, which augers well for globular cluster astronomy. On the observational front, the next couple of years should (with luck) see the refurbishment of HST with new high-resolution cameras, while the coming decade may see the construction of one or several 30-50 m-class ELTs. New high-quality observations will doubtlessly shed new light on presently open questions, as well as, in all probability, posing many new ones.

On the computational front, 2008 should see the release of the next generation GRAPE machine (GRAPE-DR), which will represent another significant leap in N-body computing power. Novel technologies, such as using commercially available graphics processing units for the force evaluation in the N-body problem (Portegies Zwart et al. 2007) are under development. N-body codes are also set to reach new levels of sophistication. At present, stellar evolution treatments are implemented via analytic formulae derived from a set of detailed models. This is relatively restrictive, since the only way to include any updates in our understanding of stellar evolution is to re-derive and re-implement these formulae. Work is therefore underway to incorporate live stellar evolution into N-body codes (e.g., Church & Tout 2004), which would immensely increase their flexibility. For example, critical parameters such as convective overshooting could be varied as necessary, while chemical inhomogeneities between clusters and even within individual clusters (cf. Section 3.) could be introduced. Incorporating new advances in stellar evolution models would also be straightforward.

On a related front, work is underway to associate spectral libraries with the evolving stars in model clusters (e.g., Borch et al. 2007), such that the integrated spectrum of a cluster may be investigated as a function of its age, and structural and dynamical state. For extra-Galactic globular clusters, being able to accurately interpret integrated spectra and colors is vitally important, since these distant objects cannot be resolved into individual stars. Finally, it may be possible to incorporate additional modules into existing N-body codes, in order to provide a more realistic treatment of certain events. A prime example of this would be the inclusion of a SPH module to accurately model the merger or collision of stars in dense clusters.

Acknowledgments. D. M. is supported by a Marie Curie Excellence Grant from the European Commission under contract MCEXT-CT-2005-025869.

References

Aarseth, S. J. 2003, Gravitational N-Body Simulations (Cambridge University Press, Cambridge)
Belokurov, V., et al. 2007, ApJ, 654, 897
Bica, E., Bonatto, C., Ortolani, S., & Barbuy, B. 2007, A&A, 472, 483
Bonatto, C., Bica, E., Ortolani, S., & Barbuy, B. 2007, MNRAS, 381, L45
Borch, A., Spurzem, R., & Hurley, J. 2007, A&A, submitted (arXiv:0704.3915)
Brodie, J. P., & Larsen, S. S. 2002, AJ, 124, 1410
Brodie, J. P., & Strader, J. 2006, ARA&A, 44, 193

Carraro, G. 2005, ApJ, 621, L61
Carraro, G., Zinn, R., & Moni Bidin, C. 2007, A&A, 466, 181
Church, R. P., & Tout, C. A. 2004, MmSAI, 75, 688
Davis, D. S. et al. 2007, MNRAS, in press (arXiv:0709.4286)
Elson, R. A. W., Freeman, K. C., & Lauer, T. R. 1989, ApJ, 347, L69
Froebrich, D., Meusinger, H., & Davis, C. J. 2007a, MNRAS, in press (arXiv:0710.2030)
Froebrich, D., Meusinger, H., & Scholz, A. 2007b, MNRAS, 377, L54
Fukushige, T., Makino, J., & Kawai, A. 2005, PASJ, 57, 1009
Gebhardt, K., Rich, R. M., & Ho, L. C. 2005, ApJ, 634, 1093
Hansen, B. M. S., et al. 2007, ApJ, in press, (astro-ph/0701738)
Harris, W. E. 1996, AJ, 112, 1487
Heggie, D., & Hut, P. 2003, The Gravitational Million-Body Problem (Cambridge University Press, Cambridge)
Hurley, J. R., Pols, O. R., & Tout, C. A. 2000, MNRAS, 315, 543
Hurley, J. R., Tout, C. A., & Pols, O. R. 2002, MNRAS, 329, 897
Huxor, A. P. et al. 2005, MNRAS, 360, 1007
Kobulnicky, H. A., et al. 2005, AJ, 129, 239
Koposov, S., et al. 2007, ApJ, 669, 337
Kroupa, P. 2001, MNRAS, 322, 231
Maccarone, T. J., Kundu, A., Zepf, S. E., & Rhode, K. L. 2007, Nature, 445, 183
Mackey, A. D., & Gilmore, G. F. 2003a, MNRAS, 338, 85
—. 2003b, MNRAS, 338, 120
Mackey, A. D., Wilkinson, M. I., Davies, M. B., & Gilmore, G. F. 2007a, MNRAS, 379, L40
—. 2007b, MNRAS, submitted
Mackey, A. D., et al. 2006, ApJ, 653, L105
Makino, J., Fukushige, T., Koga, M., & Namura, K. 2003, PASJ, 55, 1163
McLaughlin, D. E. et al. 2006, ApJS, 166, 249
Merritt, D., Piatek, S., Portegies Zwart, S., & Hemsendorf, M. 2004, ApJ, 608, L25
Meylan, G., & Heggie, D. C. 1997, A&A Rev., 8, 1
Mikkola, S., & Aarseth, S. J. 1993, Celest. Mech. Dyn. Astron., 57, 439
—. 1998, New Astron., 3, 309
Milone, A. P., et al. 2007, ApJ, in press (arXiv:0709.3762)
Ortolani, S., Bica, E., & Barbuy, B. 2006, ApJ, 646, L115
Peng, E. W., et al. 2006, ApJ, 639, 838
Piotto, G., et al. 2007, ApJ, 661, L53
Pooley, D., & Rappaport, S. 2006, ApJ, 644, L45
Portegies Zwart, S. F., Belleman, R. G., & Geldof, P. M. 2007, New Astron., 12, 641
Richer, H. B., et al. 2007, AJ, in press (arXiv:0708.4030)
Sakamoto, T., & Hasegawa, T. 2006, ApJ, 653, L29
Ulvestad, J. S., Greene, J. E., & Ho, L. C. 2007, ApJ, 661, L151
Villanova, S., et al. 2007, ApJ, 663, 296
Wilkinson, M. I., Hurley, J. R., Mackey, A. D., Gilmore, G. F., & Tout, C. A. 2003, MNRAS, 343, 1025
Willman, B., et al. 2005, AJ, 129, 2692
Zepf, S. E. et al. 2007, ApJ, 669, L69

Takashi Okamoto explains the effects of feedback on galaxy formation.

Galaxy Formation

Takashi Okamoto

Institute of Computational Cosmology, Department of Physics, Durham University, Durham, UK

Abstract. I review the current status of theoretical studies of galaxy formation. I outline the importance of the physics of baryonic component in galaxy formation by showing results obtained by using two major tools, semi-analytical approaches and cosmological simulations. In particular, I emphasize the role of feedback in galaxy formation and discuss whether apparent conflicts between the standard theory of structure formation, the cold dark matter model, and observations can be solved by the feedback. I also discuss future prospects in numerical simulations of galaxy formation.

1. Introduction

Understanding galaxy formation is a challenging problem whose solution will require a concerted approach combining observational, semi-analytical and numerical work. Perhaps the most important progress in this area is the establishment of the cold dark matter (CDM) model. The CDM model has steadily gained acceptance since it was first mooted in the early 1980s (Peebles 1982; Blumenthal et al. 1984; Davis et al. 1985). This model has a tremendous predicting power and many of these predictions have turned out to be successful. Strong support for this model comes from the measurement of temperature anisotropies in the cosmic microwave background (e.g., Smoot et al. 1992; Spergel et al. 2007) and two large galaxy surveys: the two-degree Field Galaxy Redshift Survey (2dF-GRS Colless et al. 2001) and the Sloan Digital Sky Survey (SDSS, York et al. 2000), which have permitted the most accurate measurements to date of the power spectrum of galaxy clustering (e.g., Percival et al. 2007; Tegmark et al. 2004, 2006; Pope et al. 2004; Cole et al. 2005; Padmanabhan et al. 2007). This model specifies the values of the fundamental cosmological parameters and gives the initial condition for structure formation of the Universe such as galaxies. We are now able to study galaxy formation by *ab initio* approaches based on the CDM model where the structure formation is characterized by the hierarchical clustering.

The second key advance that makes possible progress in the understanding galaxy formation is the unveiling of the high redshift Universe. Observations of galaxies over a range of redshifts allow us to compare their properties at different epochs in the history of the Universe. The unprecedented faint imaging of galaxies by the Hubble Deep Field (Williams et al. 1996), combined with the Lyman-break dropout technique to isolate high redshift galaxies (Steidel et al. 1996), was essential in making possible the first determination of the cosmic star formation history (Madau et al. 1996; Steidel et al. 1999). The detection of the emission from galaxies at sub-millimeter wavelengths (Smail et al. 1997; Barger et al. 1998; Hughes et al. 1998) offers the chance of uncovering up heavily dust obscured galaxies which are too faint to appear in optical surveys.

While it is the observations that have led the progress in this field, it is the theory that is needed to connect observed *snapshots* of evolving galaxy population. Our lack of understanding of the key physics underpinning galaxy formation, such as star formation and feedback, enforces us to introduce some phenomenological models. I will illustrate the importance of such baryonic physics and how we are beginning to understand it.

2. Formation of Dark Matter Halos

Initial tiny density fluctuations grow owing to gravitational instability and eventually form virialized objects, called dark matter halos. In a CDM universe, the process of perturbation growth is dissipationless. This means that the total kinetic energy of a system of dark matter is retained, although energy can be converted from potential to kinetic. Hence, studying the growth of dark matter structures in a CDM universe is essentially straightforward to do with cosmological N-body simulations.

2.1. The Abundance of Dark Matter Halos

The first attempt to calculate the abundance of gravitationally bound objects was made by Press & Schechter (1974) by assuming a Gaussian density field and smoothing the field on different scales. The mass function predicted by this simple calculation agrees surprisingly well with the results obtained from N-body simulations (e.g., Efstathiou et al. 1988; Lacey & Cole 1994). Sheth et al. (2001) presented a model in which they replaced the spherical collapse model with an ellipsoidal collapse. Jenkins et al. (2001) established the mass function of dark matter halos using a suite of N-body simulations and proposed a fitting formula that encapsulates the numerical results. This fitting formula was found to produce a good fit down to $\sim 10^{10}$ M$_\odot$ (Springel et al. 2005).

By utilizing the halo mass function, one can derive a mass-to-light ratio that guarantees a match between the theoretical prediction and the observed group luminosity function. A mass-to-light ratio that guarantees a match between the theoretical prediction and the observed group luminosity function is a strong function of halo mass (Yang et al. 2003; Eke et al. 2004). xx It is lowest for halos lf mass $\simeq 10^{12} h^{-1}$ M$_\odot$ and rises by a factor of $\simeq 6$ to lower and higher mass. To understand this mass dependence of galaxy formation, we should consider baryonic physics.

2.2. Density Profiles of Dark Halos

The increasing computer power and the advent of new simulation techniques enable the study of the formation of individual dark matter halos in a full cosmological context (i.e., from CDM initial conditions). Multi-mass simulation techniques, so-called *zoom simulations* (Navarro & Benz 1991; Katz & White 1993; Frenk et al. 1996), resolve individual dark halos with $\sim 10^4$ particles (Navarro et al. 1996, 1997), or even $\sim 10^6$ particles (Moore et al. 1998; Klypin et al. 1999a; Okamoto & Habe 1999).

NFW found that the density profiles of dark halos follow a universal law:

$$\rho(r) = \frac{\rho_0}{(r/r_s)(1 + r/r_s)^2} \qquad (1)$$

where r_s is a characteristic radius. The profile is characterized by the inner slope $\alpha = d\ln\rho/d\ln r = -1$ for $r \ll r_s$ and the outer slope $\alpha = -3$ for $r \gg r_s$. While many authors have confirmed that the NFW profile provides good fit, simulations with higher resolution form halos with inner slopes as steep as $\alpha = -1.5$ (Moore et al. 1999b; Fukushige & Makino 2001). Hayashi et al. (2004) has claimed that there is no universal slope for the inner density profile and the inner slopes are steeper than NFW and shallower than Moore's, i.e., $-1.5 < \alpha < -1$ for $r \ll r_s$.

The existence of such a universal density profile makes strong predictions on the shape of the rotation curve of galaxies that can be confronted with observations. Such comparisons have revealed that the inner slope of the density profiles of $\alpha < -1$, which seems too steep and inconsistent with rotation curves studies of dark matter dominated galaxies, i.e., low surface brightness galaxies (McGaugh & de Blok 1998; Flores & Primack 1994; Moore et al. 1999b; de Blok & Bosma 2002). Some authors have argued that the rotation curves are consistent with CDM models if one considers the triaxiality of dark matter halos (Hayashi & Navarro 2006; Valenzuela et al. 2007; Hayashi et al. 2007). However, we should consider effects of the baryonic component properly since the presence of a galaxy disk at the center of a dark matter halo makes the dark halo significantly rounder than dark matter halos formed in dark matter-only simulations (Bailin et al. 2005). Adiabatic contraction due to disk formation makes dark matter halos more concentrate as we shall see later. The problem is not only at the very center of low surface brightness galaxies but also seen at intermediate radii of any disk galaxies (Navarro & Steinmetz 2000b; McGaugh et al. 2007). This problem in the CDM density profile is called the *cusp-core problem*.

2.3. Substructure within Dark Matter Halos

Until the late 1990s, cosmological N-body simulations suggested that dark matter halos were smooth and featureless (Summers et al. 1995; Frenk et al. 1996). Higher resolution simulations which typically contained more than 10^6 particles in each dark matter halo showed that this phenomenon, called *over-merging*, was a numerical artifact (Ghigna et al. 1998; Moore et al. 1998; Klypin et al. 1999a; Okamoto & Habe 1999). Once a halo enters within the virial radius of a more massive halo, it is referred to as a satellite halo or substructure within the larger halo (*sub-halo*). Okamoto & Habe (1999, 2000) provided a method to construct merging histories of individual sub-halos in order to follow their formation and evolutions. Moore et al. (1999a) and Klypin et al. (1999b) pointed out that the number of satellite halos in a Milky Way-sized halo is an order of magnitude larger than the number of the satellite galaxies of the Milky Way. This problem is called the *satellite problem*.

3. Semi-Analytic Models

A powerful way of studying hierarchical galaxy formation is semi-analytic modeling of galaxy formation. White & Rees (1978) proposed that galaxy formation was a two stage process, with dark halos forming in dissipationless, gravitational collapse, and with galaxies forming inside these structures following the radiative cooling of baryons. They also argued the importance of an additional process, feedback, for avoiding the production of more faint galaxies than are observed. White & Frenk (1991) produced a galaxy formation model that included many of today's models: CDM, radiative cooling of gas, star formation, feedback, and stellar populations. Kauffmann et al. (1993) and Cole et al. (1994) presented the first models to track the formation and evolution of galaxies in the setting of evolving dark matter halos. Below I briefly describe procedures that are commonly used in semi-analytic models. A more comprehensive review on semi-analytic galaxy formation models can be found in Baugh (2006).

3.1. Basic Ingredients

Dark Halo Merger Trees   Semi-analytic models require the formation history of each dark matter halo, commonly called the merger tree, in which a galaxy forms and evolves. Merger trees can be constructed using a Monte-Carlo approach by sampling the distribution of progenitor masses predicted by the extended Press-Schechter theory (Lacey & Cole 1994; Somerville & Kolatt 1999; Cole et al. 2000). A more faithful way to construct merger trees is extracting from N-body simulations which have sufficiently frequent outputs. The first attempt to extract merger trees from N-body simulations was made by Roukema et al. (1997). Kauffmann et al. (1999) studied the galaxy distribution using merger trees generated by N-body simulations. Okamoto & Nagashima (2001) utilized merger trees of sub-halos obtained in high-resolution cosmological N-body simulations of clusters of galaxies (Okamoto & Habe 1999, 2000) to investigate the morphology-density relation of cluster galaxies (Dressler 1980). This powerful method was first used for semi-analytic studies of cluster galaxies (Okamoto & Nagashima 2001, 2003; Springel et al. 2001) and has become a standard technique for semi-analytic models combined with N-body simulations (Croton et al. 2006; Bower et al. 2006).

Gas Cooling   The basic model of how gas cools inside dark matter halos was set out in detail by White & Frenk (1991). Semi-analytic models assume that when a dark matter halo forms, gas contained in the halo is shock heated to the virial temperature of the halo:

$$T_{\rm vir} = \frac{1}{2}\frac{\mu m_{\rm p}}{k} V_{\rm c}, \qquad (2)$$

where μ is the mean molecular weight of the gas, $m_{\rm p}$ is the mass of a hydrogen atom, k is Boltzmann's constant, and $V_{\rm c} = (GM_{\rm vir}/r_{\rm vir})^{\frac{1}{2}}$ is the circular velocity of the halo with mass $M_{\rm vir}$ and virial radius $r_{\rm vir}$. This gas with $T = T_{\rm vir}$ is called the *hot gas*. It is also assumed that the hot gas distributes parallel to the dark matter. The relation between the circular velocity and halo mass is a function of redshift and cosmology. Gas can subsequently cool from the hot halo. The cooling rate of the gas is dependent on the temperature and metallicity of the gas. Hence a cooling time can be specified by dividing the thermal energy density of the gas by the cooling rate per unit volume:

$$t_{\rm cool}(r) = \left(\frac{3}{2}\frac{\rho_{\rm g}kT_{\rm vir}}{\mu m_{\rm p}}\right)/(\Lambda(\rho_{\rm g}, T_{\rm vir}, Z)) \tag{3}$$

where $\rho_{\rm g}$ is the gas density which is a function of radius and the function Λ is a cooling function. The cooling radius, $r_{\rm cool}$, is computed as the radius where the cooling time is equal to the time-step used in the merger tree. The gas within $r_{\rm cool}$ accretes to the center of the halo at a rate set by the cooling time (or dynamical time if the dynamical time is longer than the cooling time) *keeping its angular momentum*. The cooled gas is assumed to have $\simeq 10^4$ K at which the atomic cooling becomes inefficient, and this gas is called the *cold gas*.

Star Formation The lack of a theory of star formation enforces and allows a simple estimate of the global rate of star formation in a model galaxy:

$$\dot{M}_* = \frac{M_{\rm cold}}{\tau_*} \tag{4}$$

where the star formation rate, $\dot{M}_*$, depends on the amount of cold gas available, $M_{\rm cold}$ and a star formation timescale, τ_*. The star formation timescale is usually represented as

$$\tau_* = \tau_*^0 \left(\frac{V_{\rm c}}{V_*}\right)^{\alpha_*} \tag{5}$$

where τ_*^0 is either the dynamical time $\tau_{\rm dyn} = r_{\rm gal}/v_{\rm gal}$ within the galaxy, which is a function of redshift, or some constant and the dependence on the circular velocity is introduced with two free parameters, V_* and α_*, to reproduce the observed cold gas fractions in spirals as a function of luminosity (Cole et al. 1994, 2000). The star formation timescale that does not depend on the dynamical timescale (i.e., redshift) has been supported from analysis on the formation of quasars (Kauffmann & Haehnelt 2000), the evolution of damped Lyα systems (Somerville et al. 2001), and the galaxy number counts (Nagashima et al. 2001). Since the value of α_* is usually assumed to be less than -1, the star formation timescale is longer for smaller galaxies. This naturally explains the galaxy downsizing by keeping the gas in small galaxies until today, though the physical origin of the form of the star formation timescale is unclear.

Merging of Galaxies When two or more dark halos merge to form a new halo, the hot gas components in the progenitor halos are immediately merged and constitute a hot gas component in the new halo. In contrast, galaxies do not merge immediately. The central galaxy of the main progenitor halo is defined as the central galaxy of the new halo and the rest of galaxies are regarded as satellite galaxies. Satellite galaxies sink to the halo center owing to the dynamical friction and merge to the central galaxy. The satellite galaxies could merge by random collision (Somerville & Primack 1999). If the merging two galaxies have similar mass, the merger is considered as the *major merger* and a *starburst* occurs. The new galaxy becomes a pure bulge galaxy. In the case of a minor merger, the disk of the larger galaxy is not destroyed and a merging satellite is absorbed into either the disk or the bulge component of the larger galaxy. Note that some authors introduced small starbursts induced by minor mergers. For example Okamoto & Nagashima (2003) claimed that a starburst induced by a minor merger is the necessary component in order to form the galaxy population with intermediate morphology.

Feedback Feedback processes arguably have the largest impact on the form of the theoretical predictions for galaxy properties, whilst at the same time being amongst the most difficult and controversial phenomena to model. The most common form of feedback used in models is the ejection of cold gas from a galactic disk by a supernova (SN) driven wind (e.g., Larson 1974; Dekel & Silk 1986). The reheated cold gas could be blown out to the hot gas halo, from which it may subsequently recool (sometimes called *retention* feedback), or it may even be ejected from the halo (*ejection* feedback) and left unable to cool until it is incorporated into a more massive halo later. The ejection rate of the cold gas by SN driven wind is parameterized as follows:

$$\dot{M}_{\rm ej} = \left(\frac{V_{\rm c}}{V_{\rm hot}}\right)^{\alpha_{\rm hot}} \dot{M}_* \qquad (6)$$

where $V_{\rm hot}$ and $\alpha_{\rm hot}$ are free parameters. Considering the fact that the potential wells are shallower in smaller galaxies, the value of $\alpha_{\rm hot}$ should be negative. From an energy balance, $\alpha_{\rm hot} = -2$ is obtained. However many authors have used much stronger dependence on the circular velocity (i.e., $\alpha_{\rm hot} < -2$) in order to reproduce the faint end slope of luminosity functions (e.g., Cole et al. 1994; Nagashima & Yoshii 2004).

Benson et al. (2003) carried out a systematic study of the impact of various feedback mechanisms on the form of the predictions for the galaxy luminosity function. The standard SN driven winds, while helping to reduce the number of faint galaxies to match observations, were found to overproduce bright galaxies. Stronger feedback in this mode would tend to weaken the break in the predicted luminosity function rather than enhance it by wiping out galaxies around L_*. The thermal conduction and the *ejection* mode of feedback (superwinds) did the better job. However they had to assume an unphysically high conductivity in the conduction model and an implausibly high efficiency of feedback in the superwind model. Croton et al. (2006) and Bower et al. (2006) implemented a simple AGN feedback schemes into their models. By assuming the *radio mode* feedback from AGN quenches gas cooling in quasi-hydrostatic hot halos, they reproduced the break in the luminosity function.

Chemical Evolution Chemical evolution of gas and stars in a galaxy is important because (i) the cooling rate of the gas, $\Lambda(T, Z)$, strongly depends on the gas metallicity (e.g., Sutherland & Dopita 1993); (ii) the metallicity with which stars are born has an impact on the luminosity and colors of the stellar population; and (iii) the optical depth of a galaxy, which determines the extinction of starlight due to dust, scales with the metallicity of its cold gas. As stars evolve, they return material to the interstellar medium (ISM) with an enhanced metallicity in the form of stellar winds or SN explosions. The amount of gas returned to the ISM per unit mass of stars formed is therefore dependent on the shape of the initial mass function (IMF) of the stellar population.

The first attempt to follow the chemical evolution of galaxies in semi-analytic models considered type II SNe, using the instantaneous recycling approximation. While the yield of metals is determined by the chosen IMF, the effective yield of metals depends on a number of differing metallicities such as (i) the mixing of hot gas reservoirs of differing metallicities by halo mergers, (ii) accretion of gas by cooling to the disk cold gas which could have a different metallicity, and (iii) feedback processes which deliver metals in the cold gas to the hot halo gas.

Recently type Ia SNe have been included in semi-analytic models. Thomas (1999) was the first to consider the delayed enrichment due to SNe Ia by using star formation histories extracted from the model of Kauffmann et al. (1993), but neglecting any inflow or outflow of gas and metals. Nagashima & Okamoto (2006) produced the first semi-analytic model which self-consistently integrated the impact of SNe Ia, using the simplification of assuming a fixed time delay for SN Ia explosions. Nagashima et al. (2005a,b) performed the first fully consistent calculation including both type II and Ia SNe in the semi-analytic models.

3.2. Successes and Failures

Semi-analytic models now successfully reproduce the present day galaxy luminosity functions by invoking superwinds (Benson et al. 2003) or AGN feedback (Croton et al. 2006; Bower et al. 2006). Bower et al. also reproduced luminosity functions at higher redshifts. However, readers should bear in mind that any phenomenon which leads to a model successfully matching the observed break in the luminosity function uses up a dangerously high fraction of the energy released by supernova explosions. We also have to wait to see whether models that invoke AGN feedback can account for the observed AGN activities.

Semi-analytic models have difficulty matching the zero-point of the Tully-Fisher relation and the normalization of the luminosity function at the same time. The difficulty lies in the fact that the CDM halos are too centrally concentrated and therefore it might be a genuine problem of the CDM. Scaling relations such as the Tully-Fisher relation rely on the calculation of the scale lengths of galaxies. Semi-analytic models assume that hot halo gas has the same specific angular momentum as its host halo and the angular momentum is conserved when gas becomes cold and accretes to a disk. However, numerical simulations have shown that angular momentum of gas can be transferred to the dark matter. Moreover, Okamoto et al. (2005) showed that the direction of the angular momentum vector of the main progenitor halo changed significantly during galaxy formation.

Baugh et al. (2005) were able to reproduce the number counts of sub-millimeter selected galaxies and the luminosity function of Lyman-break galaxies, while retaining a fair match to the present day optical and far infrared luminosity functions. However, this was only possible with the controversial assumption of a top-heavy IMF for star formation in merger driven starbursts. This model can also account for the metallicity of the intracluster medium and elliptical galaxies (Nagashima et al. 2005a,b), though Nagashima et al. (2005b) failed to reproduce the observed trend of the α/Fe ratio that increases with the velocity dispersion of the galaxy. So far, there is no explanation for this trend in the framework of hierarchical galaxy formation.

4. Gasdynamic Simulations

Another powerful tool to study galaxy formation in a cosmological context is gasdynamic simulations. In the gasdynamic simulations, the gasdynamics are solved based on governing equations. This is the major advantage of the gasdynamic simulations over the semi-analytic models, in which gas distribution is *assumed*. There are two principal algorithm in common use to follow the hydrodynamics of gas in an expanding Universe: particle-based Lagrangian schemes that employ a technique called smoothed particle hydrodynamics (SPH) (e.g., Monaghan 1992; Couchman et al. 1995; Pearce et al. 1999; Springel & Hernquist 2003), and grid-based Eulerian schemes (e.g., Ryu et al. 1993; Cen & Ostriker 1999). To date, most of gasdynamic simulations of galaxy formation have been performed with SPH, because its Lagrangian nature is suitable to follow the collapse of gas clouds in a cosmological volume.

Prescriptions used in gasdynamic simulations, in order to treat baryonic physics such as star formation and feedback, are surprisingly similar to those used in semi-analytic models. The main difference is that such processes are formulated based on the local quantities in simulations rather than global quantities such as the circular velocity of the halo. For example, the rate of change of the internal specific energy of the i-th particle by cooling is calculated as

$$\dot{u}_i = -\frac{\Lambda(\rho_i, T_i, Z_i)}{\rho_i} \qquad (7)$$

where u_i, ρ_i, T_i, and Z_i are the specific internal energy, density, temperature, and metallicity of the i-th particle, respectively. Note that in Eq.(3), the virial temperature of the halo and the mean metallicity of the hot halo gas are used. Although many simulations followed the chemical evolution, only a handful of studies (e.g., Kawata & Gibson 2005; Okamoto et al. 2005) employed the metallicity dependent cooling function, whereas others use the cooling rate of the primordial gas (e.g., Abadi et al. 2003; Sommer-Larsen et al. 2003; Robertson et al. 2004; Governato et al. 2007) in spite of the fact that the cooling rate of the gas with the solar metallicity is more than an order of magnitude higher than that of the primordial gas. Usually, the inverse Compton cooling and effects of time evolving uniform UV background are also considered in the energy equation.

Dense ($\rho > \rho_{\rm sf} \simeq 0.1\,{\rm cm}^{-3}$) and cold ($T \sim 10^4\,{\rm K}$) gas is eligible to form stars. Star formation rate of the i-th particle is given by

$$\dot{\rho}_* = C_* \frac{\rho_i}{t_{\rm dyn}} \propto \rho_i^{1.5}, \qquad (8)$$

where C_* is the dimensionless star formation efficiency parameter. To match the observed Kennicutt relation (Kennicutt 1998), $C_* \simeq 0.05$ is often used (e.g., Navarro & Steinmetz 2000a).

Each star particle represents a single stellar population. Thus, evolved stars explode as SNe. Instantaneous recycling is often assumed. Some authors relaxed this approximation and calculated the number of SNe and returned mass as a function of the SSP's age (Okamoto et al. 2005, 2007; Governato et al. 2007). They also included SNe Ia. The feedback energy is distributed to the surrounding gas. It was found that depositing energy in the form of thermal energy had almost no effect because the energy was quickly radiated away in dense regions (Katz 1992). Note that since the cooling function depends on the temperature, not only the amount of energy given to the gas but also the total mass of gas to which the energy is given affect the thermal evolution. For example, giving all the feedback energy to the nearest particle has much stronger effects than giving it to 50 particles. To circumvent this problem, feedback energy is often given in the form of kinetic energy (e.g., Navarro & Steinmetz 2000a; Springel & Hernquist 2003) or cooling of the particles which receive the feedback energy is shutting-off for a while ($\sim 10\,\mathrm{Myr}$; Thacker & Couchman 2001).

A self-consistent AGN feedback model was first incorporated into cosmological simulations of galaxy formation by Okamoto et al. (2007) in which they discriminated the quasar mode and radio mode based on the mass accretion rates onto the central black holes. Sijacki et al. (2007) showed that such a discrimination successfully produced luminous red galaxies.

4.1. Formation of Spiral Galaxies and Effects of Feedback

The first attempts to simulate the formation of a spiral galaxy from CDM initial conditions generally failed, producing objects with overly centrally concentrated gas and stars. It became immediately apparent that the root cause of this problem was a net transfer of angular momentum from the baryons to the dark matter halo during the aggregation of the galaxy through mergers (Navarro & Benz 1991; Navarro & White 1994; Navarro et al. 1995). This is known as the angular momentum problem. It was suspected from the start that its solution was likely to involve feedback processes that would regulate the supply of gas to the galaxy.

More recent simulations within the CDM framework have produced more promising disk galaxies. Thacker & Couchman (2001) obtained a disk galaxy with a reasonable size by assuming that gas heated by SN explosions can adiabatically expand without radiative loss of energy for a while. Their simulation, however, stopped at $z = 0.5$. Steinmetz & Navarro (2002) did not assume such strong feedback and found that a broad range of galaxy morphologies could be produced by gas accretion and galaxy mergers. In a related work, Abadi et al. (2003) obtained a disk galaxy which resembled observed early-type disk galaxies. Sommer-Larsen et al. (2003) were also able to generate a variety of morphological types, including disks, by assuming high star formation efficiency and strong feedback at high redshift to prevent early collapse of baryons. Governato et al. (2007) used the feedback model similar to Thacker & Couchman (2001) in order to match the Tully-Fisher relation. Their Tully-Fisher relation was, however, calculated from rotation speed of stars younger than 4 Gyr old, which are much older populations than those observed by Hα. In fact, the rotation speed of the cold gas in their galaxy was too fast compared with observations. Robertson et al. (2004) adopted a multiphase model for the ISM which stabilized gaseous disks against the Toomre instability, and produced a galaxy with an exponential surface brightness profile but insufficient angular momentum. These simulations, however, did not take the metallicity effects on cooling into account.

Okamoto et al. (2005) also assumed a multiphase model for the ISM and a top-heavy IMF for stars formed in starbursts, as required for semi-analytic models to match the number counts of bright submillimeter galaxies (Baugh et al. 2005) and metal contents in the intracluster medium and elliptical galax-

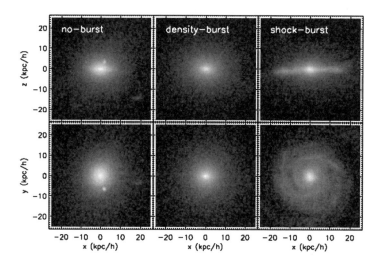

Figure 1. Stellar distributions at $z = 0$. No-burst, density-burst and shock-burst models are shown from left to right. Edge-on and face-on views of stars are given in upper and lower panels, respectively. Brightness indicates the projected mass density, and the same scaling is used for each model. All of these galaxies are obtained from the same initial conditions.

ies (Nagashima et al. 2005a,b). By varying the criteria for starbursts in their simulations, Okamoto et al. (2005) were able to produce galaxies with a variety of morphological types, from ellipticals to spirals, starting from exactly the same initial conditions (Fig. 1).

The Density Profiles In Fig. 2, density profiles of the host dark matter halos of the three galaxies in Fig. 1 are shown. The profile obtained by the simulation without a baryonic component ($\Omega_{\rm b} = 0$) is also shown as the DM-only simulation. It is clear that the dissipational nature of the baryons makes the inner density profile steeper. The no-burst galaxy which has the weakest feedback effects has the steepest profile.

The Evolution of Angular Momentum Zavala et al. (2007) analyzed two of the galaxies simulated by Okamoto et al. (2005), the no-burst galaxy (left panel in Fig. 1) and the shock-burst galaxy (right). The no-burst galaxy is referred to as the bulge-dominated galaxy and the shock-burst galaxy is called the disk-dominated galaxy. They traced the time evolution of the angular momentum of three *Lagrangian* components: the dark matter particles lie within the virial radius at $z = 0$ (halo), the 10% most bound dark matter particles within the halo (inner halo), cold gas and stars within $0.1\, r_{\rm vir}$ (galaxy). The tidal torque theory predicts that the amplitude of the angular momentum of a Lagrangian region evolves in proportion to $a^{3/2}$ until the region reaches the maximum expansion. Then the region starts to collapse and is decoupled from the tidal field, and thus the angular momentum stays constant (White 1984).

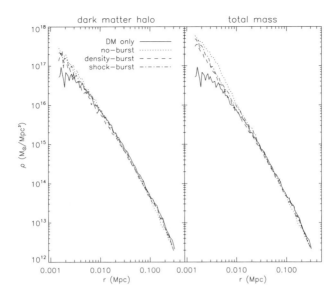

Figure 2. The left panel shows density profiles of dark halos for the DM only, *no-burst*, *density-burst*, and *shock-burst* simulations by the solid, dotted, dashed, and dot-dashed lines, respectively. The right panel is the same as the left one but for the total mass density profiles.

The angular momentum evolution of the halo component (upper left panel of Fig. 3) clearly shows two distinct phases, indicated by the dashed ($L \propto a^{3/2}$) and dotted ($L = const.$) lines. The inner halo shows quite different evolution (upper left panel). Since this subregion has a higher overdensity than the halo as a whole, it reaches maximum expansion earlier. Until $z = 0$, the inner halo loses 90% of the angular momentum by mergers. Lower panels in Fig. 3 show angular momentum evolutions of galaxies (cold baryons) for the bulge-dominated galaxy (left) and disk-dominated galaxy (right). The evolution of the bulge-dominated galaxy (lower left panel) is surprisingly similar to that of the inner halo (upper right panel). This implies that baryons that are contained in the bulge-dominated galaxy are concentrated in the dense subsystems of dark matter at high redshift and lose their angular momentum as the inner halo loses its angular momentum during the collapse. On the other hand, the evolution of the disk-dominated galaxy (lower right) resembles that of the halo as a whole (upper left panel). Strong feedback from the stellar population with a top-heavy IMF prevents early collapse of baryons and therefore baryons are diffusely distributed in the hot halo gas. When this gas cools, it accretes to the disk keeping its angular momentum. Thus, the suppression of the early collapse of baryons by some form of feedback is key to solving the angular momentum problem. Interestingly, the model used in the disk-dominated galaxy (shock-burst simulation in Okamoto et al. (2005) was also able to reproduce the luminosity function of the Milky Way satellites (Libeskind et al. 2007).

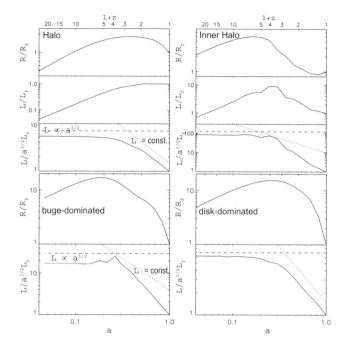

Figure 3. Upper left: the evolution of the *rms* radius (top panel) and the specific angular momentum (middle and bottom panels) of the dark mater particles that lie within $r_{\rm vir}$ at $z = 0$. In the bottom panel the dashed line indicated $L \propto a^{3/2}$ and the dotted line $L = const$. Upper right: same as the upper left but for the particles that makes the inner dark matter halo. Lower left: Evolution of the *rms* radius (upper panel) and specific angular momentum evolution (lower panel) for the baryonic component of the bulge-dominated galaxy. Lower right: same as the lower left but for the disk-dominated galaxies. All physical quantities are normalized to their values at the present day.

5. Conclusions

I have presented results on recent efforts to model the formation and evolution of galaxies in hierarchically clustering universes using two complementary techniques: semi-analytic models and gasdynamic simulations. Feedback processes are the most important ingredient in these models which have reproduced the observed luminosity function and solved the angular momentum problem and the satellite problems. It is interesting that the feedback recipes that can solve the angular momentum problem also solve the satellite problem simultaneously (Libeskind et al. 2007; Governato et al. 2007). However, in both techniques a dangerously high fraction of the energy released by SNe is used. Sometimes more than 100% of energy is used in the simulations. An excuse for doing so is that the ISM has a multiphase structure and therefore the cooling should not be as effective as in the simulations where the multiphase structure is not resolved.

This uncertainty can be easily removed if we use the resolution that is sufficiently high to resolve star forming gas with density as high as $n_{\rm H} \sim 100\,{\rm cm}^3$ and temperature as low as $\sim 100\,{\rm K}$. Simulations of the ISM on the galactic scales have recently started (Tasker & Bryan 2006; Robertson & Kravtsov 2007; Saitoh et al. 2007). Saitoh et al. (2007) found that in these high resolution simulations star formation is not controlled by the star formation time-scale defined in equation (8) but regulated by the time scale of gas supply from low density regions to high density regions, which is $\rho/\dot\rho \simeq 5 \times t_{\rm dyn}$. While there is still a long way to go, resolving multiphase structure of the ISM in cosmological simulation will be a crucial step to turn our *physically motivated* models into *models based on physics*. The radiation feedback has been largely ignored or only crudely included in our models. Murray et al. (2005) suggested that large scale galactic winds seen around local and high redshift starburst galaxies are driven by radiation pressure rather than energy from SN explosions. To test this hypothesis, we have to carry out radiation hydrodynamics in galaxy formation simulations. Semi-analytic models also need refinements. While they have already have many parameters and are well-complicated, they are still too simplified to take dynamical effects such as angular momentum transfer into account. Close interactions between simulations and semi-analytic models will be required for further improvements and better understanding of galaxy formation.

Acknowledgments. I thank the organizers for inviting me to speak on this topic. I am indebted to Takayuki Saitoh for allowing me to show results in prior to publication. I also acknowledge support from PPARC. The simulations were carried out at the Cosmology Machine at the ICC, Durham.

References

Abadi, M. G., Navarro, J. F., Steinmetz, M., & Eke, V. R. 2003, ApJ, 591, 499
Bailin, J. et al. 2005, ApJ, 627, L17
Barger, A. J., Cowie, L. L., Sanders, D. B., Fulton, E., Taniguchi, Y., Sato, Y., Kawara, K., & Okuda, H. 1998, Nature, 394, 248
Baugh, C. M. 2006, Reports of Progress in Physics, 69, 3101
Baugh, C. M., Lacey, C. G., Frenk, C. S., Granato, G. L., Silva, L., Bressan, A., Benson, A. J., & Cole, S. 2005, MNRAS, 356, 1191
Benson, A. J., Bower, R. G., Frenk, C. S., Lacey, C. G., Baugh, C. M., & Cole, S. 2003, ApJ, 599, 38
Blumenthal, G. R., Faber, S. M., Primack, J. R., & Rees, M. J. 1984, Nature, 311, 517
Bower, R. G., Benson, A. J., Malbon, R., Helly, J. C., Frenk, C. S., Baugh, C. M., Cole, S., & Lacey, C. G. 2006, MNRAS, 370, 645
Cen, R., & Ostriker, J. P. 1999, ApJ, 514, 1
Cole, S., Aragon-Salamanca, A., Frenk, C. S., Navarro, J. F., & Zepf, S. E. 1994, MNRAS, 271, 781
Cole, S., Lacey, C. G., Baugh, C. M., & Frenk, C. S. 2000, MNRAS, 319, 168
Cole, S. et al. 2005, MNRAS, 362, 505
Colless, M. et al. 2001, MNRAS, 328, 1039
Couchman, H. M. P., Thomas, P. A., & Pearce, F. R. 1995, ApJ, 452, 797
Croton, D. J. et al. 2006, MNRAS, 365, 11
Davis, M., Efstathiou, G., Frenk, C. S., & White, S. D. M. 1985, ApJ, 292, 371
de Blok, W. J. G., & Bosma, A. 2002, A&A, 385, 816
Dekel, A., & Silk, J. 1986, ApJ, 303, 39
Dressler, A. 1980, ApJ, 236, 351

Efstathiou, G., Frenk, C. S., White, S. D. M., & Davis, M. 1988, MNRAS, 235, 715
Eke, V. R. et al. 2004, MNRAS, 355, 769
Flores, R. A., & Primack, J. R. 1994, ApJ, 427, L1
Frenk, C. S., Evrard, A. E., White, S. D. M., & Summers, F. J. 1996, ApJ, 472, 460
Fukushige, T., & Makino, J. 2001, ApJ, 557, 533
Ghigna, S., Moore, B., Governato, F., Lake, G., Quinn, T., & Stadel, J. 1998, MNRAS, 300, 146
Governato, F., Willman, B., Mayer, L., Brooks, A., Stinson, G., Valenzuela, O., Wadsley, J., & Quinn, T. 2007, MNRAS, 374, 1479
Hayashi, E., & Navarro, J. F. 2006, MNRAS, 373, 1117
Hayashi, E. et al. 2004, MNRAS, 355, 794
Hayashi, E., Navarro, J. F., & Springel, V. 2007, MNRAS, 377, 50
Hughes, D. H. et al. 1998, Nature, 394, 241
Jenkins, A., Frenk, C. S., White, S. D. M., Colberg, J. M., Cole, S., Evrard, A. E., Couchman, H. M. P., & Yoshida, N. 2001, MNRAS, 321, 372
Katz, N. 1992, ApJ, 391, 502
Katz, N., & White, S. D. M. 1993, ApJ, 412, 455
Kauffmann, G., Colberg, J. M., Diaferio, A., & White, S. D. M. 1999, MNRAS, 303, 188
Kauffmann, G., & Haehnelt, M. 2000, MNRAS, 311, 576
Kauffmann, G., White, S. D. M., & Guiderdoni, B. 1993, MNRAS, 264, 201
Kawata, D., & Gibson, B. K. 2005, MNRAS, 358, L16
Kennicutt, Jr., R. C. 1998, ApJ, 498, 541
Klypin, A., Gottlöber, S., Kravtsov, A. V., & Khokhlov, A. M. 1999a, ApJ, 516, 530
Klypin, A., Kravtsov, A. V., Valenzuela, O., & Prada, F. 1999b, ApJ, 522, 82
Lacey, C., & Cole, S. 1994, MNRAS, 271, 676
Larson, R. B. 1974, MNRAS, 169, 229
Libeskind, N. I., Cole, S., Frenk, C. S., Okamoto, T., & Jenkins, A. 2007, MNRAS, 374, 16
Madau, P., Ferguson, H. C., Dickinson, M. E., Giavalisco, M., Steidel, C. C., & Fruchter, A. 1996, MNRAS, 283, 1388
McGaugh, S. S., & de Blok, W. J. G. 1998, ApJ, 499, 41
McGaugh, S. S., de Blok, W. J. G., Schombert, J. M., Kuzio de Naray, R., & Kim, J. H. 2007, ApJ, 659, 149
Monaghan, J. J. 1992, ARA&A, 30, 543
Moore, B., Ghigna, S., Governato, F., Lake, G., Quinn, T., Stadel, J., & Tozzi, P. 1999a, ApJ, 524, L19
Moore, B., Governato, F., Quinn, T., Stadel, J., & Lake, G. 1998, ApJ, 499, L5
Moore, B., Quinn, T., Governato, F., Stadel, J., & Lake, G. 1999b, MNRAS, 310, 1147
Murray, N., Quataert, E., & Thompson, T. A. 2005, ApJ, 618, 569
Nagashima, M., Lacey, C. G., Baugh, C. M., Frenk, C. S., & Cole, S. 2005a, MNRAS, 358, 1247
Nagashima, M., Lacey, C. G., Okamoto, T., Baugh, C. M., Frenk, C. S., & Cole, S. 2005b, MNRAS, 363, L31
Nagashima, M., & Okamoto, T. 2006, ApJ, 643, 863
Nagashima, M., Totani, T., Gouda, N., & Yoshii, Y. 2001, ApJ, 557, 505
Nagashima, M., & Yoshii, Y. 2004, ApJ, 610, 23
Navarro, J. F., & Benz, W. 1991, ApJ, 380, 320
Navarro, J. F., Frenk, C. S., & White, S. D. M. 1995, MNRAS, 275, 56
—. 1996, ApJ, 462, 563
—. 1997, ApJ, 490, 493
Navarro, J. F., & Steinmetz, M. 2000a, ApJ, 538, 477
—. 2000b, ApJ, 528, 607
Navarro, J. F., & White, S. D. M. 1994, MNRAS, 267, 401
Okamoto, T., Eke, V. R., Frenk, C. S., & Jenkins, A. 2005, MNRAS, 363, 1299

Okamoto, T., & Habe, A. 1999, ApJ, 516, 591
—. 2000, PASJ, 52, 457
Okamoto, T., & Nagashima, M. 2001, ApJ, 547, 109
—. 2003, ApJ, 587, 500
Okamoto, T., Nemmen, R. S., & Bower, R. G. 2007, astro-ph/0704.1218
Padmanabhan, N. et al. 2007, MNRAS, 378, 852
Pearce, F. R. et al. 1999, ApJ, 521, L99
Peebles, P. J. E. 1982, ApJ, 263, L1
Percival, W. J. et al. 2007, ApJ, 657, 645
Pope, A. C. et al. 2004, ApJ, 607, 655
Press, W. H., & Schechter, P. 1974, ApJ, 187, 425
Robertson, B., & Kravtsov, A. 2007, astro-ph/0710.2192
Robertson, B., Yoshida, N., Springel, V., & Hernquist, L. 2004, ApJ, 606, 32
Roukema, B. F., Quinn, P. J., Peterson, B. A., & Rocca-Volmerange, B. 1997, MNRAS, 292, 835
Ryu, D., Ostriker, J. P., Kang, H., & Cen, R. 1993, ApJ, 414, 1
Saitoh, T. R.and Daisaka, H., Kokubo, E., Makino, J., Okamoto, T., Tomisaka, K., Wada, K., & Yoshida, N. 2007, in preparation
Sheth, R. K., Mo, H. J., & Tormen, G. 2001, MNRAS, 323, 1
Sijacki, D., Springel, V., di Matteo, T., & Hernquist, L. 2007, MNRAS, 380, 877
Smail, I., Ivison, R. J., & Blain, A. W. 1997, ApJ, 490, L5
Smoot, G. F. et al. 1992, ApJ, 396, L1
Somerville, R. S., & Kolatt, T. S. 1999, MNRAS, 305, 1
Somerville, R. S., & Primack, J. R. 1999, MNRAS, 310, 1087
Somerville, R. S., Primack, J. R., & Faber, S. M. 2001, MNRAS, 320, 504
Sommer-Larsen, J., Götz, M., & Portinari, L. 2003, ApJ, 596, 47
Spergel, D. N. et al. 2007, ApJS, 170, 377
Springel, V., & Hernquist, L. 2003, MNRAS, 339, 289
Springel, V. et al. 2005, Nature, 435, 629
Springel, V., White, S. D. M., Tormen, G., & Kauffmann, G. 2001, MNRAS, 328, 726
Steidel, C. C., Adelberger, K. L., Giavalisco, M., Dickinson, M., & Pettini, M. 1999, ApJ, 519, 1
Steidel, C. C., Giavalisco, M., Dickinson, M., & Adelberger, K. L. 1996, AJ, 112, 352
Steinmetz, M., & Navarro, J. F. 2002, New Astronomy, 7, 155
Summers, F. J., Davis, M., & Evrard, A. E. 1995, ApJ, 454, 1
Sutherland, R. S., & Dopita, M. A. 1993, ApJS, 88, 253
Tasker, E. J., & Bryan, G. L. 2006, ApJ, 641, 878
Tegmark, M. et al. 2004, ApJ, 606, 702
—. 2006, Phys. Rev. D, 74, 123507
Thacker, R. J., & Couchman, H. M. P. 2001, ApJ, 555, L17
Thomas, D. 1999, MNRAS, 306, 655
Valenzuela, O., Rhee, G., Klypin, A., Governato, F., Stinson, G., Quinn, T., & Wadsley, J. 2007, ApJ, 657, 773
White, S. D. M. 1984, ApJ, 286, 38
White, S. D. M., & Frenk, C. S. 1991, ApJ, 379, 52
White, S. D. M., & Rees, M. J. 1978, MNRAS, 183, 341
Williams, R. E. et al. 1996, AJ, 112, 1335
Yang, X., Mo, H. J., & van den Bosch, F. C. 2003, MNRAS, 339, 1057
York, D. G. et al. 2000, AJ, 120, 1579
Zavala, J., Okamoto, T., & Frenk, C. S. 2007, astro-ph/0710.290

Masami Ouchi analyzes the epoch of reionization.

New Horizons in Astronomy: Frank N. Bash Symposium 2007
ASP Conference Series, Vol. 393, © 2008
A. Frebel, J. R. Maund, J. Shen, and M. H. Siegel, eds.

Exploring Galaxy Formation and Reionization Towards the Cosmic Dawn

Masami Ouchi[1,2]

[1] *Observatories of the Carnegie Institution of Washington, Pasadena, CA, USA*

[2] *Carnegie Fellow*

Abstract. I review observational studies of galaxies at $z \sim 4 - 10$. Recent deep surveys push the limit of large telescopes, and aim to accomplish two major goals: (i) characterizing galaxy evolution in the cosmic time and (ii) understanding the history of reionization. Luminosity functions (LFs) of dropout galaxies and Lyα emitters (LAEs) indicate that a very young Lyα emitting population emerges at $z \sim 6$, while a number of post-starburst galaxies already exist at this epoch. The combination of theoretical models, LFs, and correlation functions of LAEs places upper limits on neutral fraction of inter-galactic medium which do not prefer beginning of reionization at $z = 6 - 7$ but a moderate change of neutral hydrogen from $z = 6$ to 7. Ambitious on-going projects search for galaxies beyond the current observational limit of $z = 7$ in blank and nearby-cluster fields. These on-going searches not only extend the investigations of galaxies at $z = 3 - 7$, but also become precursors of future studies that address questions about reionization history and first-generation stars at $z \sim 10 - 20$ with the upcoming very large telescopes, i.e., JWST and ELTs, in the 2010s.

1. Introduction

Studies of high-z galaxies have shown remarkable progresses for a few years. The appearance of large format mosaic CCD imagers and spectrographs, such as HST/ACS, Subaru/Suprime-Cam, and Keck/DEIMOS, enable us to obtain data covering a large area of sky that are deep enough to detect faint galaxies at $z \gtrsim 4$. By these instrumental developments, deep wide-field surveys have identified statistical numbers of UV-bright star-forming galaxies, i.e., dropout galaxies, and Lyα emitters (LAEs) at $z \gtrsim 4$. These large data conclude evolution of luminosity function (LF), stellar-mass function, star-formation rate density, correlation function in a cosmological volume. These efforts extend the first-generation studies of high-z galaxies at $z \sim 3$ (e.g., Cowie & Hu 1998; Steidel et al. 2003) to $z \sim 4 - 6$ (e.g., Ouchi et al. 2004a; Yoshida et al. 2006; Bouwens et al. 2007).

The epoch of $z \gtrsim 4$ is very important in galaxy formation. Figure 1 presents number densities of cluster-sized ($\sim 10^{14} M_\odot$) and galaxy-sized ($\sim 10^{11} M_\odot$) dark halos as a function of redshift, which are calculated with the hierarchical structure formation of the concordance lambda cold dark matter (LCDM) model (Sheth & Tormen 1999). Figure 1 indicates that the number density of cluster-sized halos increases from $z \sim 2$ to 0 by two orders of magnitude. On the other hand, the number density of galaxy-sized halos does not show a difference between $z \sim 0$ and 3, but a significant increase from $z \sim 4$ towards higher redshifts. Since the mass of dark halos is probably the most fundamental property of galaxies, this increase of galaxy-sized halos implies that a large fraction of galaxies intensively form beyond $z \sim 4$.

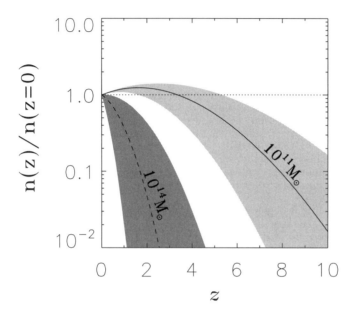

Figure 1. Number densities of dark halos as a function of redshift which are calculated with the analytic LCDM model of Sheth & Tormen (1999). Number densities at each redshift are normalized by those at $z = 0$. Solid and dashed lines represent galaxy-sized ($10^{11} M_\odot$) and cluster-sized ($10^{14} M_\odot$) dark halos. The gray area around solid (dashed) line means a mass range of $10^{10} - 10^{12} M_\odot$ ($10^{13} - 10^{15} M_\odot$).

At redshift beyond 6, studies of high-z galaxies address questions about not only the early epoch of galaxy formation but also cosmic reionization. The WMAP3 and SDSS results indicate that the Universe was reionized at around $z \sim 6-11$ (Spergel et al. 2007; Fan et al. 2006). Gunn-Peterson troughs of SDSS QSOs show that neutral fraction of inter-galactic medium (IGM) raises significantly at $z \sim 6$ (Fan et al. 2006). Since the Gunn-Peterson test cannot be applied to the IGM with a moderately large neutral fraction of $x_{\rm HI} \gtrsim 10^{-2} - 10^{-4}$ (Fan et al. 2002), an alternative probe is needed to measure neutral fraction at $z \gtrsim 6$. Lyα emission from galaxies is useful to constrain neutral fraction of IGM. A dumping wing of IGM absorbs galaxy's Lyα emission redder than 1216 Å. These absorptions are sensitive to neutral fraction of IGM, and detectability of LAEs depends on the status of IGM (e.g., Santos 2004; Furlanetto et al. 2006). Thus, Lyα LF and correlation function of LAEs can be used to investigate the ionization status of IGM at $z \gtrsim 6$ (Malhotra & Rhoads 2004; Kashikawa et al. 2006).

In this contribution, I review statistical studies of high-z galaxies and constraints of cosmic reionization based on the recent deep wide-field surveys at $z \sim 4-7$, and introduce on-going searches for galaxies at $z \sim 7-10$. Throughout this contribution, magnitudes are in the AB system. The values for the cosmological parameters adopted are: $(h, \Omega_m, \Omega_\Lambda, n, \sigma_8) = (0.7, 0.3, 0.7, 1.0, 0.9)$.

2. Luminosity Functions

Figure 2 presents UV-continuum ($\lambda \sim 1500$ Å) LFs of dropout galaxies and LAEs from $z = 3$ to $7 - 8$. Figure 3 is the same but for Lyα-emission LFs of LAEs. The UV and Lyα LFs of LAEs are derived from deep wide-field (0.2-1.0 deg^2) surveys in Subaru/XMM-Newton Deep Field (SXDF; Ouchi et al. 2007) and Subaru Deep Field (SDF; Kashikawa et al. 2006) based on large LAE samples. Although only one LAE is used to estimate the LF at $z = 7.0$ (Iye et al. 2006), the numbers of LAEs are generally $\sim 100-400$ and 60 for $z = 3 - 5.7$ and 6.6, respectively. These LFs are robust against statistical errors and field-to-field variation. The UV LFs of dropout galaxies are obtained by similar deep wide-field surveys in GOODS+HUDF (Bouwens et al. 2006) and SDF (Yoshida et al. 2006).

2.1. UV Luminosity Function

The filled circles in Figure 2 indicate that UV LF of dropout galaxies monotonically decreases from $z \sim 4$ up to, at least, $z \sim 6$ (see Section 5. for discussions of $z = 7 - 8$). If UV luminosity of dropout galaxies correlates with mass, this LF evolution of galaxies is qualitatively similar to the number density of dark halos predicted by the hierarchical structure formation model of LCDM (Figure 1).

On the other hand, the UV LFs of LAEs (the open squares in Figure 2) present a different evolutionary trend, namely, an increase in number density and/or UV luminosity from $z = 3 - 4$ up to, at least, $z \sim 5.7 - 6.6$. This implies that the ratio of LAEs to dropout galaxies increases from $z \sim 3$ to 6 (Ouchi et al. 2007). It would indicate that galaxies with Lyα emission are more common at earlier epochs. Because a large Lyα equivalent width (EW) and a flat UV continuum of LAEs suggest that LAEs are dust-poor young galaxies (e.g., Gronwall et al. 2007; Ouchi et al. 2007), the emergence of Lyα emitting population implies that a number of primeval galaxies appear at $z \sim 6$. On the other hand, recent Spitzer/IRAC studies report that about 40% of IRAC detected $z \sim 6$ dropout galaxies are post starbursts with a moderately old stellar age of 100-400 Myr (Eyles et al. 2007). The relation between the emergence of Lyα emitting population and the existence of post-starburst galaxies is not well understood.

The UV LFs of $z = 5.7$ and 6.6 LAEs are comparable to the one of $z \sim 6$ dropout galaxies within error bars (third and fourth panels of Figure 2). These comparable LFs of LAEs and dropout galaxies may imply that most of $z =$

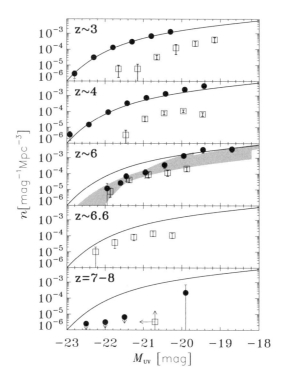

Figure 2. UV luminosity functions (LFs) at $z \sim 3, 4, 6, 6.6$, and $7-8$ from the top to bottom panels. Filled circles and open squares denote dropout galaxies and LAEs, respectively. The solid curve is the best-fit Schechter function of $z \sim 3$ dropout galaxies (Steidel et al. 1999). The data points of dropout galaxies are taken from Steidel et al. (1999) ($z \sim 3$), Yoshida et al. (2006) ($z \sim 4$), Shimasaku et al. (2005); Bouwens et al. (2006) ($z \sim 4$), and Bouwens & Illingworth (2006); Mannucci et al. (2007) ($z \sim 7-8$). The data points of LAEs are from Ouchi et al. (2007) ($z \sim 3-6$), Kashikawa et al. (2006) ($z \sim 6.6$), and Iye et al. (2006) ($z \sim 7$). The shade in the $z \sim 6$ panel indicates uncertainties of $z \sim 6$ dropout LF that correspond to diversities of various LF measurements (see Ouchi et al. 2007, for more details).

$5.7 - 6.6$ UV-bright galaxies identified by dropouts have a Lyα emission line (Shimasaku et al. 2006; Kashikawa et al. 2006), although spectroscopic follow-up results of dropout galaxies find a moderate ($\sim 30\%$) fraction of Lyα emitting galaxies (see discussions in Ouchi et al. 2007).

2.2. Lyα Luminosity Function

In Figure 3, Lyα LFs of LAEs show no significant evolution from $z = 3$ to $z = 5.7$. Note that Lyα fluxes from high-z objects are generally attenuated by

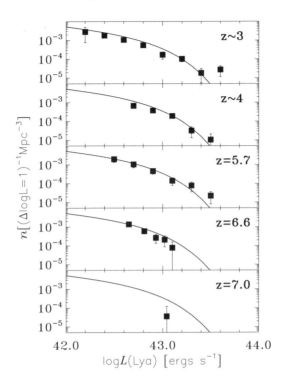

Figure 3. Lyα luminosity functions (LFs) at $z \sim 3$, 4, 5.7, 6.6, and 7.0 from the top to bottom panels. Squares denote LFs that are measured with the LAE samples. The solid curve presents the best-fit Schechter function of $z = 3.1$ LAEs (Ouchi et al. 2007). The data points are taken from Ouchi et al. (2007) ($z = 3 - 5.7$), Kashikawa et al. (2006) ($z = 6.6$), and Iye et al. (2006) ($z = 7.0$).

neutral hydrogen of intervening IGM. It is known that high-z galaxies have an asymmetric Lyα emission whose blue side of line is more strongly absorbed by neutral hydrogen of IGM than low-z galaxies (e.g., Hu et al. 2004; Shimasaku et al. 2006). Thus, the Lyα LFs in Figure 3 should be called *apparent*-Lyα LFs. Due to the absorption of IGM, we cannot directly measure *intrinsic*-Lyα fluxes that are emitted from LAEs. However, we can estimate *intrinsic*-Lyα luminosity with the average Gunn-Peterson (GP) optical depths (e.g., Madau 1995). Assuming that IGM absorbs a blue half of symmetric Lyα emission line, we estimate the ratios of intrinsic to apparent Lyα fluxes, $f_{\text{Ly}\alpha}^{int}/f_{\text{Ly}\alpha}^{app}$, to be 1.23, 1.37, and 1.85 at $z = 3.1$, 3.7, and 5.7, respectively. If Lyα LFs in Figure 3 are corrected for IGM absorption defined by this model of the GP optical depth, inferred *intrinsic* Lyα LF increases from $z = 3.1$ to 5.7. This is because the fraction of $f_{\text{Ly}\alpha}^{int}/f_{\text{Ly}\alpha}^{app}$ increases from $z = 3.1$ to 5.7 with no evolution of *apparent*-Lyα LF. This plausible evolution of *intrinsic*-Lyα LF is consistent with the evolution of the UV LF of LAEs, since UV LF of LAEs also increases from $z = 3.1$ to 5.7 with the rest-frame *intrinsic* EW of LAEs unchanged or positively evolved (Section 3.1.).

On the other hand, Figure 3 presents that Lyα LFs of LAEs would gradually decrease from $z = 5.7$ to 7.0. This decrease may be due to cosmic reionization and/or galaxy evolution. More detailed discussions are found in Section 4.

3. Have Primordial Galaxies Been Identified?

3.1. Lyα Equivalent Width Distribution

Figure 4 presents histograms of rest-frame equivalent width (EW_0) for LAEs at $z \sim 3$, 4, and 6 obtained in Ouchi et al. (2007). There is no clear evolution of EW_0 distribution from $z = 3$ to 6 within the uncertainties of measurements. These EW_0 distributions are based on the *apparent* EW. If the IGM absorption is corrected with a model of Section 2.2., the histogram of $z \sim 6$ would be shifted to large EW_0, which is systematically different from those of $z \sim 3$ and 4. Thus, $z = 5.7$ LAEs would have a large *intrinsic* EW_0 on average.

It is suggested that galaxies with a large Lyα EW are candidates of primordial galaxies. Theoretical models predict that (1) a collapse of gas clouds into a halo produces strong Lyα emission (i.e., cooling clouds, see e.g., Fardal et al. 2001), and that (2) subsequent primordial star-formation with metal poor gas also emits strong Lyα due to a top heavy IMF (i.e., population III see e.g., Schaerer 2003) whose equivalent width exceeds ~ 240 Å (Malhotra & Rhoads 2002). Based on the rest-frame EW distribution with IGM correction, the fraction of intrinsically large-EW ($EW_0 \gtrsim 240$) LAEs is 10-40% at $z = 3.1 - 5.7$, which does not significantly change from $z = 3.1 - 5.7$ within this percent range (Ouchi et al. 2007 see also Malhotra & Rhoads 2002; Dawson et al. 2004; Shimasaku et al. 2006; Saito et al. 2006).

3.2. He II Emission

He II $\lambda 1640$ emission from galaxies is a good indicator of primordial populations, i.e., cooling clouds and population III stars, since excitation of He atom requires high energy that can be produced by the gravitational collapse of gas or top-heavy IMF star-formation. Jimenez & Haiman (2006) claim that a composite spectrum of $z = 3$ Lyman break galaxies (LBGs) presents He II emission from a primordial population, although this He II emission is thought to be originated from Wolf-Rayet stars (Shapley et al. 2003). Since LAEs have younger stellar populations than LBGs (Pirzkal et al. 2007), LAEs would provide a stronger constraint on such a primordial population than LBGs with a composite spectrum. Figure 5 shows the composite spectra of LAEs whose average redshifts are $\langle z \rangle = 3.13$ and 3.68 (Ouchi et al. 2007). The inset panels of Figure 5 magnify the spectra around the wavelength of He II. No significant line is found for He II at these redshifts at the level of the upper limits of these data. The 3σ upper limits of the line ratio are $f_{\rm HeII}/f_{\rm Ly\alpha} = 0.02$ and 0.06 for $\langle z \rangle = 3.13$ and 3.68 LAEs, respectively. Since the ratio of $f_{\rm HeII}/f_{\rm Ly\alpha}$ is predicted as small as $\sim 0.1 - 0.001$ for population III star formation and cooling radiation (e.g.,

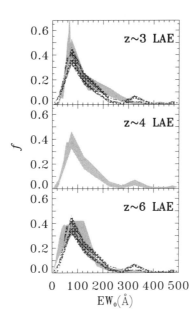

Figure 4. Histograms of the rest-frame Lyα EWs for LAEs at $z = 3.1$ (top), 3.7 (middle), and 5.7 (bottom) that are presented in Ouchi et al. (2007). At each panel, the gray region indicates the *apparent*-EW_0 histogram with uncertainties. The histogram of $z \sim 4$ LAEs is repeatedly plotted with meshes in the top and bottom panels for comparison. This figure is reproduced from Ouchi et al. (2007).

Schaerer 2003; Yang et al. 2006), it is needed to carry out future surveys whose upper limits reach at the level of $f_{\mathrm{HeII}}/f_{\mathrm{Ly}\alpha} \simeq 0.001$ to place strong constraints on primordial populations.

4. Constraints on Cosmic Reionization

4.1. Lyα Luminosity Function

Above $z = 5.7$, Lyα LF of LAEs decreases towards higher redshift, at least, up to $z = 7.0$ (Figure 3). The evolution of Lyα LF would be a signature of Lyα absorption by neutral IGM at the epoch of reionization ($z \gtrsim 6$). If no evolution of intrinsic LF is assumed at $z \sim 5.7 - 7.0$, neutral fraction of the Universe is estimated to be $x_{\mathrm{HI}} \lesssim 0.45$ at $z = 6.6$ (Kashikawa et al. 2006) with a model of Santos (2004), and an even higher upper limit of x_{HI} at $z = 7.0$ (Iye et al. 2006). Galaxy formation could be an alternative explanation for this evolution of Lyα LFs from $z = 5.7$ to 6.6. However, the UV LF of LAEs shows no evolution from $z = 5.7$ to 6.6 (open squares of the $z \sim 6$ and 6.6 panels in Figure 2). It would imply no evolution in intrinsic properties of LAEs, and suggest that the decrease

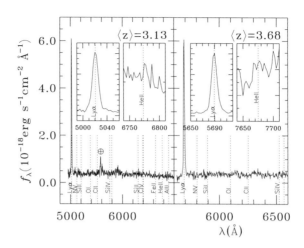

Figure 5. Composite spectra of LAEs at $\langle z \rangle = 3.13$ (left) and 3.68 (right) that are presented in Ouchi et al. (2007). The dotted lines indicate the wavelengths of interstellar absorption from star formation or highly ionized emission from AGN activities. The plots of spectrum are magnified in the inset boxes for wavelength ranges of Lyα (left) and He II (right). Note that a spectral resolution is $R \sim 500$ which is not high enough to resolve the asymmetry of Lyα line. This figure is reproduced from Ouchi et al. (2007).

of Lyα LF is originated from the evolution of IGM absorption at the epoch of cosmic reionization.

4.2. Correlation Function

Figure 6 presents angular correlation functions and biases of LAEs at $z = 5.7$ and 6.6 that are obtained from large Subaru samples in SXDF (see Ouchi et al. 2007). Note that $z = 6.6$ measurements are based on a preliminary sample of $z = 6.6$ LAEs which is made with a half of the final data. Figure 6 shows that clustering of LAEs does not significantly evolve from $z = 5.7$ to 6.6. This clustering evolution places constraints on the neutral fraction of the Universe (Section 1.). With the analytic model of Furlanetto et al. (2006), the clustering of $z = 5.7$ and 6.6 LAEs gives the upper limit of neutral fraction of $x_{\rm HI} \lesssim 0.3$ at $z = 6.6$.

This constraint from clustering, $x_{\rm HI} \lesssim 0.3$, is not only consistent with that from Lyα LF ($x_{\rm HI} \lesssim 0.45$), but also with that from GRB spectrum ($x_{\rm HI} \lesssim 0.17$ at $z = 6.3$; Totani et al. 2006). All of these complementary results indicate that the Universe is not fully neutral at $z \sim 6 - 7$ with a neutral fraction of at least $x_{\rm HI} \lesssim 0.5$. It is important to push the observational limits of galaxies to $z > 7$ and to identify very beginning of reionization.

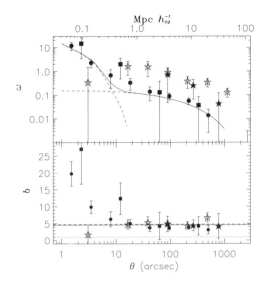

Figure 6. Angular correlation function (top) and bias (bottom) of LAEs at $z = 5.7$ and 6.6, together with those of dropout galaxies at $z = 5-6$. Filled and open stars are LAEs at $z = 6.6$ and 5.7, respectively. Squares and circles show dropout galaxies at $z \sim 6$ and 5, respectively. In the top panel, solid and dashed lines are the best-fit halo occupation distribution model of Hamana et al. (2004) for $z \sim 5$ dropout galaxies. In the bottom panel, solid and dashed lines denote the average biases of $z = 6.6$ and 5.7 LAEs, respectively. Dotted line corresponds to $b = 1$ in the bottom panel.

5. Searches for $z \gtrsim 7$ Galaxies

So far, the most distant galaxy is spectroscopically identified at $z = 6.96$ by a wide-field narrow-band imaging and spectroscopic survey of Subaru (Iye et al. 2006). Above redshift 7, there are no galaxies whose asymmetric Lyα line is confirmed by spectroscopy. However, a number of deep imaging and spectroscopic studies are currently searching for $z > 7$ galaxies. Several candidates of high-z galaxies constrain UV and Lyα LFs at $z > 7$.

Dropout galaxy searches have found one reliable and three less-reliable galaxy candidates, respectively, with $H_{AB} = 26-27$ at $z = 7-8$ in the NICMOS fields atop and around HUDF (Bouwens & Illingworth 2006). The recent study of Mannucci et al. (2007) presents a null detection of $z \gtrsim 7$ dropout candidates down to $m_{AB} = 25.5$ in the GOODS-S ISSAC field. The bottom panel of Figure 2 plots UV LF from these results with filled circles. The decrease of UV LF likely continues up to $z \sim 7-8$.

Similarly, LAE searches give constraints on Lyα LF. Figure 7 summarizes constraints of LF at $z \simeq 9$ as well as the depth and volume of on-going LAE searches. Stark et al. (2007) report some $z = 9$ LAE candidates found in nearby-cluster fields with the gravitational amplification, and claim a large increase of faint LAEs with a Lyα luminosity of $L(\text{Ly}\alpha) = 10^{40} - 10^{42.5}$ erg/s. Theoretical

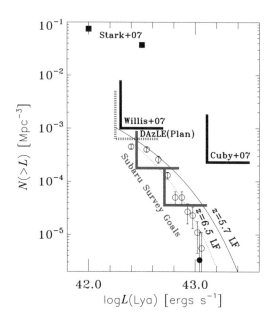

Figure 7. The survey depth and volume of recent searches for LAEs at $z \sim 9$, together with the luminosity function of LAEs at $z = 5.7$ (solid line; Ouchi et al. 2007), $z = 6.5$ (open circles and dotted line; Kashikawa et al. 2006) and $z = 7.0$ (filled circle; Iye et al. 2006). The number densities of *faint* LAE candidates are shown with squares (Stark et al. 2007). The survey limits of the previous and planned studies are presented with thick-solid lines (Willis et al. 2007; Cuby et al. 2007) and dashed lines (DAzLE; Horton et al. 2004), respectively. The survey goal of Subaru survey is presented with gray solid lines.

models show that such a faint Lyα emission line is strongly absorbed by IGM, since faint star-forming galaxies cannot make a cosmic H II region large enough for Lyα photons to escape (e.g., Haiman & Cen 2005). If the *faint* candidates of Stark et al. (2007) are real, a large number of *luminous* LAEs should be also found. Contrary to this theoretical prediction, no study has found a luminous $z \sim 9$ LAE with $L(\text{Ly}\alpha) \gtrsim 10^{42.5}$ erg/s. It may be due to strong halo evolution at massive end, or a burst of dwarf galaxy formation that is allowed in a weak UV background radiation before cosmic reionization (Wyithe & Loeb 2006). There may be some important misunderstandings of LAEs and reionization in these contradictory results between observations and theory. However, there still remains a problem in the observational side. Although both *luminous* and *faint* LAEs are key to understanding cosmic reionization, so far there are no studies which search an area large enough to constrain the luminous population of $z \sim 9$ LAEs as shown in Figure 7. Thus, ambitious wide-field narrow-band searches for $z \sim 9$ LAEs are going on (or planned) with Subaru/MOIRCS (Ouchi et al. 2007),

VLT/DAzLE (Horton et al. 2004), VLT/HAWK-I (see Willis et al. 2007), and VISTA (Nilsson et al. 2007). The Subaru/MOIRCS survey is being conducted with MOIRCS, a new wide-field near-infrared camera of Subaru (Ichikawa et al. 2007), and a narrow-band filter whose central wavelength is 1.19μm. The survey area is a ~ 100 arcmin^2 area of MOIRCS Deep Survey (MODS; Ichikawa et al. 2007) field atop GOODS-N. The goal of Subaru/MOIRCS survey is plotted in Figure 7. These wide-field imaging with a sufficient depth will reach the survey volume and sensitivity comparable to no evolution model of $z \sim 7$ LF, for the first time. The observations of Subaru/MOIRCS survey will be completed in the spring of 2008. With these observing data, the number density of luminous $z = 8.8$ LAEs from our wide-field program will clearly distinguish between evolution and no evolution models of luminous LAEs from $z \sim 7$ to 9. It will shed light on the possible misunderstandings of LAEs and reionization, which include uncertainties in galaxy/halo evolution, radiative transfer of Lyα, ionizing bubbles around galaxies, and neutral fraction at $z \sim 9$.

These on-going efforts will provide candidates of $z \sim 7 - 10$ galaxies in a few years. Because it is very difficult to take spectra of faint high-z galaxy candidates with the current technology, these candidates will be supplied to spectroscopic studies of the next-generation large telescopes, such as the James Webb Space Telescope (JWST), the Thirty Meter Telescope (TMT), the Giant Magellan Telescope (GMT), and European Extremely Large Telescope (E-ELT), which will start operations in the 2010s. Moreover, galaxies at higher redshifts of $z \sim 10 - 20$, will not be detected with the currently available telescopes, but with these next-generation large telescopes. Redshifts of $\sim 10 - 20$ are probably the epoch of the beginning of reionization and the appearance of the first-generation stars. The upcoming very large telescopes will address two major questions introduced in this contribution that are: (i) when primordial galaxies appeared and (ii) when the Universe started reionization.

Acknowledgments. I am grateful to the members of Subaru/XMM Deep Survey and MOIRCS Deep Survey teams who made invaluable efforts to obtain these data presented in this contribution. I am supported by Observatories of the Carnegie Institution of Washington through Carnegie Fellowship.

References

Bouwens, R. J., Illingworth, G. D., Blakeslee, J. P., & Franx, M. 2006, ApJ, 653, 53
Bouwens, R. J., & Illingworth, G. D. 2006, Nature, 443, 189
Bouwens, R. J., Illingworth, G. D., Franx, M., & Ford, H. 2007, ApJ, 670, 928
Cowie, L. L., & Hu, E. M. 1998, AJ, 115, 1319
Cuby, J.-G., Hibon, P., Lidman, C., Le Fèvre, O., Gilmozzi, R., Moorwood, A., & van der Werf, P. 2007, A&A, 461, 911
Dawson, S., et al. 2004, ApJ, 617, 707
Eyles, L. P., Bunker, A. J., Ellis, R. S., Lacy, M., Stanway, E. R., Stark, D. P., & Chiu, K. 2007, MNRAS, 374, 910
Fan, X., Narayanan, V. K., Strauss, M. A., White, R. L., Becker, R. H., Pentericci, L., & Rix, H.-W. 2002, AJ, 123, 1247
Fan, X., et al. 2006, AJ, 132, 117
Fardal, M. A., Katz, N., Gardner, J. P., Hernquist, L., Weinberg, D. H., & Davé, R. 2001, ApJ, 562, 605
Furlanetto, S. R., Zaldarriaga, M., & Hernquist, L. 2006, MNRAS, 365, 1012

Gronwall, C., et al. 2007, ApJ, 667, 79
Haiman, Z., & Cen, R. 2005, ApJ, 623, 627
Hamana, T., Ouchi, M., Shimasaku, K., Kayo, I., & Suto, Y. 2004, MNRAS, 347, 813
Horton, A., Parry, I., Bland-Hawthorn, J., Cianci, S., King, D., McMahon, R., & Medlen, S. 2004, SPIE, 5492, 1022
Hu, E. M., Cowie, L. L., Capak, P., McMahon, R. G., Hayashino, T., & Komiyama, Y. 2004, AJ, 127, 563
Ichikawa, T., et al. 2007, astro-ph/0701820
Iye, M., et al. 2006, Nature, 443, 186
Jimenez, R., & Haiman, Z. 2006, Nature, 440, 501
Kashikawa, N., et al. 2006, ApJ, 648, 7
Madau, P. 1995, ApJ, 441, 18
Malhotra, S., & Rhoads, J. E. 2002, ApJ, 565, L71
Malhotra, S., & Rhoads, J. E. 2004, ApJ, 617, L5
Mannucci, F., Buttery, H., Maiolino, R., Marconi, A., & Pozzetti, L. 2007, A&A, 461, 423
Nilsson, K. K., Orsi, A., Lacey, C. G., Baugh, C. M., & Thommes, E. 2007, A&A, 474, 385
Ouchi, M., et al. 2004, ApJ, 611, 660
Ouchi, M., et al. 2007, ApJ submitted, astro-ph/0707.3161
Ouchi, M., Tokoku, C., Shimasaku, K., & Ichikawa, T. 2007, Cosmic Frontiers ASP Conference Series, Vol. 379, ed. Metcalfe, N. and Shanks, T., p47
Pirzkal, N., Malhotra, S., Rhoads, J. E., & Xu, C. 2007, ApJ, 667, 49
Saito, T., Shimasaku, K., Okamura, S., Ouchi, M., Akiyama, M., & Yoshida, M. 2006, ApJ, 648, 54
Santos, M. R. 2004, MNRAS, 349, 1137
Schaerer, D. 2003, A&A, 397, 527
Shapley, A. E., Steidel, C. C., Pettini, M., & Adelberger, K. L. 2003, ApJ, 588, 65
Shimasaku, K., Ouchi, M., Furusawa, H., Yoshida, M., Kashikawa, N., & Okamura, S. 2005, PASJ, 57, 447
Shimasaku, K., et al. 2006, PASJ, 58, 313
Sheth, R. K., & Tormen, G. 1999, MNRAS, 308, 119
Spergel, D. N., et al. 2007, ApJS, 170, 377
Stark, D. P., Ellis, R. S., Richard, J., Kneib, J.-P., Smith, G. P., & Santos, M. R. 2007, ApJ, 663, 10
Steidel, C. C., Adelberger, K. L., Giavalisco, M., Dickinson, M., & Pettini, M. 1999, ApJ, 519, 1
Steidel, C. C., Adelberger, K. L., Shapley, A. E., Pettini, M., Dickinson, M., & Giavalisco, M. 2003, ApJ, 592, 728
Totani, T., Kawai, N., Kosugi, G., Aoki, K., Yamada, T., Iye, M., Ohta, K., & Hattori, T. 2006, PASJ, 58, 485
Willis, J. P., Courbin, F., Kneib, J. P., & Minniti, D. 2007, astro-ph/0709.1761
Wyithe, J. S. B., & Loeb, A. 2006, Nature, 441, 322
Yang, Y., Zabludoff, A. I., Davé, R., Eisenstein, D. J., Pinto, P. A., Katz, N., Weinberg, D. H., & Barton, E. J. 2006, ApJ, 640, 539
Yoshida, M., et al. 2006, ApJ, 653, 988

Robert Quimby discusses searches for supernovae.

Supernovae

Robert M. Quimby

Department of Astronomy, California Institute of Technology, Pasadena, CA, USA

Abstract. Supernovae are intimately entwined with virtually all areas of astronomical research from the metal content of solar system bodies, to feedback into star formation, to the production of gravitational waves. Here I will briefly review the observational properties and explanatory models of supernovae, and I will then highlight some recent results that have helped to better constrain the physical understanding of these stellar explosions. The latter portion will be strongly biased toward the Texas Supernova Search sample, which has gathered detailed observations of the normal (Type Ia SNe: 2005cg, 2005hj, 2006X; Type II SN: 2006bp) and the extraordinary (SNe 2005ap and 2006gy) among others.

1. Introduction

Supernovae are stellar explosions so catastrophic that in a brief period (seconds to days) all of a star's remaining fuel is consumed and/or expelled leaving behind a compact object, such as a neutron star or a black hole, or, in other cases, nothing but an expanding shell of gas. The glow of the cooling ashes may be augmented by the decay of radioactive species synthesized in the explosion or through interactions with circumstellar gas and last for months or years. A star's fate is essentially sealed at creation by its birth weight, metallicity, and by the presence of any companions close enough for interactions. Solitary stars that initially weigh in at $\sim$7-8 $M_\odot$ will eventually lose their extended hydrogen envelopes to expose their degenerate cores. With interior temperatures too low to continue nuclear burning and lacking external stimuli, the core will simply cool down and fade over time. The initial mass upper limit for production of such white dwarf (WD) stars is derived from stellar evolution codes, and the range reflects both uncertainties in the codes and the effects of metallicity (e.g. Eldridge & Tout 2004). If the star is initially more massive than this limit, so that it may burn carbon and later heavier elements up to iron, or if it interacts with a companion, an explosive end is possible.

Observationally, supernovae (SNe) are classified primarily by their spectral features and, in some cases, secondarily by their photometric evolution (see Filippenko 1997 for a review). In brief, SNe with (obvious) hydrogen lines are classified as Type II, and this class further breaks down into Types II-P, II-L, and IIn based on the presence of a photometric plateau (Patat et al. 1994), a linearly declining light curve, or narrow lines in the spectra, respectively. Supernovae without obvious hydrogen lines are classified as Type Ia (SNe Ia) if they show strong absorption around rest 6150 Å (usually associated with the blue shifted Si II λ 6355 doublet), or Type Ib if this feature is lacking but strong He lines are present. All remaining supernovae are classified as Type Ic. There are also hybrid objects such as the "Type IIb" SN 1993J, which initially resemble a particular class (e.g. Type II) but later transition to another (Type Ib). Given the qualitative and vague definitions delineated above, SNe are often compared to archetypal examples via χ^2 fits or cross correlation and labeled in accordance with the best match (e.g. a "1999aa-like Type Ia").

Theoretically, SNe are grouped into one of three explosion classes: thermonuclear, core-collapse, or pair-instability. The most massive stars, especially first generation stars from the early universe, are thought to explode through the pair-instability mechanism (Barkat et al. 1967; Fraley 1968), although the lower mass limit is poorly understood and highly sensitive to metallicity. In such events the interior photons, which provide the pressure necessary to balance the crushing force of gravity, attain energies greater than the rest mass of an electron-positron pair. The loss of photons to pair production then lowers the adiabatic index below the critical $\gamma = 4/3$ value, and and the core will contract. Heger & Woosley (2002) postulate that above a given mass ($\sim 260\,M_\odot$), the core will not be halted and it will form a black hole. In the range $140\,M_\odot \leq M \leq 260\,M_\odot$, thermonuclear runaway will occur and the reaction will proceed to obliterate the star; however, at lower masses the result will instead be a large amplitude pulsation that will expel a significant fraction of the outer envelope before burning in the core is halted. In this latter case, the star will survive and may explode later through core collapse (Woosley et al. 2007).

Heger et al. (2003) have calculated the end points expected for isolated stars with various initial masses greater that $\sim 8\,M_\odot$ (note that these calculations neglect physical processes such as rotation, and most massive stars are expected to possess companions). Below (metal-free) initial masses of $\sim 140\,M_\odot$, Heger et al. (2003) find that these massive stars are doomed to explode as core collapse supernovae. In these events, the progenitor forms a dense Fe core through the usual burning cycle, which collapses under its own weight. The core rebounds from this collapse and unbinds the envelope, forming a supernova. The details of this process are still poorly understood, but they may involve neutrino interactions (Janka et al. 2006), shock instabilities (Blondin & Mezzacappa 2006), acoustic instabilities (Burrows et al. 2006), and magneto-rotational effects (Akiyama et al. 2003). Heger et al. (2003) show that low metallicity progenitors with higher initial masses ($> 25\,M_\odot$) will form black holes by direct collapse or through fallback, while lighter stars will leave behind neutron stars. At metallicities near solar and higher, all massive progenitors form neutron stars. The observational classes Type II, Type Ib, and Type Ic all correspond to core collapse events. The differentiation being the extent of mass loss incurred prior to the ultimate explosion.

The remaining class of supernovae, the Type Ia events, are thought to be thermonuclear explosions of white dwarf stars (see Branch 1998; Hillebrandt & Niemeyer 2000). Except for a possible sub population of stars just below the core-collapse cutoff that may explode in solitude (the so-called "Type 1.5"; Han & Podsiadlowski 2006), low mass progenitors require interaction with an external body to explode. Possible scenarios include mass transfer from a main sequence or evolved companion onto a white dwarf via Roche Lobe overflow or a common envelope phase, and WD-WD mergers. Electron degeneracy pressure cannot support a star above the Chandrasekhar mass (about 1.4 $M_\odot$), so thermonuclear run away is triggered as accretion pushes the total mass toward this limit. The nature of the flame propagation has been debated, but delayed detonation models of C/O white dwarfs reasonably account for the observed properties of Type Ia SNe (Höflich et al. 2006).

Type Ia SNe are the best studied class of SNe owing to their bright peak luminosities (typically about -19.5 mag absolute), their small dispersion (~ 0.3 mag intrinsic, < 0.1 mag after corrections), and hence their utility as cosmological probes (Riess et al. 1998; Perlmutter et al. 1999). They comprise most, but not all, events observed to reach luminosities greater than about -19 mag. Type II supernovae show a much wider range of peak luminosities, with typical values of about -17.0 mag (Richardson et al. 2002). Type Ib and the spectroscopic default Type Ic SNe also show a range of peak magnitudes and light curve shapes, but the typical peak magnitude is $M_V \sim -18.1$ (Modjaz 2007).

2. The Texas Supernova Search

In order to obtain multi-epoch spectral observations of supernovae starting from the earliest possible phases, we began a search for optical transients with the 0.45-m ROTSE-IIIb telescope (Akerlof et al. 2003) in 2004. The Texas Supernova Search (TSS; Quimby 2006) covered the entire Virgo, Ursa Major, and Coma Galaxy clusters, and other select fields every 1-3 nights as weather and season allowed. This project was made technically feasible by ROTSE-IIIb's wide field of view ($1.85° \times 1.85°$), the large amount of telescope time available ($\sim 25\%$ of every night), and its fully robotic operation. Transient candidates were promptly uncovered with a modified version of the PSF-matched image subtraction developed by the Supernova Cosmology Project (Perlmutter et al. 1999), and image thumbnails were posted to the web for final vetting by human scanners. Candidates which passed all selection criteria (e.g. detected at $> 5\sigma$ significance, no significant motion between observations spaced by ~ 30 minutes, not matching a known variable source, etc.) were directed to the neighboring 9.2-m Hobby-Eberly Telescope (HET) for spectroscopic typing. This typically occurred the following night, although in some cases the spectra were acquired just hours after discovery (e.g. SN 2005hj). When weather or a lack of observing time precluded spectroscopic confirmation with the HET, we contacted other researchers directly or via IAU Circulars so that all supernova discovered by the TSS were spectroscopically confirmed.

From November 2004 through November 2007, we discovered or independently detected 30 supernovae, and we obtained multi-epoch spectral time series for several events. In the following sections, I will summarize some of our key findings and describe how these results and contemporary works have furthered our understanding of supernovae since the inception of the TSS.

3. Type Ia Supernovae

In this section I present three Type Ia SNe which were observed by ROTSE-III and the HET as part of the Texas Supernova Search. First reported by a Japanese amateur and the CROSS program (Ponticello et al. 2006), SN 2006X was a highly reddened event initially showing high velocity ejecta and later evolving Na I D lines. SN 2005cg is a completely typical Type Ia SN, and the

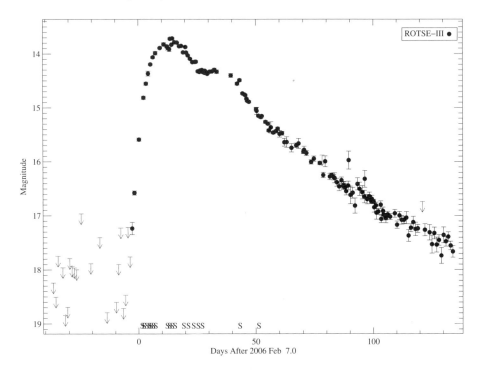

Figure 1. Preliminary ROTSE-III unfiltered light curve of the Type Ia supernova 2006X in M100. From Quimby (2006).

early spectra provide evidence for a detonation phase in such normal events. SN 2005hj may strictly be considered a normal Type Ia SN, however, multi-epoch HET observations provide evidence for shell interactions.

3.1. SN 2006X

The preliminary unfiltered ROTSE-III light curve of SN 2006X in the galaxy M100 is shown in Figure 1. We first detected the supernova on 2006 Feb 4.29 UT at about 17.2 mag. Observation upper limits from the previous nights show this was soon after explosion, and the SN brightened by more than 3 mag in the following nights to its peak of about 13.8 mag. A second maximum is seen about 25 days after this principle peak, as is commonly seen in the R-band and redder wavelengths (see Kasen 2006). Early spectral typing by the HET (Quimby et al. 2006a) revealed an unusually red continuum with rapidly expanding ejecta, and strong polarization (Wang et al. 2006).

High resolution spectral observations of SN 2006X obtained over multiple nights reveal changes in the narrow Na I D lines (Patat et al. 2007b). This is the first Type Ia SN with a published high resolution spectral time series. Such variations are not seen in other lines, such as Ca II H&K. Patat et al. (2007b) argue that the variability in the Na I D lines is a sign of circumstellar interaction with a pre-supernova wind and they use the derived wind velocity to classify the companion star as a red-giant. Similar observations have now

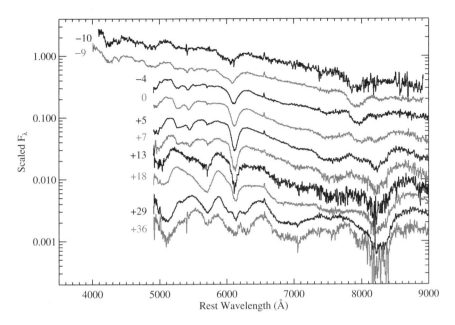

Figure 2. Spectral evolution of SN 2005cg as captured by the HET. From Quimby et al. (2006b).

been published for additional Type Ia SN. No variations are seen in the Na I D lines of SN 2007af (Simon et al. 2007) nor SN 2000cx (Patat et al. 2007a), which implies differing progenitor systems or non-spherical geometries with preferred viewing angles.

3.2. SN 2005cg

SN 2005cg eventually proved to be a completely normal Type Ia event; however, this was far from clear at the onset. The first spectrum taken by the HET revealed a Si II '6150' P-Cygni profile with what seemed to be an unusual absorption trough extending far into the blue (top spectrum in figure 2). A high velocity Ca II IR triplet feature was also clearly detected. As the supernova evolved, the light curve showed the initial spectrum was a rare, 10 day prior to maximum light observation, and the spectra taken around maximum were rather ordinary. In Quimby et al. (2006b), we suggest the presence of Si, a burning product, at high velocities is incompatible with deflagration models since the subsonic flame cannot burn through these expanding outer layers. Instead, we suggest normal Type Ia SNe, such as 2005cg, are marked by a detonation phase.

Noting that the blue wing of the Si II absorption cuts off abruptly as it meets with the continuum, and that this cutoff further corresponds to the minimum of the high velocity Ca II IR feature, we further suggest the observations are consistent with a shell interaction. Gerardy et al. (2004) claim that a solar gas mixture in the immediate vicinity of the explosion, possibly in the form of a thick accretion disk, can account for the high velocity Ca II IR feature regularly observed in Type Ia SNe at early times, and they predict such material would

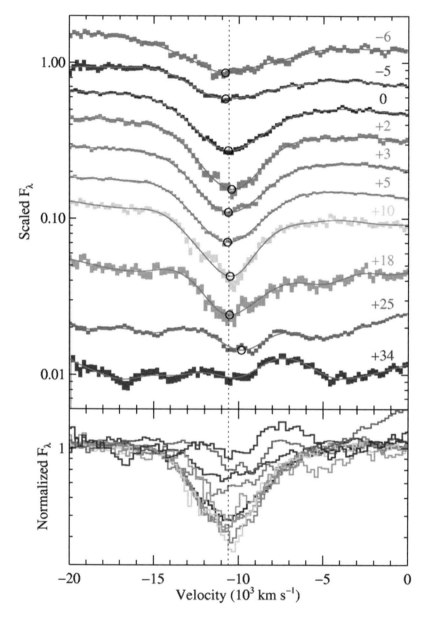

Figure 3. Spectral evolution of the Si II '6150' feature in SN 2005hj as recorded by the HET. From Quimby et al. (2007a).

truncate the velocity distribution of the highest velocity layers, much as we observe for SN 2005cg.

3.3. SN 2005hj

Much like the case of SN 2005cg, our early HET spectra of SN 2005hj showed an unusual Si II line profile and we pursued a multi-epoch spectroscopic follow-up campaign. In this case, the feature was unusually shallow and not particularly extended in the blue despite the early phase of the initial spectra. Continued monitoring over the next few weeks revealed little variation in the line until the photosphere receded below the Si layer and the Fe lines began to appear. In Quimby et al. (2007a), we find that the minimum of the Si II line, a proxy for the photospheric velocity, remains essentially constant at $10\,600 \pm 150\,\mathrm{km\,s^{-1}}$ between maximum light and +18 days (Fig. 3). We take this as a sign that the photosphere is hung up in a dense shell over this period, as could form in a pulsation delayed detonation model or in a double degenerate merger. Although SN 2005hj is technically a normal-bright (or core-normal; Branch et al. 2006) Type Ia SN, the presence of this velocity plateau is in contrast to the smooth deceleration of $1\,000\,\mathrm{km\,s^{-1}}$ or more exhibited by other normal Type Ia SNe (see Benetti et al. 2005). Thus there may be different progenitors or different explosion modes within the observed population of normal Type Ia SNe.

4. Type II Supernovae

Type II SNe, or explosions of stars with sufficient H in their envelopes to produce obvious spectral lines, are often dimmer than their Type Ia SN cousins but more prevalent in the cumulative history of the universe. In this section I will discuss the ordinary, if not unusually well observed, SN 2006bp, and the two most luminous supernovae ever identified, SNe 2006gy and 2005ap.

4.1. SN 2006bp

This Type II-P supernova was discovered by K. Itagaki of Yamagata, Japan (Nakano & Itagaki 2006). We detected SN 2006bp with ROTSE-IIIb on 2006 April 9.15 at an unfiltered magnitude of 17.75 ± 0.19 (Quimby et al. 2007c). Null detections from the previous nights and the steep initial rise (Fig. 4) indicate that the supernova was caught soon after shock breakout. Non-LTE modeling with the CMFGEN code indicates that the shock breakout date is 2005 April 7.9 ± 0.4 (Dessart et al. 2007). The light curve shows a clear plateau phase lasting ~ 100 days, defined roughly as lasting from the midpoint of the sudden rise to the midpoint of the rapid fading prior to the exponential decay tail. The average decline rate between days +121 and +335 in this unfiltered data is $0.0073\,\mathrm{mag\,day^{-1}}$, which is significantly slower than the $0.0098\,\mathrm{mag\,day^{-1}}$ decline expected from the decay of $^{56}\mathrm{Co}$, and this may suggest an additional source of energy production sustaining the luminosity.

The first spectra obtained for SN 2006bp are most notable for their mainly featureless, blue continua. Close inspection of the data, however, reveals several broad P-Cygni profiles which we identify as Hα, He I λ5876, and He II λ4686. This is the first time such broad He II has been seen in the spectra of a supernova. More intriguing still are several narrow emission features present in the April 11 data, but absent on April 12. We conclude these lines originate from highly ionized He and C atoms in the vicinity of the explosion–by April 12 the SN

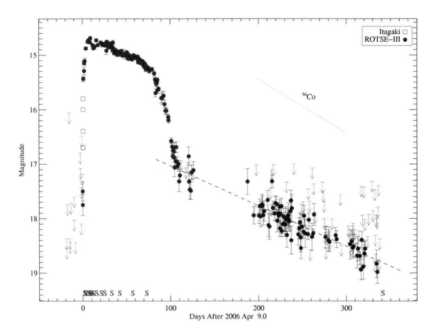

Figure 4. ROTSE-IIIb unfiltered light curve of SN 2006bp. From Quimby et al. (2007c).

ejecta have over taken this material thus squelching their narrow lines. These could be the outer layers of the progenitor itself.

At later times, the presence of "notches" to the blue of prominent P-Cygni profiles (Hα, Na I) is intriguing in its own right. Previous studies have explained these features as arising from metals such as N (Dessart & Hillier 2005). However, for SN 2006bp these features do not shift red with time in pace with other lines. In (Quimby et al. 2007c) we argue that this and other observations may signal a line forming region in higher velocity layers above the photosphere.

4.2. SN 2006gy

SN 2006gy was found near the core of NGC 1260, and there was some early speculation that it was not a supernova at all but rather an AGN. However, adaptive optic observations would eventually confirm SN 2006gy's separation from the galaxy core, and thus verify it as a supernova (Ofek et al. 2007; Smith et al. 2007). When first detected by ROTSE-IIIb, the apparent magnitude and cataloged distance to the host suggested a luminosity somewhat brighter than a Type Ia SN around maximum light. However, optical spectra indicated strong absorption along the line of sight, partially from our Galaxy but mainly from within the host (Prieto et al. 2006). After correction for this absorption the peak luminosity was corrected to an unprecedented -22 mag absolute. Smith et al. (2007) show SN 2006gy took ~ 70 days to reach its maximum brightness and that it stayed brighter than -21 mag for 100 days.

Although it is not yet clear what produced such a luminous and slowly evolving supernova as 2006gy, the available lines of evidence all point to an explosion of a massive star (Smith et al. 2007). It is at least possible that SN 2006gy was a pair instability event, which makes it the first observed supernova for which such a scenario must be considered.

4.3. SN 2005ap

Although it was found long before SN 2006gy, it was not confirmed until later that SN 2005ap was actually more luminous and in fact the brightest supernova ever observed after correcting for distance (Quimby et al. 2007b). SN 2005ap was found by ROTSE-IIIb in the Coma Cluster fields, but it is actually located well behind the cluster. A spectrum obtained by the HET shortly after discovery showed a very blue continuum with a few broad but shallow absorption features and a single narrow emission line. This narrow line was located significantly to the blue of zero-velocity Hα. A second, higher S/N spectrum taken by Keck/LRIS about 10 days later confirmed this feature and revealed a weaker emission line just to the blue, which showed this was the O III $\lambda\lambda 4959,5007$ doublet. The Keck data also showed an absorption doublet in the blue consistent with Mg II in the same rest frame. This securely places the redshift of SN 2005ap at $z = 0.283$, which sets the peak of the unfiltered light curve at a record breaking -22.7 mag absolute.

It is difficult to explain this extraordinary luminosity, which corresponds to about 10^{45} erg s^{-1} after an approximate bolometric correction, in the context of a supernova explosion. Yet the data are most consistent with a supernova. One possibility is that the engine powering SN 2005ap is similar to the engine of a gamma-ray burster, but in this case the H envelope (required by the detection of an Hα P-Cygni profile in the spectra) swallows the gamma-ray light. As suggested by Young et al. (2005), the bright subclass of Type II-L supernova may reflect such a progenitor, with the energy of the gamma-ray burst-style engine depositing a large amount of energy in the H envelope.

5. Concluding Remarks

The Texas Supernova Search was successful in its designed goal of capturing early spectra of supernovae – in fact our observations of SN 2006bp are the earliest ever obtained for a normal SNe – but the inclusion of the two most luminous supernovae ever identified in our small sample is, to say the least, unexpected. So how is it that a seemingly modest survey with a 0.45-m telescope produced such record-breaking discoveries? There are two possible avenues of explanation: 1) there is a unique aspect to the TSS that allowed these unique SNe to be found, or 2) it was complete luck. It is impossible to determine *a posteriori* if the second possibility is true especially since the detailed rates and environmental properties of SN 2005ap and/or SN 2006gy-like events are not yet known. On the other hand, there do appear to be several key differences between the TSS and other surveys that may naturally relate to the selection of such unusual SNe.

Figure 5 shows the redshift distribution of the first 30 SNe in the TSS sample including both SNe 2005ap and 2006gy. ROTSE-III is sensitive to the luminous Type Ia class out to a redshift of $z \sim 0.09$, and we found ~ 18 out to this limit. We only found one SN 2005ap-like event, however, at a redshift of $z = 0.28$, although we were sensitive to such events at closer distances. In other words our effective search volume for SN 2005ap-like events was ~ 3 times that of SNe Ia, but we found $1/18^{\mathrm{th}}$ as many, so the rate of SN 2005ap-like events cannot be much larger than $\sim 1/50^{\mathrm{th}}$ the Type Ia SNe rate. SN 2005ap-like events are thus rare, and the huge volume sampled by the TSS (comparable to all previous nearby searches) could be one key in their discovery. Also, since the spectral features of SN 2005ap are quite subtle, the high S/N spectral follow-up with the HET and Keck telescopes were critical to determining the distance and hence the intrinsic luminosity. Previous surveys that followed-up nearby SNe with smaller spectroscopic instruments or that found fainter SNe at greater distances may not have had the S/N required to make such a discovery.

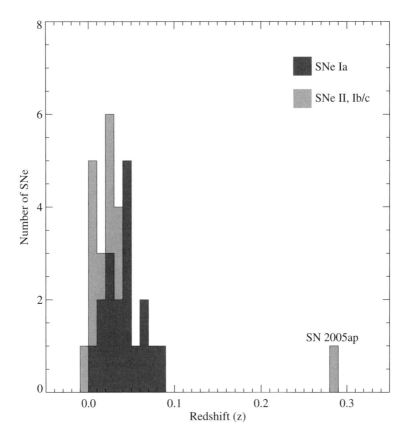

Figure 5. Redshift distribution of all supernovae in the Texas Supernova Search sample. From Quimby (2006).

For SN 2006gy, it is at least interesting that the first in a presumed class of SNe would be found near the core of a galaxy since the TSS was perhaps the first survey sensitive to so-located transients. The Lick Observatory Supernova Search, for example, observed NGC 1260 four times before ROTSE-IIIb in the summer of 2006, and it detected SN 2006gy first in the process. However the larger aperture and finer pixel scale of KAIT were rendered moot by software that automatically rejected SN candidates within 2.4″ of galaxy centers (W. Li, private communication 2007; see also Smith et al. 2007). As this is a common practice in transient searches, it is evident how SN 2006gy-like events could have been missed in the past if they prefer the cores of galaxies. There is also some physical motivation as to why the explosions of such presumably massive events may prefer the cores of galaxies. As pointed out by Portegies Zwart & van den Heuvel (2007), the deep potential well in the cores of galaxies facilitates the generation of dense star clusters, and these may include some of the most massive stars in a given galaxy. It is at least worth considering that this previously unsearched environment holds some importance to the unusual properties of SN 2006gy, although there is as yet only one example of this "class" and a statistical sample would be needed to confirm such speculation.

Acknowledgments. I thank the staff of the Hobby-Eberly Telescope and McDonald Observatory for their support of ROTSE-IIIb and for carrying out observations with the HET. I thank F. Castro, P. Mondol, and M. Sellers for their efforts in screening potential SN candidates, and J. C. Wheeler for helpful comments.

References

Akerlof, C. W., Kehoe, R. L., McKay, T. A., Rykoff, E. S., Smith, D. A., Casperson, D. E., McGowan, K. E., Vestrand, W. T., Wozniak, P. R., Wren, J. A., Ashley, M. C. B., Phillips, M. A., Marshall, S. L., Epps, H. W., & Schier, J. A. 2003, PASP, 115, 132
Akiyama, S., Wheeler, J. C., Meier, D. L., & Lichtenstadt, I. 2003, ApJ, 584, 954
Barkat, Z., Rakavy, G., & Sack, N. 1967, Physical Review Letters, 18, 379
Benetti, S., Cappellaro, E., Mazzali, P. A., Turatto, M., Altavilla, G., Bufano, F., Elias-Rosa, N., Kotak, R., Pignata, G., Salvo, M., & Stanishev, V. 2005, ApJ, 623, 1011
Blondin, J. M. & Mezzacappa, A. 2006, ApJ, 642, 401
Branch, D. 1998, ARA&A, 36, 17
Branch, D., Dang, L. C., Hall, N., Ketchum, W., Melakayil, M., Parrent, J., Troxel, M. A., Casebeer, D., Jeffery, D. J., & Baron, E. 2006, PASP, 118, 560
Burrows, A., Livne, E., Dessart, L., Ott, C. D., & Murphy, J. 2006, New Astronomy Review, 50, 487
Dessart, L., Blondin, S., Brown, P. J., Hicken, M., Hillier, D. J., Holland, S. T., Immler, S., Kirshner, R. P., Milne, P., Modjaz, M., & Roming, P. W. A. 2007, astro-ph/0711.1815
Dessart, L. & Hillier, D. J. 2005, A&A, 437, 667
Eldridge, J. J. & Tout, C. A. 2004, MNRAS, 353, 87
Filippenko, A. V. 1997, ARA&A, 35, 309
Fraley, G. S. 1968, Ap&SS, 2, 96
Gerardy, C. L., Höflich, P., Fesen, R. A., Marion, G. H., Nomoto, K., Quimby, R., Schaefer, B. E., Wang, L., & Wheeler, J. C. 2004, ApJ, 607, 391
Han, Z. & Podsiadlowski, P. 2006, MNRAS, 368, 1095
Heger, A., Fryer, C. L., Woosley, S. E., Langer, N., & Hartmann, D. H. 2003, ApJ, 591, 288
Heger, A. & Woosley, S. E. 2002, ApJ, 567, 532

Hillebrandt, W. & Niemeyer, J. C. 2000, ARA&A, 38, 191
Höflich, P., Gerardy, C. L., Marion, H., & Quimby, R. 2006, New Astronomy Review, 50, 470
Janka, H. ., Langanke, K., Marek, A., Martinez-Pinedo, G., & Mueller, B. 2007, Phys. Rep., 442, 38
Kasen, D. 2006, ApJ, 649, 939
Modjaz, M. 2007, in American Institute of Physics Conference Series, Vol. 924, American Institute of Physics Conference Series, 285–290
Nakano, S. & Itagaki, K. 2006, IAU Circ., 8700, 4
Ofek, E. O., Cameron, P. B., Kasliwal, M. M., Gal-Yam, A., Rau, A., Kulkarni, S. R., Frail, D. A., Chandra, P., Cenko, S. B., Soderberg, A. M., & Immler, S. 2007, ApJ, 659, L13
Patat, F., Barbon, R., Cappellaro, E., & Turatto, M. 1994, A&A, 282, 731
Patat, F., Benetti, S., Justham, S., Mazzali, P. A., Pasquini, L., Cappellaro, E., Della Valle, M., -Podsiadlowski, P., Turatto, M., Gal-Yam, A., & Simon, J. D. 2007a, A&A, 474, 931
Patat, F., Chandra, P., Chevalier, R., Justham, S., Podsiadlowski, P., Wolf, C., Gal-Yam, A., Pasquini, L., Crawford, I. A., Mazzali, P. A., Pauldrach, A. W. A., Nomoto, K., Benetti, S., Cappellaro, E., Elias-Rosa, N., Hillebrandt, W., Leonard, D. C., Pastorello, A., Renzini, A., Sabbadin, F., Simon, J. D., & Turatto, M. 2007b, Science, 317, 924
Perlmutter, S., Aldering, G., Goldhaber, G., Knop, R. A., Nugent, P., Castro, P. G., Deustua, S., Fabbro, S., Goobar, A., Groom, D. E., Hook, I. M., Kim, A. G., Kim, M. Y., Lee, J. C., Nunes, N. J., Pain, R., Pennypacker, C. R., Quimby, R., Lidman, C., Ellis, R. S., Irwin, M., McMahon, R. G., Ruiz-Lapuente, P., Walton, N., Schaefer, B., Boyle, B. J., Filippenko, A. V., Matheson, T., Fruchter, A. S., Panagia, N., Newberg, H. J. M., Couch, W. J., & The Supernova Cosmology Project. 1999, ApJ, 517, 565
Ponticello, N. J., Burket, J., Li, W., Chen, Y.-T., Yang, M., Lin, C.-S., Soma, M., Migliardi, M., & Dimai, A. 2006, IAU Circ., 8667, 1
Portegies Zwart, S. & van den Heuvel, E. P. J. 2007, Nature, 450, 388
Prieto, J. L., Garnavich, P., Chronister, A., & Connick, P. 2006, Central Bureau Electronic Telegrams, 648, 1
Quimby, R., Brown, P., Gerardy, C., Odewahn, S. C., & Rostopchin, S. 2006a, Central Bureau Electronic Telegrams, 393, 1
Quimby, R., Höflich, P., Kannappan, S. J., Rykoff, E., Rujopakarn, W., Akerlof, C. W., Gerardy, C. L., & Wheeler, J. C. 2006b, ApJ, 636, 400
Quimby, R., Höflich, P., & Wheeler, J. C. 2007a, ApJ, 666, 1083
Quimby, R. M. 2006, PhD thesis, University of Texas, U.S.A.
Quimby, R. M., Aldering, G., Wheeler, J. C., Höflich, P., Akerlof, C. W., & Rykoff, E. S. 2007b, ApJ, 668, L99
Quimby, R. M., Wheeler, J. C., Höflich, P., Akerlof, C. W., Brown, P. J., & Rykoff, E. S. 2007c, ApJ, 666, 1093
Richardson, D., Branch, D., Casebeer, D., Millard, J., Thomas, R. C., & Baron, E. 2002, AJ, 123, 745
Riess, A. G., Filippenko, A. V., Challis, P., Clocchiatti, A., Diercks, A., Garnavich, P. M., Gilliland, R. L., Hogan, C. J., Jha, S., Kirshner, R. P., Leibundgut, B., Phillips, M. M., Reiss, D., Schmidt, B. P., Schommer, R. A., Smith, R. C., Spyromilio, J., Stubbs, C., Suntzeff, N. B., & Tonry, J. 1998, AJ, 116, 1009
Simon, J. D., Gal-Yam, A., Penprase, B. E., Li, W., Quimby, R. M., Silverman, J. M., Prieto, C. A., Wheeler, J. C., Filippenko, A. V., Martinez, I. T., Beeler, D. J., & Patat, F. 2007, ApJ, 671, L25
Smith, N., Li, W., Foley, R. J., Wheeler, J. C., Pooley, D., Chornock, R., Filippenko, A. V., Silverman, J. M., Quimby, R., Bloom, J. S., & Hansen, C. 2007, ApJ, 666, 1116

Wang, L., Baade, D., Patat, F., & Wheeler, J. C. 2006, Central Bureau Electronic Telegrams, 396, 2
Woosley, S. E., Blinnikov, S., & Heger, A. 2007, Nature, 450, 390
Young, T. R., Smith, D., & Johnson, T. A. 2005, ApJ, 625, L87

Jane Rigby presents an infrared view of AGN and starburst galaxies.

Recent Progress on the Evolution of Rapidly Star–forming Galaxies and Active Galactic Nuclei

J. R. Rigby

Observatories of the Carnegie Institution of Washington, Pasadena, CA, USA

Abstract. I highlight recent results from the Spitzer Space Telescope that have advanced our understanding of rapidly star–forming galaxies and active galactic nuclei, and discuss what advances we should expect in the next decade with coming instrumentation.

1. Historical Perspective

The IR emission of star–forming galaxies and AGN have been characterized for more than three decades now. Very quickly it was realized that a basic characteristic of AGN is that they are bright in the infrared (Oke et al. 1970 for QSOs; Rieke 1978 for Seyferts). The dust temperatures are high, indicating that AGN contain dust close to the engine, up to the sublimation radius. This informs a picture where AGN generically keep substantial amounts of dust close to the central engine, which informs the unification paradigm.

While all galaxies produce some IR emission, the surprising early result was that some star–forming galaxies emit almost all their power in the IR. The must–read Annual Review articles are Rieke & Lebofsky (1979) and Low et al. (2007). The state–of–M82 summary in Rieke & Lebofsky (1979) is particularly instructive: M82 emits almost all its considerable luminosity in the IR; its mid–IR spectrum boasts prominent broad emission features (now recognized as from aromatic hydrocarbons, see for example Draine & Li 2007), strong $10\,\mu m$ silicate absorption, Brackett recombination lines, and fine structure lines of [Ar II], [Ne II], and [S III]. That's a perfect summary of the spectral diagnostics now used to study star–forming galaxies and AGN out to high redshift. Thus, the one–sentence summary of progress is that, by 1979 the major mid–IR spectral features were known (in M82), and that the last 30 years have built the sample size from one (M82) to tens of thousands, and moved out from the local universe to high redshift.

The next major progress in the understanding of IR–bright star–forming galaxies came with IRAS's discovery that ultraluminous infrared galaxies (ULIRGS) exist (Neugebauer et al. 1984; Aaronson & Olszewski 1984). ISO, launched in 1995, detected ULIRGs out to $z = 1$, enough to conclusively demonstrate that the IR luminosity function strongly evolves (Elbaz et al. 1999; Genzel & Cesarsky 2000; Rowan-Robinson et al. 2004). Today, the sensitive detectors onboard Spitzer are detecting mid–IR emission from star–forming galaxies out to $z \sim 3$, and from the highest–redshift AGN known (Jiang et al. 2006).

156   Rigby

Given that star–forming galaxies are rare today, but a thousand times more common in the past, how does this rapid star formation take place? Is there any trigger, or is it the inevitable result of gas–rich galaxies? What drives the evolution in the luminosity function? These are the foci of current investigations, selected results of which I will now summarize.

2. Spitzer–derived Star Formation Rates in Galaxies

The 24 μm band of Spitzer is commonly used to estimate star formation rates in galaxies (see calibrations in Genzel & Cesarsky 2000; Alonso-Herrero et al. 2006; Calzetti et al. 2007). This technique works because warm dust continuum and aromatic feature[1] luminosities scale with total infrared luminosity, which in turn scales with star formation rate. This requires screening out the galaxies with AGN first (since their dust is partially heated by non–stellar means). Le Floc'h et al. (2005) use such a conversion of $f_\nu(24)\,\mu$m–to– star formation rate to plot the star formation rate density as a function of redshift, reprinted here in Figure 1.

At $z \sim 0$ low–luminosity galaxies dominate the star formation rate density. As redshift increases, the contribution from more luminous galaxies dramatically increases, such that luminous infrared galaxies[2] dominate at $z > 0.7$. Even more luminous galaxies[3] are increasingly important as redshift increases, and are seen to take over for $z > 2$ (Caputi et al. 2006; Pérez-González et al. 2005). Thus, the past was a much more exciting time — not only was the star formation rate density much higher, but a much larger fraction of the Universe's star formation occurred in large bursts.

3. How Reliable Are Mid–IR–derived Star Formation Rates?

While the basic evolution of the star formation rate density is now known, the detailed picture is complicated by several sources of potential systematic error: namely, the use of photometric redshifts; the assumption that AGN contamination rates are low; and the assumption that $z = 0$ spectral templates accurately describe high–z galaxies, which are generally only faint photometric detections. In Rigby et al. (2008), we present evidence that the spectra of gravitationally lensed high–z star–forming galaxies are indeed different than local analogues, such that high–z galaxies create preferentially more aromatic emission for a given total infrared luminosity. Figure 2 illustrates the evolution. Therefore, high–redshift galaxies are well–fit by local templates of much lower luminosity, and are poorly fit by comparable-luminosity local templates. To test and (if

---

[1]The aromatic features are commonly identified in the literature as arising from polycyclic aromatic hydrocarbons (PAHs), though this identification is not yet definitive. In daily life, we encounter PAH molecules as carcinogens associated with diesel exhaust and grill–charred brisket.

[2]Luminous IR galaxies or LIRGs are defined as having $10^{11} < L(8\text{–}1000\,\mu\text{m}) < 10^{12}\ L_\odot$.

[3]Ultraluminous infrared galaxies or ULIRGS are defined as having $10^{12} < L(8\text{–}1000\,\mu\text{m}) < 10^{13}\ L_\odot$.

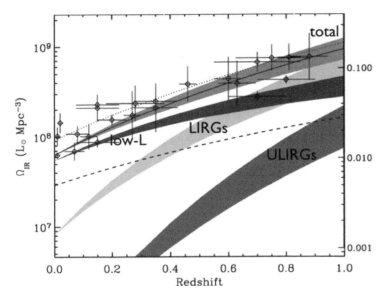

Figure 1. Evolution of the comoving IR luminosity density out to $z = 1$, taken from Le Floc'h et al. (2005). The total measurements are shown in the top shaded region. This is then broken up into contributions from the following constituents: low–luminosity galaxies (L(IR) $< 10^{11}$ L$_\odot$) (darkest shaded region); LIRGs ($10^{11} <$ L(IR) $< 10^{12}$ L$_\odot$) (lightest shaded region) and ULIRGs (L(IR) $> 10^{12}$ L$_\odot$) (bottom shaded region). Estimates are translated into a star formation rate density on the right vertical axis.

necessary) recalibrate star formation rate diagnostics in the distant universe, I am now working to measure recombination lines in a small number of lensed star–forming galaxies at $z = 1$ and $z = 2.5$. At lower redshifts, the mid-IR fine structure lines can also be used to test SFR diagnostics.

4. The New Observable at High Redshift: Stellar Mass

One of the key advances Spitzer has brought to extragalactic studies is the ability of the 3.5–8 μm IRAC bands to photometer the peak and Rayleigh–Jeans tail of the redshifted integrated stellar light of galaxies; this allows measurement of stellar mass out to $z \sim 4$ and beyond. I count 42 papers to date in the literature doing this. One example is Pérez-González et al. (2008), where the authors plot the evolution of stellar mass density with redshift, which we reprint as Figure 3. They find that half the stellar mass is in place by $z < 1$, large spheroids already exist by $z = 3$; and the galaxies doing most of the star formation are not the galaxies that contain most of the mass. While similar results have been reported previously, what's remarkable is how good the data are (their sample contains 3×10^4 galaxies); that samples can now be sub-divided by stellar mass; that the data are not confusion limited and thus will improve with deeper integrations;

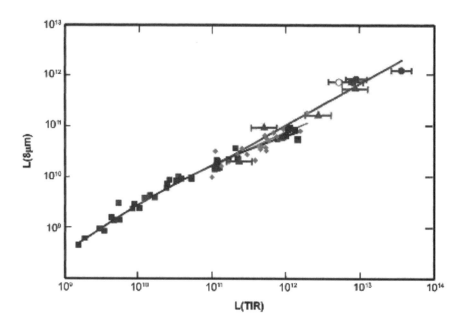

Figure 2. Comparison of L(8μm) and L(TIR) for star–forming galaxies at $z = 0$ and $z = 2$, taken from Rigby et al. (2008). Local galaxies with above average L(8μm)/L(TIR) are indicated as squares. The lower line is a fit to them. High-z galaxies from Yan et al. (2007) are shown as circles. Triangles are lensed galaxies from Rigby et al. (2008). The upper line is a fit to the $z \sim 2$ points, constrained to agree with the local fit at 10^{11} L$_\odot$. The diamonds are individual galaxies at $z \sim 0.85$, from Marcillac et al. (2006).

and that both star formation rate and its integral are well–measured, such that it's not silly to test whether the past IMF grossly differed from its present shape. The proposed no–cryogen, IRAC–only Spitzer "Warm Mission" would extend these studies to lower masses and higher redshift. Already IRAC has detected a lensed galaxy at $z \sim 6$ (Egami et al. 2005); the warm mission would bring unlensed galaxies from the recombination epoch into view.

5. Selected AGN Science with Spitzer

An obvious application of Spitzer for AGN science is to test the unification paradigm by measuring the reprocessed luminosity (e.g., Siebenmorgen et al. 2005; Shi et al. 2006). Another is to use the aromatic features to estimate star formation rates in the host galaxies of AGN (Schweitzer et al. 2006; Shi et al. 2007; Lutz et al. 2007; Lacy et al. 2007), to test the long–standing suspicion that obscured AGN may have more star formation than unobscured AGN, in conflict with strict unification.

Another productive avenue for Spitzer has been the search for highly–obscured AGN in the distant universe. Locally, half of AGN are highly obscured (Risaliti et al. 1999; Maiolino & Rieke 1995). This may be true for the distant

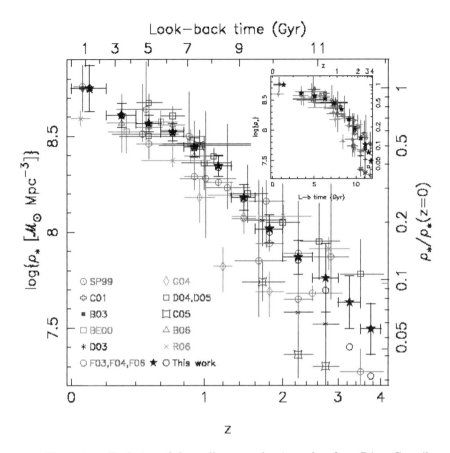

Figure 3. Evolution of the stellar mass density, taken from Pérez-González et al. (2008). Black points show the IRAC measurements; grey points show data from the literature. The IRAC points have small errorbars out to $z = 2.5$; the data at higher redshift will improve dramatically with deeper surveys from a no-cryogen Spitzer Warm mission.

universe as well, but it's been challenging to test since buried AGN are hard to find with current X-ray telescopes. As illustrated in Figure 4, the softer (easier-to–detect) X-rays are absorbed, leaving only the hard (and harder–to–detect) X-rays. To address the limitations of X-ray selection, several IR–based detection methods are being tested (Cutri et al. 2001; Glikman et al. 2007; Lacy et al. 2004; Martínez-Sansigre et al. 2005; Stern et al. 2005; Donley et al. 2005; Alonso-Herrero et al. 2006; Donley et al. 2007) and progress is being made, but the definitive census of highly–obscured AGN may require a future high–energy X-ray telescope like Nu-Star.

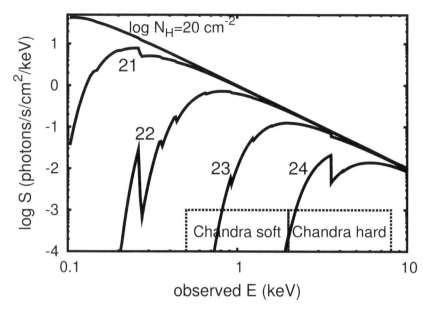

Figure 4. Effect of photoelectric absorption and Compton scattering by intervening gas on a typical AGN spectrum. In this example, the AGN and obscuring gas are both at $z = 1$; boxes mark the observed Chandra 0.5–2 and 2–8 keV bands, which at this redshift correspond to rest-frame energies of 1–4 4–16 keV. Note that Compton-thickness ($\log N_H = 24$ cm^{-2}) completely suppresses the soft band and significantly suppresses the hard band. As a result, Chandra can detect very few distant, Compton-thick AGN. The absorption is even worse for lower-z sources.

6. Jet-packs, Flying Cars, and New Telescopes

The organizers asked us to highlight how future instrumentation will impact the science we've discussed. Here's my list with example applications.

- Multi-object near-IR spectrographs, to get redshifts, Hα luminosities, and AGN diagnostics for large samples of Spitzer–selected galaxies, many of which are maddenly faint for optical spectroscopy.

- Adaptive–optic near-IR spectrographs for following up individual Spitzer–selected galaxies at very high spatial resolution.

- Herschel photometry at 60–670 μm (covering the bolometric dust emission) and cooling–line spectroscopy.

- The Spitzer Warm mission at 3.5 and 4.5 μm, which will do very deep stellar mass surveys out to high redshift.

- The Spitzer archive, especially the wealth of mid-IR diagnostics in IRS spectra.

- The WISE mission, which will survey the whole sky at 4–24 μm.

- The Nu-Star X-ray telescope, which will image from 6 to 80 keV, and thus should definitively determine what fraction of AGN are highly obscured.

- JWST, which will detect Hα star formation rates of 1 $M_\odot$ yr^{-1} out to $z = 6$, will resolve stellar populations and morphologies, and can detect mid–IR aromatic features for thousand of galaxies at moderate luminosities out to $z = 6$. It's exciting to imagine having high–quality JWST spectra of any galaxy we can currently detect photometrically with Spitzer.

Acknowledgments. Thanks to the organizers, especially Kurtis Williams and Justyn Maund, for making this year's Bashfest possible. Thanks to the other participants for making the conference so enjoyable.

References

Aaronson, M., & Olszewski, E. W. 1984, Nature, 309, 414
Alonso-Herrero, A., et al. 2006, ApJ, 640, 167
Alonso-Herrero, A., Rieke, G. H., Rieke, M. J., Colina, L., Pérez-González, P. G., & Ryder, S. D. 2006, ApJ, 650, 835
Calzetti, D., et al. 2007, ApJ, 666, 870
Caputi, K. I., et al. 2006, ApJ, 637, 727
Cutri, R. M., Nelson, B. O., Kirkpatrick, J. D., Huchra, J. P., & Smith, P. S. 2001, The New Era of Wide Field Astronomy, 232, 78
Donley, J. L., Rieke, G. H., Rigby, J. R., & Pérez-González, P. G. 2005, ApJ, 634, 169
Donley, J. L., Rieke, G. H., Pérez-González, P. G., Rigby, J. R., & Alonso-Herrero, A. 2007, ApJ, 660, 167
Draine, B. T., & Li, A. 2007, ApJ, 657, 810
Egami, E., et al. 2005, ApJ, 618, L5
Elbaz, D., et al. 1999, A&A, 351, L37
Genzel, R., & Cesarsky, C. J. 2000, ARA&A, 38, 761
Glikman, E., Helfand, D. J., White, R. L., Becker, R. H., Gregg, M. D., & Lacy, M. 2007, ApJ, 667, 673
Jiang, L., et al. 2006, AJ, 132, 2127
Lacy, M., et al. 2004, ApJS, 154, 166
Lacy, M., Sajina, A., Petric, A. O., Seymour, N., Canalizo, G., Ridgway, S. E., Armus, L., & Storrie-Lombardi, L. J. 2007, ApJ, 669, L61
Le Floc'h, E., et al. 2005, ApJ, 632, 169
Lutz, D., et al. 2007, ApJ, 661, L25
Low, F. J., Rieke, G. H., & Gehrz, R. D. 2007, ARA&A, 45, 43
Marcillac, D., Elbaz, D., Chary, R. R., Dickinson, M. E., Galliano, F., & Morrison, G. 2006, A&A, 451, 57
Martínez-Sansigre, A., Rawlings, S., Lacy, M., Fadda, D., Marleau, F. R., Simpson, C., Willott, C. J., & Jarvis, M. J. 2005, Nature, 436, 666
Maiolino, R., & Rieke, G. H. 1995, ApJ, 454, 95
Neugebauer, G., et al. 1984, Science, 224, 14
Oke, J. B., Neugebauer, G., & Becklin, E. E. 1970, ApJ, 159, 341
Pérez-González, P. G., et al. 2005, ApJ, 630, 82
Pérez-González, P. G., et al. 2008, ApJ, in press, astro-ph/0709.135
Rieke, G. H. 1978, ApJ, 226, 550
Rieke, G. H., & Lebofsky, M. J. 1979, ARA&A, 17, 477
Rigby, J. et al. 2008, ApJ, in press, astro-ph/0711.1902
Risaliti, G., Maiolino, R., & Salvati, M. 1999, ApJ, 522, 157

Rowan-Robinson, M., et al. 2004, MNRAS, 351, 1290
Schweitzer, M., et al. 2006, ApJ, 649, 79
Shi, Y., et al. 2006, ApJ, 653, 127
Shi, Y., et al. 2007, ApJ, 669, 841
Siebenmorgen, R., Haas, M., Krügel, E., & Schulz, B. 2005, A&A, 436, L5
Stern, D., et al. 2005, ApJ, 631, 163
Yan, L., et al. 2007, ApJ, 658, 778

Enrico Ramirez-Ruiz and Dougal Mackey dispense wisdom from graduate student Q&A lunch.

Chris Sneden enjoys Casey Deen's poster.

Ian Roederer and Mike Dunham discuss their hike to the Bash symposium.

Juntai Shen listens to Ross Falcon's explanation
of density measurements of the local ISM.

Julie Krugler explains Be abundances to Ivan Ramirez.

Sean Couch teaches the breaststroke to Jane Rigby, Enrico Ramirez-Ruiz and Craig Wheeler.

Frank Bash and Bill Guest at lunch.

Masami Ouchi and Dougal Mackey contribute to the panel discussion.

Casey Deen and Amanda Bayless at their posters.

At the graduate student Q&A lunch, Ian Roederer listens in awe as Avi Mandell explains how to karate chop a sunflower.

Eichiro Komatsu and Beth Fernandez talk about galaxies at high redshift.

Part II
Research Highlights

Five-minute talks were presented by several undergraduate and graduate students about their posters.
Top row, left to right: Athena Stacy, Amanda Bayless, Bi-Qing For. Middle row, left to right: Casey Deen, Julie Krugler, Lee Powell. Bottom row: Sean Couch.

Distances to the High Velocity Clouds: A Forty-Year Mystery on the Way to Solution

J. C. Barentine,[1,2] B. P. Wakker,[3] D. G. York,[4] J. C. Howk,[5] R. Wilhelm,[6] H. van Woerden,[7] R. F. Peletier,[7] T. C. Beers,[8] P. Richter,[9] Ž. Ivezić,[10] and U. J. Schwarz[11]

Abstract. A full understanding of the High Velocity Clouds (HVCs) is still lacking more than four decades after their discovery. Determining the clouds' locations in relation to the Galaxy is an important constraint on hierarchical assembly models of galaxies in a Λ-CDM universe. However, quantifying physical cloud properties such as mass and size have been difficult because of their unknown distances. We report the first definitive distance determinations for three HVCs (Complex C, 6.4-11.3 kpc; the Cohen Stream, 5.0-11.7 kpc; Complex GCP, 9.8-15.1 kpc) and two intermediate-velocity clouds (IVCs: IV-South, 1.0-2.7 kpc; cloud g1, 1.8-3.8 kpc) using the absorption line bracketing method. From the distances to the larger cloud complexes we calculate H I masses ranging from $1.5 \times 10^6 \, M_\odot$ to $8 \times 10^6 \, M_\odot$. In the case of Complex C this implies a mass inflow rate of $\sim 0.1 \, M_\odot \, \text{yr}^{-1}$. The measured distances place the HVCs in the hot Galactic Corona and the IVCs in the lowest reaches of the Halo. With this information we may now positively identify at least two roles played by HVCs in the internal and external dynamics of the Milky Way: (1) the infall of low-metallicity gas and (2) the supply and return streams of the Galactic Fountain.

---

[1] Department of Astronomy, University of Texas, Austin, TX, USA

[2] Apache Point Observatory, Sunspot, NM, USA

[3] Department of Astronomy, University of Wisconsin, Madison, WI, USA

[4] Astronomy and Astrophysics Center, University of Chicago, Chicago, IL, USA

[5] Department of Physics, University of Notre Dame, Notre Dame, IN, USA

[6] Department of Physics and Astronomy, Texas Tech University, Lubbock, TX, USA

[7] Kapteyn Astronomical Institute, University of Groningen, Groningen, Netherlands

[8] Dept. of Physics and Astronomy, Center for the Study of Cosmic Evolution (CSCE) and the Joint Institute for Nuclear Astrophysics (JINA), Michigan State University, East Lansing, MI 48824, USA

[9] Institut für Physik, Universität Potsdam, Potsdam, Germany

[10] Department of Astronomy, University of Washington, Seattle, WA, USA

[11] Department of Astrophysics, Radboud University Nijmegen, Nijmegen, Netherlands

1. Introduction

Of the outstanding problems in the formation and evolution of galaxies, an important example is the nature of the Galactic high velocity clouds (HVCs). These are clouds of H I gas at observed velocities not predicted by simple galactic differential rotation models; see the review by Wakker & van Woerden (1997). Interpretations of the HVCs include infalling, low-metallicity gas from the intergalactic medium; material left over from the original assembly of the Milky Way; cycling of gas between the Disk and Halo in a Galactic Fountain; and tidally-stripped debris from passing dwarf galaxies.

Testing these hypotheses requires knowing the basic parameters of the clouds, such as their masses and sizes. However, empirical measurements of these quantities have been frustrated for decades by a more fundamental unknown: their distances. HVC distances allow us to (1) quantify the inflow rate of low-metallicity gas into the Galaxy, possibly solving the G-dwarf problem (Pagel 1989) and explaining why the ISM was not completely turned into stars long ago; (2) place meaningful constraints on the strength of the metagalactic ionizing radiation field; (3) probe the shape of the dark matter potential of the Milky Way; and (4) understand observed clouds of H I with anomalous velocities around external galaxies in the context of galaxy formation and evolution

2. Experimental Method

We use the absorption line bracketing method to constrain HVC/IVC distances. This involves searching for interstellar (IS) absorption at cloud velocities toward probe stars in the direction of cloud complexes. Upper distance limits are set by detection of IS lines. Lower limits are established by "significant" non-detections in which the ratio of the expected equivalent width (EW) of an IS line to the observed EW upper limit is at least ten (Wakker 2001). Expected EWs come from the empirical relation between N(H I) and N(Ca II) of Wakker & Mathis (2000).

Our probe star selection criteria require candidates to be (1) observable at large distances; (2) relatively metal-poor, to avoid contamination of the IS metal line regions of the spectrum with stellar photospheric lines; and (3) hot enough to ensure sufficient flux in the near-UV near the Ca II H&K lines. The blue horizontal branch (BHB) stars meet these requirements but until recently have been difficult to identify in large quantities. We select probe candidates from the HK survey, SDSS and 2MASS. Candidates are correlated with H I maps of the HVCs, noting N(H I) toward each star. Photometry and medium-resolution spectroscopy are conducted to derive T_{eff}, $\log g$, [Fe/H] and M_V (Wilhelm et al. 1999). A final check compares the medium-resolution spectroscopy against the H I spectra to ensure the stellar lines do not obscure the IS lines. This sample is then observed at high spectral resolution to detect the IS lines.

3. Observations

Probe stars are correlated with the 21 cm maps of Hulsbosch & Wakker (1988) and Morras et al. (2000). These provide a list of HVCs on a $1° \times 1°$ grid at

16 km s^{-1} resolution. The H I spectrum is checked using the Leiden-Argentina-Bonn (LAB) survey (Kalberla et al. 2005), which covers the sky on a 0.5°×0.5° grid with 1 km s^{-1} resolution. For SDSS stars we used SDSS photometry and spectroscopy when available. Photometry of other candidates was performed at the ESO/Danish 1.5-m, WIYN 0.9-m, CTIO 0.9-m, Yale 1.0-m and MDM 2.4-m. We obtained spectroscopy with the ARC 3.5-m at Apache Point for HK and 2MASS stars and some SDSS stars. We observed some relatively nearby HK stars toward complex C at high-resolution in 1997 with the 4.2-m William Herschel Telescope at a resolution of 6 km s^{-1}.

High-resolution spectroscopy of 25 probe stars in the direction of HVC complexes GCP and IVCs IV-South and cloud g1 was obtained with the VLT 8.4 m telescope and UV-Visual Echelle Spectrograph in 37 hours between April-September 2006. Observations of three stars toward complex C at a resolution of 8.8 km s^{-1} were obtained in April 2007 with the 10-m Keck I telescope. The VLT observations were highly successful while the night of Keck time was plagued with poor seeing and foggy conditions. However, we obtained spectra of three probe stars toward complex C sufficient to determine a meaningful distance bracket.

4. Results

Distances for the clouds we measured are given in Table 1. We specify heliocentric distances to the clouds, and where available, the z distance above (positive) or below (negative) the plane of the Milky Way (MW). Using the H I masses derived from 21 cm maps, we calculated the mass inflow rates associated with complex C and the Cohen Stream. We conclude by summarizing the implications of our distance measurements for each cloud or complex.

Table 1. Distance brackets and derived physical parameters of HVCs and IVCs observed.

HVC/IVC	Distance (kpc)	z (kpc)	H I Mass ($M_\odot$)	$\dot{M}$ ($M_\odot$ yr^{-1})
complex C	3.7-11.2	3-9	0.7-6×10^6	0.1-2.5
Cohen Stream (CS)	9.3-11.4	(−6.4)-(−8.4)	2.5-3.9×10^5	4×10^{-3}
g1	1.8-3.8	(−0.8)-(−1.7)	< 1×10^4	−
complex GCP	9.8-15.1	(−2.5)-(−3.9)	1×10^6	−
IV-South	1.0-4.5	−	1×10^5	−

Complex C: Located high above the Disk of the MW, this subsolar-metallicity complex represents infalling gas that has not been previously cycled through the Galaxy. The mass infall rate associated with complex C is 0.1-0.25 $M_\odot$ yr^{-1}, a substantial fraction of the estimated ∼1 $M_\odot$ yr^{-1} (depending on some assumptions) required to solve the G-dwarf problem.

Cohen Stream: This cloud complex is falling into the MW from high above the Disk, but at present its metallicity is not constrained.

Cloud g1: The location of this cloud suggests that it is in an outflow of the Galactic Fountain. Cloud g1 may be condensing out of this flow.
Complex GCP: A cloud complex of still unknown origin, and with still unknown metallicity. Its location and motion suggests a close association with the Disk, but modeling is required.
IV-South: This IVC represents the return flow of the Galactic Fountain and is identified specifically with the Perseus Arm of the MW.

References

Hulsbosch, A. N. M. & Wakker, B. P. 1988, A&AS, 75, 191
Kalberla, P. M. W., Burton, W. B., Hartmann, D., Arnal, E. M., Bajaja, E., Morras, R., & Pöppel, W. G. L. 2005, A&A, 440, 775
Morras, R., Bajaja, E., Arnal, E. M., & Pöppel, W. G. L. 2000, A&AS, 142, 25
Pagel, B. E. J. 1989, in Evolutionary Phenomena in Galaxies, ed. J. E. Beckman & B. E. J. Pagel, 201
Wakker, B. P. 2001, ApJS, 136, 463
Wakker, B. P. & Mathis, J. S. 2000, ApJ, 544, L107
Wakker, B. P. & van Woerden, H. 1997, ARA&A, 35, 217
Wilhelm, R., Beers, T. C., & Gray, R. O. 1999, AJ, 117, 2308

Hydrodynamic Instabilities in Jet-Induced Supernovae: Results of 2D Simulations

Sean M. Couch, J. Craig Wheeler and Miloš Milosavljević

Department of Astronomy, The University of Texas, Austin, TX, USA

Abstract. The structure of supernovae and young supernova remnants, such as SN 1987A and Cas A, suggests hydrodynamic instabilities during the course of the explosion. Various instabilities have been studied in detail for the case of spherical, neutrino shock-revival models, but a variety of studies, including spectropolarimetry, suggest that the intrinsic explosions are far from spherical. We study the intrinsically aspherical jet-induced model, wherein bipolar (and possibly unipolar) jets drive the explosion. We present preliminary results of numerical simulations of jet-induced explosions in a realistic stellar model. We find a growth of Rayleigh-Taylor instabilities at the contact discontinuity between the jet fluid and the post-shock stellar medium. The equatorial flow established by the obliquely colliding bipolar bow shocks is subject to shear-driven Kelvin-Helmholtz instabilities. These results are part of a larger study characterizing the large-scale structure of aspherical jet-induced supernovae.

1. Introduction

The physical mechanism of core-collapse supernovae (SNe) is an outstanding mystery in modern astrophysics. This complex process is crucial to our understanding of numerous phenomena such as the formation of neutron stars, pulsars and black holes, nucleosynthesis, feedback on galactic evolution and the origin of gamma-ray bursts. From observations, we know that SNe are ubiquitously asymmetric events. The richness of this problem is discussed in detail for the specific case of Cas A by Wheeler, Maund & Couch (2007). For distant SNe, spectropolarimetric measurements indicate that these catastrophic events are generally bi-polar in nature while showing distinct departures from axisymmetry. To account for asymmetries, hydrodynamic instabilities have been studied in detail for the case of spherical, neutrino-driven explosions (Kifonidis et al. 2006), but these calculations have not explained the routinely observed polarization. Explaining the observed structure of SNe, near and far, is a major challenge for theories of stellar core collapse.

The jet-induced model of core collapse SNe presents a natural explanation for the elongated structure inferred from spectropolarimetry. In this picture, non-relativistic bi-polar jets launched from a central engine drive bow shocks through the progenitor star. The two expanding bow shock cocoons collide along the equator of the star and drive an equatorial, lateral outflow. Khokhlov et al. (1999) showed that this model leads to a jet/torus structure in which the torus is the result of the equatorial flow. The 3D calculations of Khokhlov et al. did not resolve any fluid instabilities that may naturally arise in jet-driven SNe. This problem requires sufficient resolution to capture the fastest

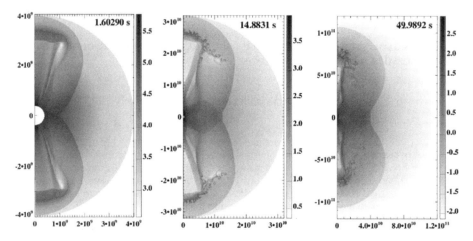

Figure 1. Results of our simulations at various times. The scale is the logarithmic density in units of g cm^{-3}. All axes are in units of cm.

growing wavelengths of relevant instabilities. Among important hydrodynamical instabilities that are likely to occur within SNe are the Kelvin-Helmholtz (KH), Rayleigh-Taylor (RT), and Richtmyer-Meshkov (RM) instabilities.

2. Simulations

For the present study, we have employed the FLASH code, developed and maintained by the ASC FLASH Center at the University of Chicago (Fryxell et al. 2000). FLASH uses the adaptive mesh refinement (AMR) package PARAMESH (MacNeice et al. 2000) and a direct Eulerian piecewise parabolic method (Colella & Woodward 1984). All simulations were run on the dual-core Linux cluster, *Lonestar*, maintained by the Texas Advanced Computing Center. The initial conditions of our simulation are taken from a 15 M$_\odot$ red supergiant SN progenitor model of Woosley et al. (2002).

The present calculation was performed in 2D spherical geometry. The grid covers π radians in the angle and stretches from 3.82×10^8 cm to 4×10^{12} cm in radius. Bipolar jets were injected at the inner spherical boundary. The physical parameters of the jets were chosen to correspond to those of Khokhlov et al. (1999), who in turn based their parameters on the magnetohydrodynamics calculations of LeBlanc & Wilson (1970). The jet opening half-angle is 15°, the density of the jet fluid is 6.5×10^5 g cm^{-3}, the pressure of the jet fluid is 10^{23} erg cm^{-3}, and the injection velocity is 3.22×10^9 cm s^{-1} in the radial direction. The upper and lower jets are identical. As in Khokhlov et al. (1999), the jet injection velocity is held constant for a small time and then linearly reduced to zero at about $t = 2.5$ s. The time scales were chosen to give a total input energy between the two jets of 10^{51} erg, sufficient to unbind the progenitor (binding energy $\sim 2 \times 10^{48}$ erg).

One aspect of 2D spherical geometry that must be considered is that the spatial resolution decreases in the angular dimension with increasing radius. As

Hydrodynamic Instabilities in Jet-Induced Supernovae

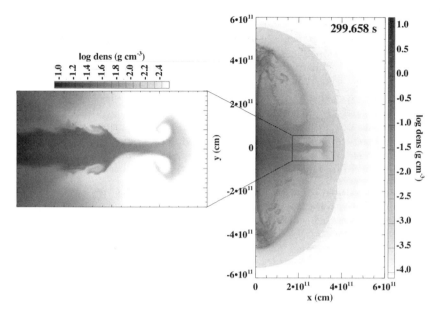

Figure 2. Density at $t = 300$ s. The equatorial outflow has developed a significant KH cap.

the explosion expands, the radial resolution will remain the same. This becomes very computationally inefficient. FLASH has no built-in means of altering the refinement and grid parameters after a simulation has started. To overcome this hurdle, we have implemented a routine to re-map simulation data to a new, arbitrary grid. This method of periodically re-mapping the simulation data to a grid of lower refinement has reduced computation time by about a factor of ten.

3. Results and Discussion

Our simulations show that hydrodynamic instabilities develop and grow in jet-induced supernovae. Instabilities in the jet heads are evident from early times. The equatorial outflow is also subject to instabilities, particularly KH. Figure 1 shows density snapshots at different times in our calculation. As early as a few seconds, instabilities start to grow at the contact discontinuity between the jet and star. By $t = 15$ s (middle pane), these instabilities have already grown to significant proportions, and continue to grow thereafter. The nature of these instabilities is likely to be both RM and RT: the shocks establish opposed density and pressure gradients (RT unstable) at the contact discontinuity.

As the bow shocks cross at the equator of the star, the post-shock media collide obliquely, resulting in a net outward velocity. This 'pancake'-like structure propagates outward at more than the escape velocity ($v_{\rm esc} = \sqrt{2Gm_r/r}$). For example, at $r = 2 \times 10^{11}$ cm the mass enclosed is $m_r = 8.9 \times 10^{34}$ g, giving an escape velocity $v_{\rm esc} = 2 \times 10^8$ cm/s. The velocity along the equator at

$r = 2 \times 10^{11}$ cm is 1.2×10^9 cm/s. Significant shear is present on the top and bottom surfaces of the pancake and clear KH instability sets in. Figure 2 shows the late time evolution of our calculation. The prominent 'mushroom' cap on the equator is the result of the growth of this KH instability.

The work presented here is the preliminary outcome of an ongoing effort to explore the effect the jet-induced Ansatz may have on observable features of core-collapse supernovae. Forthcoming calculations will incorporate composition tracking and a more sophisticated equation of state. Our future work will include investigating the influence the detailed jet parameters have on the nature of the SN evolution, as well as explosions in different progenitor models, such as blue supergiants and stars lacking hydrogen envelopes.

Acknowledgments. The authors would like to thank the Texas Advanced Computing Center for providing and maintaining excellent computational resources that made this work possible. This research is supported in part by NSF Grant 0707769 and NASA Grant NNG04GL00G. The software used in this work was in part developed by the DOE-supported ASC / Alliance Center for Astrophysical Thermonuclear Flashes at the University of Chicago.

References

Colella, P., & Woodward, P.R. 1984, JCoPh, 54, 174
Fryxell, B., Olson, K., Ricker, P., Timmes, F. X. , Zingale, M., Lamb, D. Q., MacNeice, P., Rosner, R., Truran, J. W. and Tufo, H. 2000, ApJS, 131, 273
Khokhlov, A. M., Hoflich, P. A., Oran, E. S., Wheeler, J. C., Wang, L., & Chtchelkanova, A. Y. 1999, ApJ, 524, L107
Kifonidis, K., Plewa, T., Scheck, L., Janka, H.T., & Muller, E., 2006, A&A, 453, 661
LeBlanc, J.M. & Wilson, J.R. 1970, ApJ, 161, 541
MacNeice, P., Olson, K., Mobarry, C., deFainchtein, R., & Packer, C., 2000, CPC, 126, 330
Wheeler, J. C., Maund, J. R. and Couch, S. M. 2007, ApJ, in press (astro-ph/0711.3925)
Woosley, S.E., Heger, A., & Weaver, T.A., 2002, RvMP, 74, 1015

The Hobby-Eberly Telescope Chemical Abundances of Stars in the Halo (*CASH*) Project III. Abundance Analysis of Three Bright Hamburg/ESO Survey Stars

L. A. Davies,[1] A. Frebel,[2] J. J. Cowan,[1] C. Allende Prieto[3] and C. Sneden[3]

Abstract. We present an abundance analysis of three newly discovered stars from the Hamburg/ESO survey for which HET observations have been obtained as part of the *CASH* project. Light elemental abundances of all three stars agree with those of other metal-poor stars. This means that they likely formed from well-mixed gas. Upper limits on the heavier neutron-capture abundances have not eliminated the possibility that these stars are r-process enhanced. However, the measured barium abundances are rather low.

1. Introduction

The abundances and locations of metal-poor stars can give insight into the chemical elements present in the early Universe and the evolutionary history of the Galaxy. The Hamburg/ESO objective-prism survey (HES; Christlieb (2003)) has sought to study these oldest stars in the halo and contains two main samples: the original *faint* ($14 < B < 17.5$) sample and the *bright* ($10 < B < 14$) sample (Frebel et al. 2006) for which the survey data partially suffers from saturation effects. It is from the bright sample that these three stars, HE 0049+0005, HE 0124+0119 and HE 0433−1008 are chosen.

2. Observations and Measurements

High-resolution data for HE 0049+0005, HE 0124+0119 and HE 0433−1008 were obtained with the High-Resolution Spectrograph at the Hobby-Eberly Telescope on December 26, 2006 as part of the Chemical Abundances of Stars in the Halo (*CASH*) Project (Frebel et al. 2008). The wavelength coverage of the blue spectrum used in this analysis is $\sim 4100 - 5800$ Å.

Equivalent Width Measurements

To spectroscopically obtain the metallicity, a comprehensive iron line list was created that included information on wavelength, excitation potential and transition probabilities. The Kurucz atomic line database (Kurucz & Bell 1995) was

---

[1] Homer L. Dodge Department of Physics & Astronomy, University of Oklahoma, Norman, OK, USA

[2] McDonald Observatory, University of Texas, Austin, TX, USA

[3] Department of Astronomy, University of Texas, Austin, TX, USA

utilized paying special care to lines regularly utilized in the analysis of metal-poor stars. For elements other than iron, Barklem et al. (2005) and Aoki et al. (2002) were utilized to locate lines of calcium, magnesium, scandium, titanium, barium, and some lanthanides. Again, the Kurucz atomic line database was used to complete the line list data. Equivalent width measurements were made by utilizing the ESO-MIDAS software. Custom written MIDAS scripts allow the user to visually inspect features and semi-automatically fit Gaussian profiles to the chosen absorption feature.

3. Stellar Parameters

In order to carry out an abundance analysis, effective temperature, gravity, metallicity, and microturbulent velocity must be determined. Optical B, V, R_C, I_C (where 'C' indicates the Cousins system; Beers et al. 2007) and near-infrared 2MASS J, H, K (Skrutskie et al. 2006) photometry is available for HE 0433−1008. For the other two stars only J, H, K and the HES V magnitudes are available. Based on these data, and using the Alonso et al. (1999) calibration for giants, we derive effective temperatures (T_{eff}) for all three stars. The iron abundance, gravity and microturbulent velocity are calculated simultaneously by inputting a model atmosphere and iron equivalent widths list into the LTE stellar line analysis program MOOG (Sneden 1973) utilizing the abfind driver. Table 1 summarizes the four parameters found for each star.

Table 1. Stellar parameters

Star Name	T_{eff} K	$\log g$ dex	[Fe/H] dex	v_{micr} km s^{-1}
HE 0049+0005	4750	1.53	−2.81	2.07
HE 0124+0119	4480	1.59	−3.38	2.25
HE 0433−1008	4710	1.18	−2.49	2.55

4. Abundance Patterns

Once the stellar parameters are established, the abundances of light elements can be found rather quickly. By inputting the correct model atmosphere and new line list containing equivalent widths for elements other than iron, the abundances of each element are found.

The spectra used in this study, though taken with a high-resolution spectrograph, have a relatively low S/N of 30-50 per resolution element at ~ 5000 Å. We were able to measure the most common lighter elements ($Z < 30$) as well as the neutron-capture element Ba. Figures 1 and 2 compare our findings to the abundances found in similar stars by Cayrel et al. (2004) and Barklem et al. (2005) for light element abundances and Ba abundances, respectively. Upper limits can be placed on abundances by creating synthetic spectra and noting the abundance at which the element could have been seen if it were stronger than the noise level of the continuum. Table 2 summarizes these upper limits.

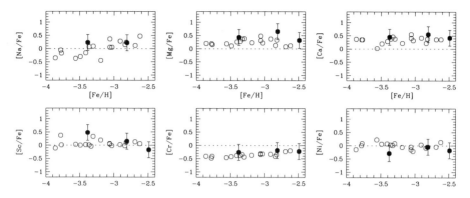

Figure 1. Selected abundances of our three stars in comparison with the metal-poor giants of Cayrel et al. (2004).

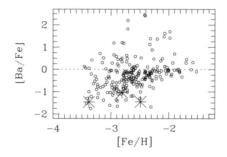

Figure 2. Ba abundance of our three stars in comparison with the metal-poor giants of Barklem et al. (2005).

5. Conclusions

Spectra of three stars from the bright sample of the HES were analyzed for stellar parameters and elemental abundances. The abundances of Na, Mg, Ca, Sc, Cr, Ni and Ba were found to be consistent with previous studies. Upper limits were places on abundances of seven neutron-capture elements.

Further photometric observations will refine the temperature and, consequently, the other stellar parameters and the abundances. Future work will include the red HET spectra for improved abundance measures. This study will also serve as a template for subsequent abundance analyzes to be performed for the other stars scheduled for high-resolution observations in the *CASH* project.

Acknowledgments. Based on observations that were obtained with the Hobby-Eberly Telescope. The HET is named in honor of its principal benefactors, William P. Hobby and Robert E. Eberly. A. F. acknowledges support through the W. J. McDonald Fellowship of the McDonald Observatory. Funding

Table 2. Elemental Abundances and Upper Limits

elm	HE 0049+0005 log ϵ	HE 0049+0005 [A/Fe]	HE 0124+0119 log ϵ	HE 0124+0119 [A/Fe]	HE 0433−1008 log ϵ	HE 0433−1008 [A/Fe]
Y	< −0.41	< 0.20	< −0.76	< 0.42	< −0.36	< −0.07
La	< −0.78	< 0.85	< −0.98	< 1.22	< −0.78	< 0.53
Ce	< 0.30	< 1.50	...	...	...	...
Pr	< −0.79	< 1.24	...	...	...	...
Nd	< −0.25	< 1.10	< −0.50	< 1.42	< −0.40	< 0.63
Sm	< −0.25	< 1.61	< 0.00	< 2.43	< 0.25	< 1.79
Eu	< −1.49	< 0.80	...	...	...	...

for this project has also been provided by the US National Science Foundation (grants AST 06-07708 to C. S. and AST 07-07447 to J. C.).

References

Alonso, A., Arribas, S., & Martínez-Roger, C. 1999, A&AS, 140, 261
Aoki, W., Norris, J. E., Ryan, S. G., Beers, T. C., & Ando, H. 2002, PASJ, 54, 933
Barklem, P. S., Christlieb, N., Beers, T. C., Hill, V., Bessell, M. S., Holmberg, J., Marsteller, B., Rossi, S., Zickgraf, F.-J., & Reimers, D. 2005, A&A, 439, 129
Beers, T. C., Flynn, C., Rossi, S., Sommer-Larsen, J., Wilhelm, R., Marsteller, M., Lee, Y. S., De Lee, N., Krugler, J., Deliyannis, C. P., Zickgraf, F.-J., Holmberg, J., Onehag, A., Eriksson, A., Terndrup, D. M., Salim, S., Christlieb, N., & Frebel, A. 2007, ApJS, 168, 128
Cayrel, R., Depagne, E., Spite, M., Hill, V., Spite, F., François, P., Plez, B., Beers, T., Primas, F., Andersen, J., Barbuy, B., Bonifacio, P., Molaro, P., & Nordström, B. 2004, A&A, 416, 1117
Christlieb, N. 2003, in Reviews in Modern Astronomy, Vol. 16, Reviews in Modern Astronomy, ed. R. E. Schielicke, 191
Frebel, A., Allende Prieto, C., Roederer, I., Shetrone, M., Rhee, J., Sneden, C., Beers, T. C., & Cowan, J. J. 2008, in New Horizons in Astronomy: Frank N. Bash Symposium, ed. A. Frebel, J. Maund, J. Shen, & M. Siegel, Astronomical Society of the Pacific Conference Series
Frebel, A., Christlieb, N., Norris, J. E., Beers, T. C., Bessell, M. S., Rhee, J., Fechner, C., Marsteller, B., Rossi, S., Thom, C., Wisotzki, L., & Reimers, D. 2006, ApJ, 652, 1585
Kurucz, R. & Bell, B. 1995, Atomic Line Data (R.L. Kurucz and B. Bell) Kurucz CD-ROM No. 23. Cambridge: Smithsonian Astrophysical Observatory, 23
Skrutskie, M. et al. 2006, AJ, 131, 1163
Sneden, C. A. 1973, PhD thesis, The University of Austin at Texas

Using Stellar Photospheres as Chronometers for Studies of Disk Evolution

Casey P. Deen and Daniel T. Jaffe

Department of Astronomy, University of Texas, Austin, TX, USA

Abstract. Studies of young stellar objects and their associated disks can provide constraints on theories of planet formation. Disk lifetimes can determine the amount of time available for the formation of planets. Accurate stellar ages are important for the study of circumstellar disks and can be determined from the photospheric properties of the associated young stellar objects. We present the results of a survey of the photospheres of young stellar objects in the ρ-Ophiuchus star formation region. Using near-infrared spectra, we assign spectral types to the heavily reddened objects, determine extinctions, and measure surface gravities. We then look for a correlation between surface gravities and luminosities.

1. Introduction

Circumstellar disks serve an important role in the process of star and planet formation. They regulate the angular momentum evolution of their associated young stellar object while providing the material for planet formation. However, the factors driving the evolution of the disks are not well understood. If age were the only factor involved in the evolution of circumstellar disks, one would expect a smooth transition from young objects mostly with disks, to older objects mostly without disks. While fractions of objects with observable disks do decrease with age, suggesting that age is an important factor, there exist a significant fraction of objects which do not follow this evolutionary paradigm. There are examples of extremely young stars without disks which should not exist if disks have a well defined lifetime. The reason for this dispersion is currently unknown. Previous studies have suffered from numerous theoretical and observational difficulties. One method of discerning a subtler picture of disk evolution is to determine detailed disk parameters (such as disk mass, flaring angle, inner disk size) from theoretical disk models and examine trends in disk properties with age. Accurate ages for the disks are important for this study and can be determined from the photospheric properties of the associated young stellar objects. As young stellar objects age, it is expected that their surface gravity increases as they contract while decreasing in luminosity as they approach the main sequence (Baraffe et al. 1998). Previous studies of stellar photospheres have relied on the H-R diagram and photometrically determined luminosities to determine stellar age. Other studies have used high resolution spectra to determine photospheric properties, such as surface gravity and effective temperature (Doppmann & Jaffe 2003). We present the results of a comparison between derived surface gravities with luminosities.

2. Observations

We have obtained near-infrared spectra and photometry of 66 YSOs in the ρ-Ophiuchus cloud core. The sample was selected from the mid-infrared ISOCAM survey of Bontemps et al. (2001) to be K-band magnitude limited to 10.0. Moderate resolution ($R \sim 2000$) near-infrared spectra were obtained using SpeX on the NASA IRTF in 2002 and 2003. Spectra were reduced using SpeXtool. Photometry for the objects was retrieved from the 2MASS archive[1].

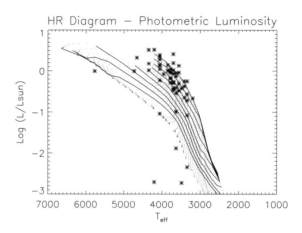

Figure 1. H-R diagram using photometrically determined luminosities. Isochrones are from Baraffe et al. (1998). The earliest isochrone is $\log t$ (yr) = 6.0 and each subsequent isochrone is 0.2 dex older.

3. Analysis

Spectral types were determined by comparing flattened spectra of sources to flattened spectra of optically typed standard stars. The spectral typing scheme depends on ratios of gravity-insensitive absorption lines, and is independent of veiling, reddening, and variations in surface gravity. Extinctions were determined by dereddening the source spectra to match the J-band spectral shape of the corresponding spectral type template. The template for a given spectral type was constructed by averaging together normalized (at $\lambda = 1.2 \mu$m) dwarf and giant spectra of the same spectral type (Luhman 1999). Extinctions, J-band photometry, and J-band bolometric corrections (Kenyon & Hartmann 1995) were used to determine source luminosities. Effective temperatures were determined

---

[1]This publication makes use of data products from the Two Micron All Sky Survey, which is a joint project of the University of Massachusetts and the Infrared Processing and Analysis Center/California Institute of Technology, funded by the National Aeronautics and Space Administration and the National Science Foundation.

using the spectral type-effective temperature relation of Luhman (1999). The sources were then placed on an H-R diagram (see Figure 1).

3.1. Surface Gravities

Following the prescription of Doppmann & Jaffe (2003), we look for ratios of absorption lines which are sensitive to surface gravity. By comparing spectra of optically typed giant and dwarf standard stars (Cushing et al. 2005; Rayner et al. 2007), we chose the ratio of the flux contained in the Ca feature at 1.98 μm to the flux contained in the first CO bandhead at 2.3 μm as a preliminary gravity-sensitive ratio. As shown in Figure 2, at temperatures of K and M stars, the dwarfs (objects with high surface gravities) and giants (objects with low surface gravities) are well separated. Reassuringly, computing the ratio for the Ophiuchus sources places them in between the dwarf and giant populations.

In order to derive surface gravities for our objects, we calculate the ratio for a grid of synthetic spectra generated from NexGen model atmospheres (Hauschildt et al. 1997). The synthetic spectra produce iso-gravity contours in Figure 2, from which we can interpolate the surface gravity of the sample of Ophiuchus stars. Once we have surface gravities, we then place the objects in a surface gravity evolutionary diagram (see Figure 3). The surface gravity evolutionary tracks shown are also from Baraffe et al. (1998).

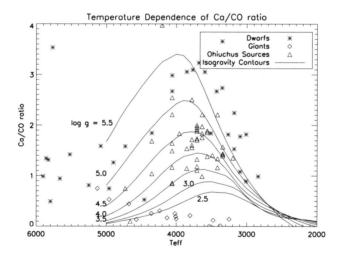

Figure 2. Dependence of Ca/CO ratio on effective temperature, luminosity class, and surface gravity

4. Results

Comparing the derived surface gravities with the photometric luminosities, no clear trend emerges, even within objects of the same spectral type. There are many possible reasons for this, including poor telluric cancellation, difficulties

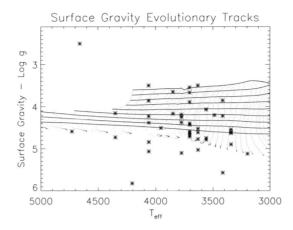

Figure 3. Surface gravity evolutionary tracks from Baraffe et al. (1998)

in assigning the continuum level, and possible emission/absorption from circumstellar material.

5. Conclusions

There is no obvious correlation between surface gravity and photospheric luminosity. While uncertainties due to noise in the spectra certainly contribute to the scatter of the surface gravity measurements, one would hope some correlation would be apparent. The lack of correlation may in fact be evidence for either emission or absorption from circumstellar material. Future studies will focus on different ratios of gravity sensitive lines.

References

Baraffe, I., Chabrier, G., Allard, F., & Hauschildt, P. H. 1998, A&A, 337, 403
Bontemps, S., André, P., Kaas, A. A., Nordh, L., Olofsson, G., Huldtgren, M., Abergel, A., Blommaert, J., Boulanger, F., Burgdorf, M., Cesarsky, C. J., Cesarsky, D., Copet, E., Davies, J., Falgarone, E., Lagache, G., Montmerle, T., Pérault, M., Persi, P., Prusti, T., Puget, J. L., & Sibille, F. 2001, A&A, 372, 173
Cushing, M. C., Rayner, J. T., & Vacca, W. D. 2005, ApJ, 623, 1115
Doppmann, G. W. & Jaffe, D. T. 2003, AJ, 126, 3030
Hauschildt, P. H., Baron, E., & Allard, F. 1997, ApJ, 483, 390
Kenyon, S. J. & Hartmann, L. 1995, ApJS, 101, 117
Luhman, K. L. 1999, ApJ, 525, 466
Rayner, J. T., Cushing, M. C., & Vacca, W. D. 2007, in prep.

Through the Thick and Thin: Electron Density Measurements of the Local Interstellar Medium

Ross E. Falcon and Seth Redfield[1]

Department of Astronomy and McDonald Observatory, University of Texas, Austin, TX, USA

Abstract. In a comprehensive survey of observations from the high resolution spectrographs on board the *Hubble Space Telescope* (*HST*), the Goddard High Resolution Spectrograph (GHRS) and the Space Telescope Imaging Spectrograph (STIS), we have detected interstellar C II absorption for the 1334.5 Å ground-state transition and the 1335.7 Å (doublet) collisionally populated excited-state transition for 13 sight lines toward stars within 100 parsecs. We derive column densities of C II in each of the states and use methods implemented by Wood & Linsky (1997) to measure the electron densities along the lines of sight. Electron densities in the local interstellar medium (LISM) are particularly difficult to ascertain, but are essential for understanding the ionization structure of the LISM and density variations which impact the morphology of the heliosphere. To combat systematic error due to the saturation of the C II resonance line, we also use S II measurements as a proxy to determine column densities. Our sample consists of 13 stars, a significant improvement in the number of measurements that allows us to search for variations in density for the first time. The sample of electron densities appears consistent with a log-normal distribution, with an unweighted mean density of $n_e = 0.13^{+0.15}_{-0.07}$ cm^{-3} and, based on S II as a proxy, $n_e(\text{C II}_{\text{SII}}) = 0.11^{+0.10}_{-0.05}$ cm^{-3}.

1. Introduction

The interstellar medium (ISM) consists of all the "stuff" between the stars. It includes atomic, ionized and molecular gas, and dust. Most of it is quite diffuse, having densities on the order of 10^{-4} to 1 particle cm^{-3}. The most dense and therefore the most magnificent examples of the ISM include nebulae like that of Orion and consist of densities on the order of $10^3 - 10^4$ particles cm^{-3}. None of this is very easily imagined for us here on Earth, though, where the air we breathe has a density of roughly 10^{19} particles cm^{-3}.

Our solar system lies within a region of warm, partially ionized gas moving with a coherent velocity (Lallement & Bertin 1992; Frisch 1995). This Local Interstellar Cloud (LIC) of ∼93 pc^3 with ∼0.32 M$_\odot$ of material (Redfield & Linsky 2000) is one of many distinct warm "clouds" located within the Local Bubble, a larger cavity, which extends to a radius of about 100 pc (Lallement et al. 2003) and is filled with low density, highly ionized gas. It is this Local Bubble that defines the *local* interstellar medium (LISM).

---

[1] Hubble Fellow

How the LIC interacts with the heliosphere, the interface between the LISM and the solar wind, is largely determined by the cloud's density (Zank 1999) as well as the ionization structure of the surrounding ISM (Müller et al. 2006). Both of these quantities are related to electron density, which has few previous measurements in the LISM (Frisch et al. 2006; Oliveira et al. 2003; Vennes et al. 2000; Wood & Linsky 1997). This survey greatly expands the total number of electron density measurements.

2. Data Acquisition and Analysis

We compiled all moderate to high resolution observations of nearby stars with the *HST* spectrographs: GHRS and STIS. From the complete sample of 417 unique targets within 100 pc, we identify 13 sight lines that show interstellar absorption in both C II and C II*.

We analyze LISM absorption in the S II triplet near 1250 Å in order to better constrain the C II column density. Ten of the 13 targets have spectra that include the S II lines. In addition, we fit the LISM absorption in Mg I and a Mg II doublet in order to further constrain the electron density. Only three of the 13 targets have spectra that include both Mg I and Mg II.

We calculate electron densities (see Figure 1) by making use of the knowledge that collisions with electrons are responsible for populating the C II* excited state. Hence, the ratio of the column densities of the C II resonance and excited lines is proportional to the electron density. Our use of this method is similar to that implemented by Spitzer & Fitzpatrick (1993) and Oliveira et al. (2003) but most parallels that of Wood & Linsky (1997).

From thermal equilibrium between collisional excitation of the excited state and radiative de-excitation, one can derive the following relation:

$$\frac{N(\text{C II}^*)}{N(\text{C II})} = \frac{n_e C_{12}(T)}{A_{21}}. \tag{1}$$

$N(\text{C II})$ and $N(\text{C II}^*)$ are the column densities of the resonance and excited lines, respectively. n_e is the electron density. The radiative de-excitation rate coefficient $A_{21} = 2.29 \times 10^{-6}$ s^{-1} as listed by Nussbaumer & Storey (1981), and $C_{12}(T)$ is the collision rate coefficient, which weakly depends on temperature.

We are motivated to find an alternative means of estimating the C II column density, due to the difficulty of obtaining it directly from the strongly saturated resonance line. We use the optically thin S II triplet as a proxy for C II due to their similarity in ionization potential (Oliveira et al. 2003).

The conversion from $N(\text{S II})$ to $N(\text{C II})$ takes into account the different abundances and depletion levels of sulfur and carbon in the LISM:

$$N(\text{C II}_{\text{SII}}) = N(\text{S II}) \times 10^{[C_\odot + D(C)] - [S_\odot + D(S)]}, \tag{2}$$

where $N(\text{C II}_{\text{SII}})$ is the estimated C II column density based on S II as a proxy, $N(\text{S II})$ is our measured column density of S II, $C_\odot = 8.39 \pm 0.05$ and $S_\odot = 7.14 \pm 0.05$ are the solar abundances of carbon and sulfur (Asplund et al. 2005), and $D(C)$ and $D(S)$ are the depletion levels (Jenkins 2004). The calculation of electron density using S II as a proxy for C II simply replaces $N(\text{C II})$ with $N(\text{C II}_{\text{SII}})$ in Equation 1.

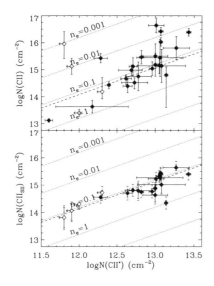
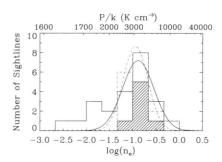

Figure 1. *Left:* Excited line column density versus resonance line column density (top) and versus C II column density using S II as a proxy (bottom). The systematic errors due to the conversion from $N(\text{S II})$ (solid lines) extend beyond the random S II fitting errors. The dotted lines indicate lines of equal electron density, assuming the LISM average temperature of 6680 K (Redfield & Linsky 2004). The unweighted mean electron densities are $n_e(\text{C II}) = 0.13^{+0.15}_{-0.07}\,\text{cm}^{-3}$ and $n_e(\text{C II}_{\text{SII}}) = 0.11^{+0.10}_{-0.05}\,\text{cm}^{-3}$ (thick dashed lines), as calculated from the histograms of n_e and $n_e(\text{C II}_{\text{SII}})$ in logarithm. *Right:* Histograms of measured electron densities n_e (solid curve) and electron densities based on S II as a proxy for C II (dashed curve) in logarithm with Gaussian fits. The unweighted centroid of the solid log-normal distribution is -0.88 ± 0.33 log (cm^{-3}) and of the dashed distribution, -0.94 ± 0.26 log (cm^{-3}). The shaded histogram indicates the electron densities of sight lines that are kinematically and spatially identified with the LIC. The top axis gives the estimated pressure $P/k = nT$.

3. Results

We analyze high spectral resolution observations of LISM absorption in order to survey the electron density in nearby interstellar material. A summary of our results is as follows:

1. We searched the entire *HST* spectroscopic database of nearby stars ($< 100\,\text{pc}$) for detections of C II*. Of these 417 sight lines, we find 13 that show C II* absorption in 23 different velocity components. The majority of these detections are new.

2. Using the C II* to C II ratio, we infer the electron density. To increase our accuracy in measuring the column density of the saturated C II resonance line, we employ multiple strategies, including using measured S II column densities as a proxy for C II column density.

3. The distribution of electron densities based on using S II as a proxy for C II is similar to the distribution based on carbon alone, while significantly tighter. This is a promising technique to avoid grossly overestimating the C II column density based on the saturated line profile.

4. We find the distribution of measured LISM electron densities (n_e) is consistent with a log-normal profile with an unweighted mean value of $n_e = 0.13^{+0.15}_{-0.07}\,\mathrm{cm}^{-3}$ and, based on S II as a proxy, $n_e(\mathrm{C\,II_{SII}}) = 0.11^{+0.10}_{-0.05}\,\mathrm{cm}^{-3}$.

5. We assign individual velocity components to specific LISM clouds, based on kinematical and spatial properties (Redfield & Linsky 2008). In particular, the LIC is probed by seven different sight lines which all give roughly identical electron density measurements. The weighted mean value for the LIC is $n_e = 0.12 \pm 0.04\,\mathrm{cm}^{-3}$.

6. The range in electron density is used to estimate the range of pressures that may be found in warm LISM clouds. Assuming all else is equal, the measured electron densities correspond to an unweighted mean pressure $P/k = 3300^{+1880}_{-900}\,\mathrm{K\,cm}^{-3}$.

Acknowledgments. S. R. would like to acknowledge support provided by NASA through Hubble Fellowship grant HST-HF-01190.01 awarded by the STScI, which is operated by the AURA, Inc., for NASA, under contract NAS 5-26555.

References

Asplund, M., Grevesse, N., & Sauval, A. J. 2005, in ASP Conf. Ser., Vol. 336, Cosmic Abundances as Records of Stellar Evolution and Nucleosynthesis, ed. T. G. Barnes, III & F. N. Bash, 25
Frisch, P. C. 1995, Space Science Reviews, 72, 499
Frisch, P. C., Jenkins, E. B., Aufdenberg, J., Sofia, U. J., York, D. G., Slavin, J. D., & Johns-Krull, C. M. 2006, in Bulletin of the AAS, Vol. 38, 922
Jenkins, E. B. 2004, in Origin and Evolution of the Elements, ed. A. McWilliam & M. Rauch, 336
Lallement, R., & Bertin, P. 1992, A&A, 266, 479
Lallement, R., Welsh, B. Y., Vergely, J. L., Crifo, F., & Sfeir, D. 2003, A&A, 411, 447
Müller, H.-R., Frisch, P. C., Florinski, V., & Zank, G. P. 2006, ApJ, 647, 1491
Nussbaumer, H., & Storey, P. J. 1981, A&A, 96, 91
Oliveira, C. M., Hébrard, G., Howk, J. C., Kruk, J. W., Chayer, P., & Moos, H. W. 2003, ApJ, 587, 235
Redfield, S., & Linsky, J. L. 2000, ApJ, 534, 825
Redfield, S., & Linsky, J. L. 2004, ApJ, 613, 1004
Redfield, S., & Linsky, J. L. 2008, ApJ, in press
Spitzer, L. J., & Fitzpatrick, E. L. 1993, ApJ, 409, 299
Vennes, S., Polomski, E. F., Lanz, T., Thorstensen, J. R., Chayer, P., & Gull, T. R. 2000, ApJ, 544, 423
Wood, B. E., & Linsky, J. L. 1997, ApJ, 474, L39
Zank, G. P. 1999, Space Science Reviews, 89, 413

Mass-to-light Ratio of Lyα Emitters: Implications of Lyα Surveys at Redshifts $z = 5.7$, 6.5, 7, and 8.8

Elizabeth Fernandez and Eiichiro Komatsu

Department of Astronomy, University of Texas, Austin, TX, USA

Abstract. Using a simple method to interpret the luminosity function of Lyα emitters, we explore properties of Lyα emitters from $5.7 \leq z \leq 8.8$ with various assumptions about metallicity and stellar mass spectra. We constrain a mass-to-"observed light" ratio, M_h/L_{band}. The mass-to-"bolometric light", M_h/L_{bol}, can also be deduced, once the metallicity and stellar mass spectrum are given. Lyα emitters are consistent with either starburst galaxies ($M_h/L_{bol} \sim 0.1 - 1$) with a smaller Lyα survival fraction, $\alpha_{esc}\epsilon^{1/\gamma} \sim 0.01 - 0.05$, or normal populations ($M_h/L_{bol} \sim 10$) if a good fraction of Lyα photons survived, $\alpha_{esc}\epsilon^{1/\gamma} \sim 0.5 - 1$. We find no evidence for the end of reionization in the luminosity functions of Lyα emitters discovered in the current Lyα surveys. The data are consistent with no evolution of intrinsic properties of Lyα emitters or neutral fraction in the intergalactic medium up to $z = 7$. No detection of sources at $z = 8.8$ does not yield a significant constraint yet.

1. Introduction

It is very likely that there was significant star formation above $z > 6$. With the introduction of new, more powerful telescopes and deep field searches, an interesting question arises: do these first stars form galaxies that are bright enough to be seen today? Several deep field Lyα searches have been performed at redshifts of six and higher. In this paper, we present a simple method to calculate the luminosity function of high-z galaxies, and compare this with the results of Lyα searches to constrain properties of Lyα emitters.

2. A Simple Model of Galaxy Counts

The simplest way to predict the cumulative luminosity function of galaxies is to count the number of halos available in the universe above a certain mass,

$$N(>L) = V(z) \int_{M_h(L)}^{\infty} \frac{dn}{dM_h} dM_h, \qquad (1)$$

where $V(z)$ is the survey volume, dn/dM_h the comoving number density of halos per unit mass range, and M_h the total mass of a halo.

From equation (1) we can derive the number of galaxies observed above a certain flux density as:

$$\int_{F_{limit}}^{\infty} \frac{d^2N}{dFd\Omega} dF = 4\pi d_L^2(z) \frac{dV}{dzd\Omega} \Delta z \Delta \nu_{obs} \frac{M_h}{L_{band}} \int_{F_{limit}}^{\infty} \frac{dn(M_h(F))}{dM_h} dF$$
$$\times \vartheta(M_h - M_{min}(z)). \qquad (2)$$

Not all dark matter halos will be forming stars - only halos with a mass above some critical minimum mass (M_{min}). We model the galaxy number counts by placing one galaxy per halo with only a fraction of photons escaping from galaxies and the IGM.

Several telescopes are now powerful enough to attempt to locate galaxies at $z \gtrsim 6$. One effective method of locating high redshift galaxies is to use narrow-band filters to detect Lyα emission from a small range of redshifts.

Two surveys at $z = 5.7$ and $z = 6.56$ were taken at the Subaru telescope using the Suprime-Cam. At $z = 5.7$, 34 Lyα emitters have been confirmed (Shimasaku et al. 2006). At $z = 6.56$, 17 Lyα emitters have been confirmed (Taniguchi et al. 2005; Kashikawa et al. 2006). From these detections, they were able to fit a Schechter function. The Large Area Lyman Alpha (LALA) survey searched for galaxies at a redshift of around 6.55 (Rhoads et al. 2004). There was one confirmed Lyα emitter. Another survey using the Subaru telescope at $z = 7.025$ found one Lyα emitter at $z = 6.96$ (Iye et al. 2006). ZEN, which stands for z equals nine, is a narrow J-band mission using the ISAAC on the VLT. Its central redshift is 8.76. No galaxies were found (Willis & Courbin 2005; Willis et al. 2006). Cuby et al. (2007) did a followup narrow-band search, using the ISAAC at the VLT, with a larger field of view (hereafter referred to as the ISAAC ext). They also detected no galaxies within the fields.

3. Properties of Lyα Emitters

In equation (2), the only free parameter was the mass-to-"observed light" ratio, M_h/L_{band}, which was varied to fit observations. The number of galaxies drops as M_h/L_{band} increases (See Figure 1). As the mass-to-light ratio increases, the star formation is spread out over a longer period of time.

M_h/L_{band} only describes the light observed over the narrow band. To proceed further and understand physical properties of Lyα emitters better, however, we must relate M_h/L_{band} to M_h/L_{bol} (where L_{bol} is the bolometric luminosity). In order to get the actual mass to light ratio, the spectra of a stellar population of galaxies (dependent on metallicity and mass function of the stars) was modeled and integrated first over all frequencies and then compared to the light that is observed in the narrow band. We use analytical formulas for these spectra given in section 2 of Fernandez & Komatsu (2006), paired with a line profile of Lyα emission from Loeb & Rybicki (1999); Santos et al. (2002).

The outcome of our analysis is $(M_h/L_{bol})(\alpha_{esc}\epsilon^{1/\gamma})^{-1}$, where M_h/L_{bol} is the mass-to-bolometric light ratio, L_{bol} is the *intrinsic* luminosity of galaxies before absorption or extinction of Lyα photons, α_{esc} is the Lyα survival fraction, and $\epsilon^{1/\gamma}$ gives the effect of the duty cycle (ϵ is the "duty cycle" of Lyα emitters, and $\gamma \sim 2$ is a local slope of the cumulative luminosity function, $N(>L) \propto L^{-\gamma}$, to which the current data are sensitive). We tabulate this quantity inferred from various narrow-band searches in Table 1.

We conclude from these results that the Lyα emitters detected in these narrow-band surveys are either normal galaxy populations with $M_h/L_{bol} \sim 10$ and having a fair fraction of Lyα photons escape, $\alpha_{esc}\epsilon^{1/\gamma} \sim 0.5-1$, or starburst galaxies with $M_h/L_{bol} \sim 0.1-1$ and a smaller fraction of the Lyα photons escaped from the galaxies themselves *and* the surrounding IGM,

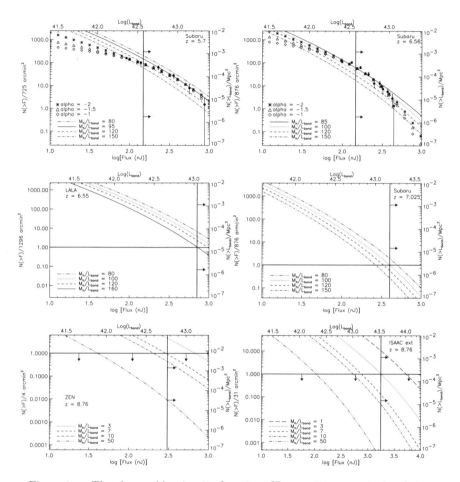

Figure 1. The observed luminosity function of Lyα emitters constrains their mass-to-"observed light" ratio. Each panel shows the cumulative number of sources detected in each field above a certain flux density, $N(>F)$. (The flux density limits of each survey are indicated by the vertical lines with right arrows). This figure is reproduced from Fernandez & Komatsu (2008).

Table 1. The mass (total halo mass) to light (bolometric luminosity) ratio times $1/(\alpha_{esc}\, \epsilon^{1/\gamma})$.

Field	Redshift	$\frac{M_h}{L_{bol}} \frac{1}{\alpha_{esc}\, \epsilon^{1/\gamma}}$ ($Z=0$)	$\frac{M_h}{L_{bol}} \frac{1}{\alpha_{esc}\, \epsilon^{1/\gamma}}$ ($Z=1/50\ Z_\odot$)	$\frac{M_h}{L_{bol}} \frac{1}{\alpha_{esc}\, \epsilon^{1/\gamma}}$ ($Z=1\ Z_\odot$)
Subaru	5.7	$28-38$	$19-26$	$12-17$
LALA	6.55	$\sim 32-34$	$\sim 21-23$	$\sim 13-15$
Subaru	6.56	$21-26$	$14-19$	$8.9-12$
Subaru	7.025	$\sim 28-30$	$\sim 19-21$	$\sim 12-14$
ZEN	8.76	$> 0.79-0.84$	$> 0.53-0.59$	$> 0.34-0.39$
ISAAC ext	8.76	$> 0.34-0.36$	$> 0.23-0.25$	$> 0.14-0.17$

$\alpha_{esc}\epsilon^{1/\gamma} \sim 0.01-0.05$. Consistency across redshifts ($z = 5.7$, 6.5, and 7.0) as well as across different observations is striking.

4. Conclusions

A simple model based upon the halo mass function coupled with a constant mass-to-light ratio fits the luminosity functions measured and constrained by the current generation of narrow-band Lyα surveys from $5.7 \leq z \leq 8.8$. We have explored various metallicities and stellar mass spectra.

The inferred mass-to-light ratios are consistent with no evolution in the properties of Lyα emitters or opacity in the IGM from $5.7 \leq z \leq 7$. Therefore, the current data of the luminosity functions do not provide evidence for the end of reionization. The data at $z = 8.8$ do not yield a significant constraint yet.

These mass-to-light ratios suggest that the Lyα emitters discovered in the current surveys are either starburst galaxies with only a smaller fraction of Lyα photons escaped from galaxies themselves and the IGM, $\alpha_{esc}\epsilon^{1/\gamma} \sim 0.01-0.05$, or normal populations with a fair fraction of Lyα photons escaped, $\alpha_{esc}\epsilon^{1/\gamma} \sim 0.5-1$. The luminosity function alone cannot distinguish between these two possibilities.

References

Cuby, J.-G. et al. 2007, A&A, 461, 911
Fernandez, E., & Komatsu, E. 2006, ApJ, 646, 703
Fernandez, E., & Komatsu, E. 2008, MNRAS, in press, astro-ph/0706.1801
Iye, M., et al. 2006, Nature, 443,186
Kashikawa, N., et al. 2006, ApJ, 648, 7
Loeb, A., & Rybicki, G. 1999, ApJ, 524, L527
Rhoads, J. E., et al. 2004, ApJ, 611, 59
Santos, M. R., Bromm, V., & Kamionkowski, M. 2002, MNRAS, 336, 1082
Shimasaku, K., et al. 2006, PASJ, 58, 313
Taniguchi, Y., et al. 2005, PASJ, 57, 165
Willis, J. P., & Courbin, F. 2005, MNRAS, 357,1348
Willis, J. P., et al. 2006, New Astronomy Reviews, 50, 70

The Hobby-Eberly Telescope Chemical Abundances of Stars in the Halo (*CASH*) Project I. Observations of the First Year

A. Frebel,[1] C. Allende Prieto,[2] I. U. Roederer,[2] M. D. Shetrone,[1]
J. Rhee,[3] C. Sneden,[2] T. C. Beers,[4] and J. J. Cowan[5]

Abstract. We present preliminary results obtained from the first year of observations of a new, long-term project of the University of Texas, the Hobby-Eberly Telescope Chemical Abundances of Stars in the Halo (*CASH*) Project.

1. Scientific Goals

The story of early Galactic nucleosynthesis is written in the chemical compositions of metal-poor halo stars. The abundances in these stars reflect only a few chemical enrichment events and hence, this fossil record can be used to trace the chemical and dynamical evolution of the early Galaxy. While the individual abundances of most metals in these stars will remain unchanged throughout the stellar lifetime, close examination is necessary to discern exceptions to this rule.

We have recently started the Chemical Abundance of Stars in the Halo (*CASH*) Project with the Hobby-Eberly-Telescope (HET) located at McDonald Observatory. The project aims to characterize the chemical composition of the Galactic halo through abundance analyses of large numbers of metal-poor stars. Our goal is to build up the largest high-resolution database available for these objects over several years. The *CASH* Project has among its primary goals the identification of large numbers of metal-poor stars that are 1. α-element rich (and -poor), 2. carbon-rich, with and without accompanying s-process overabundances, or 3. highly r-process-enhanced. For the first time, the absolute frequencies of abundance anomalies based on statistically significant samples of metal-poor stars will be obtained. Such information is required for an improvement in our understanding of the nature and interplay of the nucleosynthetic processes, and to help in identifying their astrophysical sites in the early Galaxy.

The second major scientific question we seek to investigate is the recent the claim by Carollo et al. (2007) that there is a chemical difference between the so-

---

[1] McDonald Observatory, University of Texas, Austin, TX, USA

[2] Department of Astronomy, University of Texas, Austin, TX, USA

[3] Department of Physics, Purdue University, IN, USA

[4] Department of Physics and Astronomy, Center for the Study of Cosmic Evolution, and Joint Institute for Nuclear Astrophysics, Michigan State University, East Lansing, MI, USA

[5] Homer L. Dodge Department of Physics & Astronomy, University of Oklahoma, Norman, OK, USA

called "inner" and "outer" halo populations. By means of chemical analyses of large numbers of stars associated with both populations we are able to trace the chemical evolution of the Galactic halo. This, in turn, will lead to an improved understanding of Galaxy formation.

2. Target Selection

We have access to several sources of yet unstudied northern-hemisphere targets that have (or will be acquired during the course of this project) preliminary low resolution ($R \sim 2,000$) spectroscopic data: Re-processed HK-I survey (Beers et al. in prep.) and HK-II survey (Rhee 2001) Bright Hamburg/ESO survey (Frebel et al. 2006) SDSS and SEGUE surveys

An effort has been made over the course of the past few years to obtain low-resolution spectroscopic confirmation of their metal-poor status using 2-4 m telescopes at the McDonald Observatory and elsewhere. The HK-II candidates and northern bright targets from the Hamburg/ESO survey are of particular interest for our proposed work, as they range in brightness from $B = 12 - 15$.

3. Observing and Data Analysis Strategy

Our new comprehensive, multi-year, survey aims at obtaining moderately high-resolution spectroscopic data of ~ 1000 very metal-poor stars ([Fe/H] < -2.0) discovered during the course of previous low-resolution spectroscopic searches, but not yet observed at high spectral resolution. We can meet our intended S/N for HK-II stars with total integration times from 10 to 30 minutes. Over 100 such candidates are immediately available that are observable from the HET. Further targets will be identified in winter 2007 using the 2.1m telescope at McDonald Observatory (West Texas) and the 2.3 m at Siding Spring Observatory (Australia). In total, we plan to target a sample of ~ 1000 such stars with [Fe/H] < -2.0 and $4500 < T_{\rm eff} < 7000$ K, over a period of several years, in order to obtain "snapshot" ($R \sim 15,000$, $S/N \sim 30 - 50$) spectra with the HRS on HET. This will be the largest high-resolution database for halo objects.

The REDUCE (Piskunov & Valenti 2002) pipeline has been tailored to our HRS setup. The data obtained in the first year (as of Sep. 2007) are reduced, and stellar parameters were obtained. The reduction was validated with an independent IRAF reduction. We are in the process of obtaining abundances other than Fe for our detailed chemical abundance analyses.

4. Results from the First Year

During the first year (as of Sep 2007), we have obtained data for the anticipated ~ 200 objects per year. We have found ~ 60 stars with [Fe/H] < -2.5 of which 6 have [Fe/H] < -3. Among the data taken in UT07-01, we have found a very particular object – a Li self-enriched red giant with r+s process enhancement (Frebel et al. 2008). An abundance analysis based on newly obtained HET/*CASH* data was carried out for three bright stars from the Hamburg/ESO survey with [Fe/H] ~ -2.9 (Davies et al. 2008).

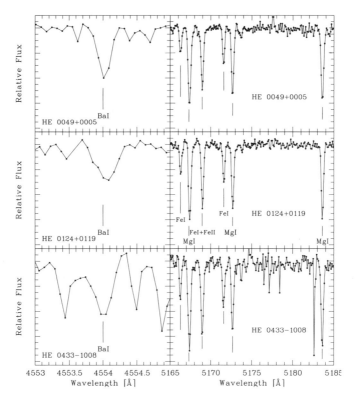

Figure 1. Example spectra of three metal-poor stars ([Fe/H] ~ -2.8, T_{eff} 4500 K, $\log g$ 1.0) observed with the proposed HET/HRS setup in 2007. Left panel: The Ba line at 4554 Å a neutron-capture abundance indicator that is detectable in our proposed spectra. Right panel: Mg b lines at ~ 5170 Å from which properties about the alpha-elements can be inferred.

5. Outlook

The *CASH* project will represent the largest high-resolution survey of metal-poor halo stars ever conducted, and thus will be one of the "legacy results" of the Hobby-Eberly Telescope. We expect that this survey will provide the fundamental spectroscopic database capable of resolving numerous fundamental questions concerning the origin of elements in the first generations of stars.

More details and further results will be posted on our new project website http://www.as.utexas.edu/cash.html

Acknowledgments. Based on observations that were obtained with the Hobby-Eberly Telescope. The HET is named in honor of its principal benefactors, William P. Hobby and Robert E. Eberly. A. F. acknowledges support through the W. J. McDonald Fellowship of the McDonald Observatory. Funding for this project has also been generously provided by the US NSF (grants AST 06-07708 to C. S. and AST 07-07447 to JJC). T. C. B. acknowledges funding

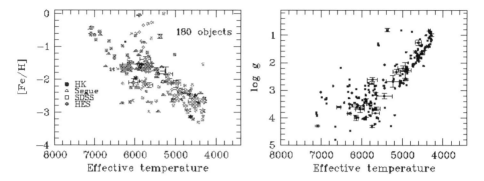

Figure 2. Preliminary stellar parameters for ~ 180 stars out of the ~ 220 observed objects. These stars have been observed in one or several nights, and the stellar parameters determined for the observations of several nights agree well (small or no error bars indicate agreement between different nights of observations). For the remaining stars, we are in the process of combining the data before the stellar parameters are determined, to achieve the minimum S/N necessary for the parameter determination. Right panel: Hertzsprung-Russell-Diagram of our first year sample. As can be seen we cover a large parameter space with our targets.

from grants AST 04-06784, AST 07-07776 as well as PHY 02-15783; Physics Frontiers Centers/ JINA: Joint Institute for Nuclear Astrophysics, awarded by the US NSF.

References

Carollo, D., Beers, T. C., Lee, Y. S., Chiba, M., Norris, J. E., Wilhelm, R., Sivarani, T., Marsteller, B., Munn, J. A., Bailer-Jones, C. A. L., Re Fiorentin, P., & York, D. G. 2007, astro-ph/0706.3005
Davies, L. A., Frebel, A., Cowan, J. J., Allende Prieto, C., & Sneden, C. 2008, in New Horizons in Astronomy: Frank N. Bash Symposium, ed. A. Frebel, J. Maund, J. Shen, & M. Siegel, Astronomical Society of the Pacific Conference Series
Frebel, A., Christlieb, N., Norris, J. E., Beers, T. C., Bessell, M. S., Rhee, J., Fechner, C., Marsteller, B., Rossi, S., Thom, C., Wisotzki, L., & Reimers, D. 2006, ApJ, 652, 1585
Frebel, A., Roederer, I., Shetrone, M., Allende Prieto, C., Rhee, J., Gallino, R., Bisterzo, S., Sneden, C., Beers, T. C., & Cowan, J. J. 2008, in New Horizons in Astronomy: Frank N. Bash Symposium, ed. A. Frebel, J. Maund, J. Shen, & M. Siegel, Astronomical Society of the Pacific Conference Series
Piskunov, N. E. & Valenti, J. A. 2002, A&A, 385, 1095
Rhee, J. 2001, PASP, 113, 1569

The Hobby-Eberly Telescope Chemical Abundances of Stars in the Halo (*CASH*) Project II. The Li-, *r*- and *s*-Enhanced Metal-Poor Giant HK-II 17435-00532

A. Frebel,[1] I. U. Roederer,[2] M. Shetrone,[1] C. Allende Prieto,[2] J. Rhee,[3] R. Gallino,[4] S. Bisterzo,[4] C. Sneden,[2] T. C. Beers,[5] and J. J. Cowan[6]

Abstract. We present the first detailed abundance analysis of the metal-poor giant HK-II 17435-00532. This star was observed as part of the University of Texas Long-Term *Chemical Abundances of Stars in the Halo* (*CASH*) Project. We find that this metal-poor ([Fe/H] = −2.2) star has an unusually high lithium abundance (log ε (Li) = +2.1), mild carbon ([C/Fe] = +0.7) and sodium ([Na/Fe] = +0.6) enhancement, as well as enhancement of both *s*-process ([Ba/Fe] = +0.8) and *r*-process ([Eu/Fe] = +0.5) material. The high Li abundance can be explained by self-enrichment through extra mixing mechanisms. If so, HK-II 17435-00532 is the most metal-poor star in which this short-lived phase of Li enrichment has been observed. The *r*- and *s*-process material was not produced in this star but was either present in the gas from which HK-II 17435-00532 formed or was transferred to it from a more massive binary companion. Despite the current non-detection of radial velocity variations (over a time span of ∼ 180 days), it is possible that HK-II 17435-00532 is in a long-period binary system, similar to other stars with both *r* and *s* enrichment.

1. Introduction

The story of early nucleosynthesis is written in the chemical compositions of the most metal-poor Galactic halo stars. Li, the only metal produced during the Big Bang, is often observed in unevolved metal-poor stars. Information on the primordial Li abundance can be inferred from, e.g., the Spite Plateau Li abundance that most metal-poor main-sequence stars tend to possess, log ε (Li) = 2.21 ± 0.09 (e.g., Charbonnel & Primas 2005), although the stellar result does not agree with the WMAP estimate of the primordial Li abundance, log ε (Li) = 2.64 ± 0.03 (Spergel et al. 2007). Some evolved stars, however, appear to have *overabundances* of Li—in contrast to the sparse galactic produc-

---

[1] McDonald Observatory, University of Texas, Austin, TX, USA

[2] Department of Astronomy, University of Texas, Austin, TX, USA

[3] Department of Physics, Purdue University, IN, USA

[4] Dipartimento di Fisica Generale, Universitàdi, Torino, Italy

[5] Department of Physics and Astronomy, Center for the Study of Cosmic Evolution, and Joint Institute for Nuclear Astrophysics, Michigan State University, East Lansing, MI, USA

[6] Homer L. Dodge Department of Physics & Astronomy, University of Oklahoma, Norman, OK, USA

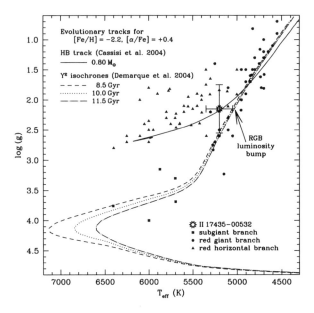

Figure 1. Spectroscopic gravities are shown as a function of effective temperature for HK-II 17435-00532 and a sample of other evolved metal-poor stars from previous studies (Behr 2003; Cayrel et al. 2004; Preston et al. 2006). Three sets of Y^2 isochrones (Demarque et al. 2004) are displayed, along with a synthetic HB track (Cassisi et al. 2004).

tion mechanisms of Li through cosmic ray spallation, and in spite of the fact that Li is completely diluted and destroyed in the stellar atmosphere by the time the star reaches the giant branch.

We have recently started the *Chemical Abundances of Stars in the Halo* (*CASH*) Project with the Hobby-Eberly Telescope (HET) located at McDonald Observatory (Frebel et al. 2008). This project aims to characterize the chemical composition of the Galactic halo by means of abundance analyses of metal-poor stars. HK-II 17435-00532 was observed as part of *CASH* project The $R \sim 15,000$ spectra covering the range from 4120 to 7850 Å.

Based on several radial velocity measurements we currently find no evidence that HK-II 17435-00532 is in a binary system. We derive the effective temperature of HK-II 17435-00532 from the color-T_{eff} calibrations (Alonso et al. 1999). The surface gravity and microturbulence were derived with the usual spectroscopic methods.

We find $T_{\text{eff}}/\log(g)/v_t/[\text{Fe}/\text{H}] = 5200\,\text{K}/2.15/2.0\,\text{km s}^{-1}/-2.23$. We perform our equivalent width and spectrum synthesis LTE analyses with MOOG (Sneden 1973). We use the Grevesse & Sauval (2002) solar abundances.

2. Abundance pattern and Origin of HK-II 17435-00532

The abundances are displayed in Figure 2. The Li abundance of $\log \varepsilon\,(\text{Li}) = 2.1$, is at odds with the evolutionary status of HK-II 17435-00532, i.e., it is far higher than expected from standard stellar evolution models. If this star is evolving up the red giant branch (RGB), it is located near the RGB luminosity bump, where extra mixing has been found to cause a short-lived phase of Li enrichment in more massive, metal-rich stars (Charbonnel & Balachandran 2000). See Figure 1. It is possible that the present Li abundance has already been reduced from its maximum abundance. If HK-II 17435-00532 is evolving through the RGB luminosity bump, it is the most metal-poor star for which an extra mixing mechanism has been shown to produce Li enrichment in the stellar envelope. Enrichment of Li is not expected to occur on the RHB or early-AGB; if HK-II 17435-00532 is on the RHB or early AGB, we are left to postulate that a previously-unidentified efficient extra mixing episode may be operating during this stage of evolution in low-mass, low-metallicity stars.

C and O are overabundant, [C/Fe] = +0.7 and [O/Fe] = +1.1. A comparison of HK-II 17435-00532 with the ten most metal-rich stars ($-2.8 \leq$ [Fe/H] ≤ -2.0) in the sample of McWilliam et al. (1995) suggests that all α- and Fe-peak abundances are consistent with "typical" metal-poor stars. The exception is Na, which is clearly enhanced, [Na/Fe] = +0.6. HK-II 17435-00532 exhibits overabundances of all n-capture elements, such as [Ba/Fe] = +0.86 and [Eu/Fe] = +0.48. These ratios suggest that HK-II 17435-00532 is enriched to some degree in both s- and r-process material, although neither the scaled-solar s- nor r-process abundances alone provide a satisfactory fit to the observed abundances. HK-II 17435-00532 has a similar n-capture abundance pattern compared with other $r+s$ stars. The s and r nucleosynthesis reactions are not expected to operate in stars of sub-solar mass such as HK-II 17435-00532, so the n-capture elements must have been either present in the material from which the star formed or were transferred to it from an undetected binary companion.

Like HK-II 17435-00532, several other $r+s$ stars also have enhanced Na abundances. The C, Na, and s-process enhancements can all be explained by assuming material was transferred to HK-II 17435-00532 from an undetected binary companion that passed through the AGB phase. In Figure 2 we display the measured [X/Fe] abundance ratios for $6 \leq Z \leq 63$ in HK-II 17435-00532, along with the predicted abundance ratios (based on the FRANEC stellar evolution models; e.g., Straniero et al. 2003) from a binary mass transfer event. The best-fit is obtained with the a 1.5 $M_\odot$ companion. We derive a dilution factor of 1.8 dex (i.e., a factor of ≈ 63), which means that one part of accreted material is present in the stellar envelope of HK-II 17435-00532 for every 63 parts of material originally present in the star. If we assume an initial [r/Fe] = +0.3 for this system, then only ~ 0.2 dex of Eu needs to be acquired from the s-process material of the AGB companion. In contrast to the Eu, which mostly reflects the initial composition of the ISM from which the system formed, the Ba (and other s-process species) was dominantly produced by the companion during its AGB phase.

Acknowledgments. A. F. gratefully acknowledges support through the W. J. McDonald Fellowship of the McDonald Observatory. Funding has been provided by the U. S. National Science Foundation (grants AST 06-07708 to C. S.; AST 04-06784, AST 07-07776, and PHY 02-15783 to T. C. B.; AST 07-07447 to J. J. C.). J. R. acknowledges partial support by NASA through the AAS Small Research Grant Program and GALEX GI grant 05-GALEX05-27.

210 Frebel et al.

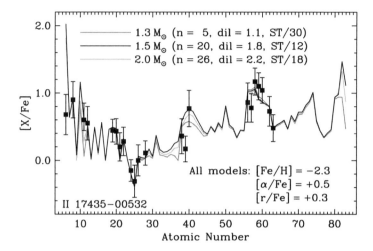

Figure 2. Predicted [X/Fe] ratios in HK-II 17435-00532 assuming pollution from a companion star that passed through the AGB phase. The initial mass of the AGB star is the primary variable between the different sets of abundance predictions, although changing the mass also changes the number of thermal pulses ("n") and necessitates altering the logarithmic dilution factor ("dil") and ^{13}C pocket efficiency ("ST/"). The black curve reflects our best fit model.

R. G. acknowledges the Italian MIUR-PRIN06 Project, and T. C. B. the Physics Frontier Center/Joint Institute for Nuclear Astrophysics (JINA) for partial support.

References

Alonso, A., Arribas, S., & Martínez-Roger, C. 1999, A&AS, 140, 261
Behr, B. B. 2003, ApJS, 149, 101
Cassisi, S., Castellani, M., Caputo, F., & Castellani, V. 2004, A&A, 426, 641
Cayrel, R., et al. 2004, A&A, 416, 1117
Charbonnel, C., & Balachandran, S. C. 2000, A&A, 359, 563
Charbonnel, C., & Primas, F. 2005, A&A, 442, 961
Demarque, P., Woo, J.-H., Kim, Y.-C., & Yi, S. K. 2004, ApJS, 155, 667
Frebel, A., Allende Prieto, C., Roederer, I., Shetrone, M., Rhee, J., Sneden, C., Beers, T. C., & Cowan, J. J. 2008, in New Horizons in Astronomy: Frank N. Bash Symposium, ed. A. Frebel, J. Maund, J. Shen, & M. Siegel, Astronomical Society of the Pacific Conference Series
Grevesse, N., & Sauval, A. J. 2002, Adv. Space Res., 30, 3
McWilliam, A., Preston, G. W., Sneden, C., & Searle, L. 1995b, AJ, 109, 2757
Preston, G. W., Sneden, C., Thompson, I. B., Shectman, S. A., & Burley, G. S. 2006, AJ, 132, 85
Sneden, C. A. 1973, Ph.D. Thesis, Univ. of Texas at Austin
Spergel, D. N., et al. 2007, ApJS, 170, 377
Straniero, O., Domínguez, I., Cristallo, R., & Gallino, R. 2003, Publications of the Astronomical Society of Australia, 20, 389

New Horizons in Astronomy: Frank N. Bash Symposium 2007
ASP Conference Series, Vol. 393, © 2008
A. Frebel, J. R. Maund, J. Shen, and M. H. Siegel, eds.

Morphological Transformations of Galaxies in the A901/02 Supercluster from STAGES

Amanda Heiderman,[1] S. Jogee,[1] D. Bacon,[9] M. Balogh,[10] M. Barden,[11]
F. D. Barazza,[4] E. F. Bell,[6] A. Böhm,[12] J. A. R. Caldwell,[18]
M. E. Gray,[3] B. Häußler,[3] C. Heymans,[2,5] K. Jahnke,[6]
E. van Kampen,[11] S. Koposov,[6] K. Lane,[3] I. Marinova,[1] D. McIntosh,[15]
K. Meisenheimer,[6] C. Y. Peng,[7,8] H.-W. Rix,[6] S. F. Sánchez,[13]
R. Somerville,[6] A. Taylor,[14] L. Wisotzki,[12] C. Wolf,[16] and X. Zheng[17]

[1] *Dept. of Astronomy, University of Texas at Austin, Austin, TX, USA*
[2] *Dept. of Physics and Astron., Univ. of Brit. Col., Vancouver, CAN*
[3] *School of Physics and Astronomy, Univ. of Nottingham, GBR*
[4] *Laboratoire d'Astrophysique, EPFL, Sauverny, SUI*
[5] *Institut d'Astrophysique de Paris, Paris, FRA*
[6] *Max-Planck-Institut für Astronomie, Heidelberg, GER*
[7] *NRC Herzberg Institute of Astrophysics, Victoria, CAN*
[8] *Space Telescope Science Institute, Baltimore, MD, USA*
[9] *Inst. of Cosmology and Gravitation, Univ. of Portsmouth, GBR*
[10] *Dept. of Physics and Astronomy, Univ. of Waterloo, Ontario, CAN*
[11] *Inst. for Astro- and Particle Physics, University of Innsbruck, AUT*
[12] *Astrophysikalisches Insitut Potsdam, Potsdam, GER*
[13] *Centro Hispano Aleman de Calar Alto, Almeria, ESP*
[14] *Scottish Universities Physics Alliance, Inst. for Astron., Univ. of Edinburgh, GBR*
[15] *Dept. of Astronomy, Univ. of Massachusetts, Amherst, MA, USA*
[16] *Department of Astrophysics, University of Oxford, Oxford, GBR*
[17] *Purple Mountain Obs., Chinese Academy of Sciences, Nanjing, CHN*
[18] *University of Texas, McDonald Observatory, Fort Davis, TX, USA*

Abstract. We present a study of galaxies in the Abell 901/902 Supercluster at $z \sim 0.165$, based on *HST* ACS F606W, COMBO-17, *Spitzer* 24 μm, XMM-Newton X-ray, and gravitational lensing maps, as part of the STAGES survey. We characterize galaxies with strong externally-triggered morphological distortions and normal relatively undisturbed galaxies, using visual classification and quantitative CAS parameters. We compare normal and distorted galaxies in terms of their frequency, distribution within the cluster, star formation properties, and relationship to dark matter (DM) or surface mass density, and intra-cluster medium (ICM) density. We revisit the morphology density relation, which postulates a higher fraction of early type galaxies in dense environments, by considering separately galaxies with a low bulge-to-disk (B/D) ratio and a low gas content as these two parameters may not be correlated in clusters. We report here on our preliminary analysis.

1. Introduction

The systematic quest to understand how galaxies evolve as a function of epoch and environment remains in its infancy. Galaxies in cluster environments may differ from field galaxies due to high initial densities leading to early collapse. Furthermore, the relative importance of galaxy-galaxy interactions (e.g., galaxy harassment, tidal interactions, minor mergers, and major mergers) and galaxy-ICM interactions (e.g., ram pressure stripping and compression) are likely to differ between cluster and field environments due to the different number density of galaxies, galaxy velocity dispersions, ICM density, and DM density. In order to constrain how galaxies evolve in cluster environments, we present a study based on the STAGES survey of the Abell 901/902 supercluster (M. Gray et al. 2008, in prep.). The survey covers 0.5×0.5 degrees on the sky, and includes high resolution (0.1″, corresponding to 280 pc at $z \sim 0.165$[1]) *HST* ACS F606W images, along with COMBO-17, *Spitzer* 24 µm, XMM-Newton X-ray data, and gravitational lensing maps (Gray et al. 2002; C. Heymans et al. 2008, submitted). The STAGES survey is complemented with accurate spectrophotometric redshifts with errors $\delta_z/(1+z) \sim 0.02$ down to $R_{\text{Vega}} = 24$ from the COMBO-17 survey (Wolf et al. 2004), and stellar masses (Borch et al. 2006). Star formation rates (SFRs) are derived from the COMBO-17 UV and *Spitzer* 24 µm data (Bell et al. 2007). The supercluster sample of 2309 galaxies covers a broad range of luminosities, encompassing both dwarf and larger galaxies (E to Sd). We present many of our results separately for bright ($M_V \leq -18$; 798 galaxies), and faint galaxies ($-18 < M_V \leq -15.5$; 1286 galaxies) because the morphological characterization is less robust for small dwarf galaxies, than for normal galaxies, due to surface brightness, spatial resolution (280 pc), and contamination from field galaxies.

2. Methodology and Preliminary Results

Using the CAS code (Conselice et al. 2000), the concentration (C), asymmetry (A), and clumpiness (S) parameters were derived from the *HST* ACS F606W images. Given that the CAS merger criteria (A>S and A>0.35) tend to capture only a fraction of interacting/merging galaxies (e.g., Conselice 2006; Jogee et al. 2007), we also characterize the morphological properties via visual classification. In order to identify galaxies that are distorted due to a recent merger or tidal interaction, we take special care to classify galaxies into three distinct visual classes: (1) Galaxies with *externally-triggered* distortions: These distortions, triggered by tidal interactions or mergers, include double or multiple nuclei inside a common body, tidal tails, arcs, shells, ripples, or tidal debris in body of galaxy, warps, offset rings, and extremely asymmetric SF or spiral arms on one side of the disk. Galaxies are classified as strongly or weakly distorted, according to whether the distortions occupy a large or small fraction of the total light. (2) Galaxies with *internally-triggered* asymmetries (classified as Irr-1): Internally-triggered asymmetries are due to stochastic star-forming regions or the low ratio of ordered to random motions, common in irregular galaxies. These asymmetries

---

[1] We assume in this paper a flat cosmology with $\Omega_M = 1 - \Omega_\Lambda = 0.3$ and $H_0 = 70 \text{ km s}^{-1} \text{ Mpc}^{-1}$.

tend to be correlated on scales of a few hundred parsecs, rather than on scales of a few kpc. (3) Relatively *undistorted symmetric* galaxies (classified as Normal).

The assumed correlation implicit in the Hubble sequence, between a high B/D and a low gas and dust content, can often fail in cluster environments, where systems of low B/D may have smooth featureless disks (e.g., Koopmann & Kenney 1998). This begs the question of whether the classical morphology-density relation, which claims a larger fraction of early-type galaxies in dense environments, is driven by a larger fraction of gas-poor galaxies in such environments, or a larger fraction of bulge-dominated systems, or both. In order to address this question, we visually classify the clumpiness and B/D ratio of galaxies separately, classifying galaxies into five categories: pure bulge (pB), bulge plus disk (B+D) clumpy or smooth, and pure disk (pD) clumpy or smooth. We present below extracts of our preliminary findings.

1. Fraction of strongly distorted galaxies: Visual classification shows that 3% and 0.4% of bright ($M_V \leq -18$) and faint ($-18 < M_V \leq -15.5$) galaxies are strongly distorted, respectively. (The results for faint galaxies are highly uncertain.). The distortion fraction for bright galaxies in Abell 901/902 is lower than the values of 7% to 9% reported in the field for bright ($M_B \leq -20$; Lotz et al. 2007) or massive ($M \geq 2.5 \times 10^{10} M_\odot$; Jogee et al. 2007) galaxies.

2. CAS recovery rate: The CAS merger criteria (A>S and A>0.35) capture 54% (bright sample) and 40% (faint sample) of the strongly distorted galaxies.

3. Distribution of strongly distorted vs. normal galaxies: Most strongly distorted galaxies lie outside the cluster cores, avoid the peaks in DM surface mass density ($\kappa \geq 0.1$) and ICM density (Figure 1A). These results are consistent with high galaxy velocity dispersions in the core being unfavorable to mergers and strong tidal interactions. It may also be due, at least in part, to the predominance in the core of gas-poor systems (see point 6 below; Figure 1B), which tend to show shorter-lived tidal signatures.

4. SFR of strongly distorted vs. normal galaxies: The UV-based SFR ranges primarily from 0.001 to $14 M_\odot \text{yr}^{-1}$ and rises with stellar mass (Fig. 1D). The average SFR_{UV} is enhanced only by a modest factor of ~ 4 in strongly distorted galaxies compared to normal galaxies (Fig. 1C). A similar result is reported for distorted galaxies in field galaxies at $z \sim 0.2$ (e.g., Jogee et al. 2007). For 12% of the 798 bright galaxies having $Spitzer$ 24 μm detection, the UV+IR-based SFR ranges primarily from 0.01 to $50 M_\odot \text{yr}^{-1}$. The median ratio of ($\text{SFR}_{\text{UV+IR}}/\text{SFR}_{\text{UV}}$) is ~ 3, indicating significant amounts of obscured star formation.

5. Specific SFR (SSFR) of strongly distorted vs. normal galaxies: The SSFR or SFR per unit stellar mass is on average lower at higher stellar mass, consistent with a large fractional growth happening in lower mass galaxies at later times (Figure 1C). Only a modest enhancement in SSFR_{UV} is seen in distorted galaxies compared to normal galaxies.

6. The Morphology-Density relation in A901/02: Smooth (gas-poor) galaxies cluster in regions of high ICM density while clumpy (gas-rich) systems dominate at low ICM densities (Fig 1B). The ratio of smooth (gas-poor) galaxies to very clumpy (gas-rich) galaxies within the central 0.3 Mpc is (89%:11%), (75%:25%), and (86%:14%), respectively, for A901a, A901b, and A902. These values are comparable to the ratio of (80%:20%) for early-type to late-type galaxies in the classical morphology-density relation (Dressler 1980). Future work will

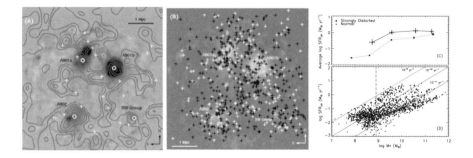

Figure 1. (**A**) Strongly distorted galaxies (white crosses) are overplotted on the ICM density (greyscale) and DM surface mass density κ (contours ranging from 0.2 to 0.12, in steps of 0.2). Most strongly distorted galaxies lie outside region of high ICM density and DM peaks ($\kappa \sim 0.1$). (**B**) We revisit the morphology-density relation by showing the distribution of bright ($M_{\rm V} \leq -18$) smooth (gas-poor) galaxies (black crosses) and clumpy (gas-rich) galaxies (white crosses) on the ICM density. Smooth (gas-poor) galaxies populate the highest density regions. (**C**) The average SFR$_{\rm UV}$ is shown as a function of stellar mass. It is enhanced by a modest factor of 4 in strongly distorted galaxies compared to normal systems. (**D**) Strongly distorted (black stars) and normal relatively undisturbed (open circles) galaxies are plotted on the SFR$_{\rm UV}$ *versus* stellar mass plane. Loci of constant specific SFR$_{\rm UV}$ are marked in units of yr^{-1}. The dashed vertical line denotes the COMBO-17 mass completeness limit for the red sequence. Note that SFR$_{\rm UV}$ for red sequence galaxies are upper limits.

explore whether the ratio of bulge-dominated galaxies to disk-dominated galaxies also shows a similar trend with galaxy number density.

Acknowledgments. AH and SJ acknowledge support from NSF grant AST-0607748, LTSA grant NAG5-13063, and HST-GO-10861 from STScI, which is operated by AURA, Inc., for NASA, under NAS5-26555.

References

Bell, E., et al. 2007, ApJ, 663, 834
Borch, A., et al. 2006, A&A, 453, 869
Conselice, C., Bershady, M. A., & Jangren, A. 2000, 529, 886
Conselice, C. J. 2006, ApJ, 638, 686
Dressler, A. 1980, ApJ, 236, 351
Gray, M., et al. 2002, ApJ, 568, 141
Jogee, S. et al. 2007, to appear in proceedings of "Formation and Evolution of Galaxy Disks", Rome, October 2007, ed. J.G. Funes, S.J. and E.M. Corsini.
Koopmann, R., & Kenney, J. 1998, ApJ, 497, 75
Wolf, C., et al. 2004, A&A, 421, 913

Radiative Feedback in the Formation of the First Protogalaxies

Jarrett L. Johnson

Department of Astronomy, University of Texas, Austin, TX, USA

Thomas H. Greif

Institut für Theoretische Astrophysik, Universität Heidelberg, Heidelberg, Germany

Volker Bromm

Department of Astronomy, University of Texas, Austin, TX, USA

Abstract. The formation of the first galaxies is influenced by the radiative feedback from the first generations of stars. Using a ray-tracing method in three-dimensional cosmological simulations, we self-consistently track the formation of, and radiative feedback from, individual stars in the course of the formation of a protogalaxy in the early universe. We follow the thermal, chemical, and dynamical evolution of the primordial gas as it becomes incorporated into the protogalaxy. While the IGM is, in general, optically thin to Lyman-Werner (LW) photons over cosmological distances, the high molecule fraction that is built up in relic H II regions and their increasing volume-filling factor renders even the local IGM optically thick to LW photons over physical distances of the order of a few kiloparsecs. Overall, we find that the local radiative feedback from the first generations of stars suppresses the star formation rate by only a factor of, at most, a few.

1. Introduction

The first galaxies form under the influence of the radiative feedback from the first generations of stars. This feedback acts to heat and ionize the gas within the H II regions surrounding the first stars, as well as to photodissociate hydrogen molecules within the larger Lyman-Werner (LW) bubbles that surround these sources. However, because the first stars (Population III; Pop III), believed to have had masses of the order of 100 $M_\odot$, lived for only ~ 3 Myr, the local radiative feedback from these stars is intermittent. Thus, atoms and molecules are often able to reform, with molecule formation being strongly promoted within the prevalent relic H II regions that the first stars leave behind, perhaps leading to subsequent star formation. We investigate this complex interaction between the radiation emitted by the first stars and the processes that characterize subsequent star formation in the assembly of the first protogalaxies.

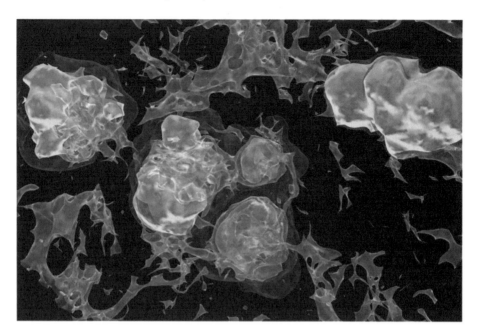

Figure 1. The chemical interplay in relic H II regions. While all molecules are destroyed in and around active H II regions, within the relic H II regions of extinct Population III stars, the high residual electron fraction catalyzes the formation of an abundance of H_2 and HD molecules. Here, plotted over the density field, the light and dark shaded bubbles denote relic H II regions, where the free electron fraction is 5×10^{-3} and 5×10^{-4}, respectively. In these regions, with radii of the order of $\sim 5\,\mathrm{kpc}$ (physical), the H_2 fraction can rise to values greater than 10^{-4}. Relic H II regions thus play an important role in subsequent star formation, allowing molecules to become shielded from photodissociating radiation and altering the cooling properties of the primordial gas.

2. Methodology

We employ the parallel version of the smoothed particle hydrodynamics (SPH) code GADGET (Springel et al. 2001) to carry out three-dimensional cosmological simulations, in which we self-consistently track the formation of, and radiative feedback from, individual stars in the formation of a protogalaxy. Our simulation evolves both the dark matter and baryonic components, initialized according to the ΛCDM model at $z = 100$. We adopt cosmological parameters close to those measured by WMAP in its first year to initialize our periodic box, which has a comoving size $L = 460\,\mathrm{kpc}\,h^{-1}$. We include an extensive chemical network for deuterium, along with the hydrogen and helium chemistry, allowing us to account for the important coolant HD.

We follow the evolution of the primordial gas until the gas collapses to our resolution limit of $n_{\mathrm{res}} \sim 20\,\mathrm{cm}^{-3}$. At this point we photoheat the high density gas, from a point source, using rates computed for a 100 $M_\odot$ primordial star. This allows us to capture the correct density profile for the gas within the minihalo hosting the star. Using a ray-tracing algorithm, we then compute in detail the H II regions of the star, as well as the regions affected by its molecule-dissociating radiation. With the thermal and chemical state of the gas in these regions adjusted accordingly, we continue the simulation, stopping to carry out this routine for each of the first eight stars that form in our box.

3. Results

In this section, we discuss a selection of our results on the evolution of the primordial gas under the influence of the radiative feedback from the first stars in the formation of a protogalaxy.

3.1. Local Radiative Feedback on Pop III Star Formation

While the formation of each Pop III star in our box is accompanied by strong photoheating and the photodissociation of molecules, each of these stars lives for only ~ 3 Myr, and upon the deaths of these stars the primordial gas cools and molecule formation begins anew. Thus, while in principle, the strong radiative feedback from Pop III stars could strongly suppress local star formation by destroying the molecules required to cool the primordial gas, we find that the intermittent nature of this local feedback leads to only a slight decrease in the Pop III star formation rate (Johnson et al. 2007a).

3.2. Shielding from Relic H II Regions

In principle, molecular hydrogen (H_2 and HD) can be easily destroyed by the LW radiation emitted by the first stars. Because these molecules are the coolants in the primordial gas which allow for star formation to take place, it is crucial to track the effects of the LW radiation from the first stars in order to follow the evolution of the primordial gas as it is incorporated into a protogalaxy. We find in our simulation that a high molecule fraction is generated within the relic H II regions left by the first stars, owing to the high residual electron fraction which persists in these regions, as shown in Figure 1. This can lead to significant self-shielding of H_2 molecules, owing to the high optical depth to LW photons through relic H II regions, as shown in Figure 2 (Johnson et al. 2007a). In turn, this hinders the build-up of a strong molecule-dissociating global background radiation field, suggesting that LW radiation from the first stars may not greatly suppress Pop III star formation (Johnson et al. 2007b).

Overall, our findings suggest that both the local and global radiative feedback from the first stars may have only a weak effect on the Pop III star formation rate during the assembly of the first galaxies.

Acknowledgments. The simulations presented here were carried out at the Texas Advanced Computing Center (TACC). We are grateful to Paul Navrátil for producing visualizations of our simulations. We acknowledge support from NSF grant AST-0708795.

References

Johnson, J. L., Greif, T. H., & Bromm, V. 2007a, ApJ, 665, 85

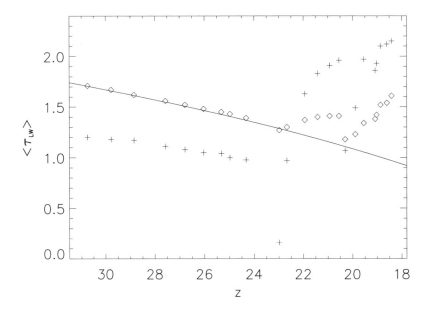

Figure 2. Optical depth to LW photons averaged over two volumes in our box, as a function of redshift z. The diamonds denote the optical depth averaged over the entire cosmological box of comoving length $460\,{\rm kpc}\,h^{-1}$, while the plus signs denote the optical depth averaged only over a cube centered on the middle of the box and containing the inner comoving $153\,{\rm kpc}\,h^{-1}$ of the box. The solid line denotes the optical depth to LW photons, averaged over the whole box, that would be expected for the case in which the H_2 fraction does not change from the primordial value of 2×10^{-6}; for this case, the optical depth changes owing only to cosmic expansion. This figure is reproduced from Johnson et al. (2007a).

Johnson, J. L., Greif, T. H., & Bromm, V. 2007b, MNRAS, submitted, astro-ph/0711.4622

Springel, V., Yoshida, N., & White, S. D. M. 2001, New Astronomy, 6, 79

Radioactive ^{30}S Beam to Study X-Ray Bursts

D. Kahl,[1] A. A. Chen,[1] J. Chen,[1] S. Hayakawa,[2] A. Kim,[3] S. Kubono,[2] S. Michimasa,[2] K. Setoodehnia,[1] Y. Wakabayashi,[2] and H. Yamaguchi[2]

Abstract. The understanding of energy generation in Type I X-ray bursts will be enhanced by a direct measurement of the nuclear reaction cross section ^{30}S(α,p)^{33}Cl. Models of this explosive stellar nucleosynthesis are constrained by the observations of bursts with multiple peaks in their bolometric luminosity. A recent model by Fisker, Thielemann, & Wiescher (2004) gives a theoretical basis for a nuclear waiting point at ^{30}S in order to explain these observations. In order to experimentally test the predictions of this model, the authors undertook two days of ^{30}S radioactive ion beam (RIB) development in December 2006 in preparation for a 2008 experiment directly measuring the $\alpha(^{30}$S,p) cross section.

1. Astrophysical Motivation

X-ray bursters (XRBs) are binary stars with an accreting neutron star. Thermal instabilities in the neutron star atmosphere lead to recurrent runaway thermonuclear explosions ($T \sim 10^9$ K) which are observed as a sharp rise in the bolometric luminosity followed by a thermal decay. Several observed bursts have exhibited multi-peaked bolometric luminosity profiles (Sztajno et al. 1985; van Paradijs et al. 1986; Penninx et al. 1989; Kuulkers et al. 2002), and none of the numerous explanations have yielded irrefutable experimental predictions.

Recently, Fisker, Thielemann, & Wiescher (2004), using state-of-the-art 1D hydrodynamic modeling, proposed a theory which accounts for these double-peaked bursts and relies only on the nuclear physics involved. It was inferred that a bottleneck may impede the nuclear burning pathway at ^{30}S, resulting in the observed drop in luminosity until the waiting point may be bypassed (see Figure 1). The Fisker et al. calculations indicate that by varying the Hauser-Feshbach ^{30}S(α,p) reaction rate by a factor of 100, the double-peaked structure of modeled XRBs is significantly diminished. The aim of our research is to experimentally test the predictions of the Fisker et al. theory by performing the first direct measurement of the $\alpha(^{30}$S,p) reaction and comparing it to the theoretical model reaction rate.

---

[1]Department of Physics & Astronomy, McMaster University, Hamilton, Ontario, Canada

[2]Center for Nuclear Study, University of Tokyo, Wako Branch at RIKEN, Wako, Saitama, Japan

[3]Department of Physics, Ewha Womans University, Seoul, Korea

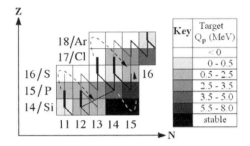

Figure 1. N vs. Z diagram near ^{30}S, for Z=16, N=14, $t_{\frac{1}{2}} \simeq 1.2\,$s (Fisker, Thielemann, & Wiescher 2004). Black lines indicate strength of reaction flow for (p,γ), (γ,p), β^{+} and (α,p), except for thick vertical lines indicating (p,γ)-(γ,p) equilibrium. Grey-scale indicates the energy released by proton capture (Q_p). Models indicate that XRB nucleosynthesis must pass through ^{30}S.

2. Beam Development and Experimental Setup

The experimenters undertook two days of ^{30}S radioactive ion beam development in 2006 at the Center for Nuclear Study RIB (CRIB) facility of the University of Tokyo (Kubono et al. 2002) using the ^{3}He(^{28}Si,n)^{30}S reaction (Bohne et al. 1982) to produce a ^{30}S beam for a future measurement of the $\alpha(^{30}$S,p)^{33}Cl reaction cross section. A ^{28}Si beam impinged on a cryogenic ^{3}He gas target, and the reaction products were separated by the CRIB (see Figure 2). Particles were grouped based on charge-to-mass ratio ($\frac{A}{Q}$) after they traversed the magnetic fields in the dipoles. Thus, by considering time-of-flight versus energy, we extracted particle identification (PID) (see Figure 3). We optimized magnetic field settings of D1 & D2 for the nucleus and charge state of interest, and we inserted a slit at F1 to block particles with differing $\frac{A}{Q}$. We then measured the beam purities and intensities for two charge states (^{30}S$^{16+,15+}$) in the experimental scattering chamber (F3).

The future experiment will measure the $\alpha(^{30}$S,p)^{33}Cl cross section using the thick-target method by bombarding a havar foil windowed He-filled gas cell with the ^{30}S RIB. Detectors in F3 will give event-by-event trajectory and PID for beam particles upstream from the He gas target using two microchannel plates (MCP). We will extract energy and angular data for reaction protons downstream of the He gas target with an array of three ΔE-E silicon telescopes. Reactions of interest will be reconstructed by coincidence of upstream ^{30}S nuclei with downstream protons, allowing us to veto background events. The He gas cell pressure will be set based on the mean energy of incident ^{30}S particles to scan energies in the center-of-mass system corresponding to the Gamow window at $T = 0.4$-$1.2 \cdot 10^{9}$ K (E_{beam}=6.8-23 MeV) for this reaction; the cross section measurement will correspond exactly with XRB temperatures during the rp-process. Kinematic calculations using the theoretical reaction rate of ^{30}S(α,p) indicate that 10^{5} particles per second (pps) at 50% purity are required to see 200 events over the course of 250 hours of beam-time. In order to determine the

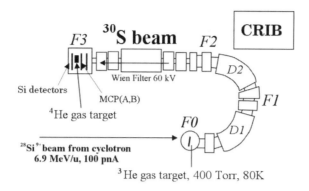

Figure 2. The ^{3}He(^{28}Si,n)^{30}S reaction products are sent through a dispersive magnetic dipole (D1), followed by a doubly achromatic dipole (D2). The beam is then sent through a Wien (velocity) filter before entering F3, the experimental scattering chamber.

efficiency of our detector system, we are planning to do a calibration run using the previously measured $\alpha(^{28}$Si,p)^{31}P reaction (Buckby & King 1984).

3. Results and Future Plans

A ^{30}S^{16+} beam of 1.8 x 10^3 pps at 10% primary beam intensity (10 particle nano-ampere) arrived at the He gas target location with an energy of 11.3 ± 1.8 MeV and a purity of 13% (Kahl et al. 2007). Although the prospects to increase the purity of the RIB appeared very promising, we chose to scan the charge state ^{30}S^{15+} due to time constraints and the consideration that the beam intensity was two orders of magnitude lower than desired. In the remaining beam-time, we developed a ^{30}S^{15+} beam of 3 x 10^3 pps at lower primary beam intensities (2 particle nano-ampere) with an energy 9 MeV, but due to contamination from ^{28}Si^{14+}, which has the same $\frac{A}{Q}$ value as ^{30}S^{15+}, the purity did not exceed 2.1%.

During the beam development run, the ^{30}S beam purity was measured at F3 (Kumagai et al. 2001) using consecutive parallel plate avalanche counters (PPAC). Since the run, MCP detectors have been tested and installed at F3, and ^{30}S particles will lose less energy traversing the new detection system, increasing the beam energy on target. The Wien filter has since attained ~50% higher voltages, which will increase the RIB purity if these conditions can be maintained throughout the experiment. Finally, we are investigating the use of a higher energy ^{28}Si primary beam, which would increase the energy and yield of ^{30}S particles. A second RIB development run is planned for early 2008 to verify the effectiveness of these proposed changes. Even without a substantial increase in beam intensities, in 2008 our group looks forward to collecting the first direct experimental data of the $\alpha(^{30}$S,p) cross section. If the ^{30}S(α,p) cross section is 10^2 above the Hauser-Feshbach prediction, it is inconsistent with the rate adopted in the Fisker et al. model, and our present sensitivity allows us to directly measure such a cross section.

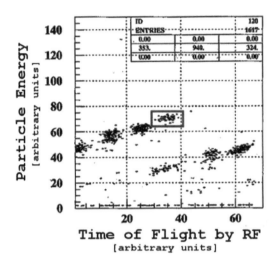

Figure 3. Sample on-line PID spectrum at F2. ^{30}S^{16+} ions are enclosed in the solid box, with a purity of 7.4%. The abscissa data is obtained by comparing the cyclotron RF signal with the triggering of the PPAC at F2, and the ordinate data is from the SSD at F2. After PID, these detectors are removed from the beam-line.

Acknowledgments. This experiment was made possible through the CNS and RIKEN collaboration. The McMaster University group is appreciative of funding from the National Science and Engineering Research Council of Canada.

References

Bohne, W. et al. 1999, Nuc. Inst. & Meth. A, 425, 1
Buckby, M. A. & King, J. D. 1984, Can. J. Phys., 62, 134
Fisker, J. L., Thielemann, F.-K., & Wiescher, M. 2004, ApJ, 608, L61
Kahl, D. et al. 2007, CNS Annual Report 2006, 1, 1
Kubono, S. et al. 2002, Eur. Phys. J. A, 13, 217
Kumagai, H. et al. 2001, Nuc. Inst. & Meth. A, 470, 562
Kuulkers, E., Homan, J., van der Klis, M., Lewin, W.H.G., & Méndez, M. 2002, A&A, 382, 947
Penninx, W., Damen, E., van Paradijs, J., & Lewin, W. H. G. 1989, A&A, 208, 146
Sztajno, M., van Paradijs J., Lewin, W. H. C., & Trümper, J. 1985, ApJ, 299, 487
van Paradijs, J. et al. 1986, MNRAS, 221, 617

The Cluster-Merger Shock in 1E 0657-56: Faster than a Speeding Bullet?

Jun Koda,[1] Miloš Milosavljević,[2] Paul R. Shapiro,[2] Daisuke Nagai[3] and Ehud Nakar[3]

Abstract. The merging galaxy cluster 1E 0657-56, known as the "bullet cluster," is one of the hottest clusters known. The X-ray emitting plasma exhibits bow-shock-like temperature and density jumps. The segregation of this plasma from the peaks of the mass distribution determined by gravitational lensing has been interpreted as a direct proof of collisionless dark matter. If the high shock speed inferred from the shock jump conditions equals the relative speed of the merging CDM halos, however, this merger is predicted to be such a rare event in a ΛCDM universe that observing it presents a possible conflict with the ΛCDM model.

We examined this question using high resolution, 2D simulations of gas dynamics in cluster collisions to analyze the relative motion of the clusters, the bow shock, and the contact discontinuity, and relate these to the X-ray data for the bullet cluster. We find that the velocity of the fluid shock need not equal the relative velocity of the CDM components. An illustrative simulation finds that the present relative velocity of the CDM halos is 16% lower than that of the shock. While this conclusion is sensitive to the detailed initial mass and gas density profiles of the colliding clusters, such a decrease of the inferred halo relative velocity would significantly increase the likelihood of finding 1E 0657-56 in a ΛCDM universe.

1. Introduction

We use gas dynamical simulations of the nearly head-on collision of two unequal-mass clusters to model the "bullet cluster" 1E 0657-56, to show that the high shock velocity of $4700 \, \text{km s}^{-1}$ inferred from X-ray observations (Markevitch 2006; Markevitch & Vikhlinin 2007) may exceed the relative velocity of the merging halos. Previous estimates of the extremely small probability of finding such a high-velocity cluster-cluster collision in a ΛCDM universe assumed that the shock and the halo-collision velocities were the same (Hayashi & White 2006; Farrar & Rosen 2007). Our results, first described in Milosavljević et al. (2007), significantly increase the likelihood of observing such a merger event in ΛCDM.

---

[1]Department of Physics, University of Texas, Austin, TX, USA

[2]Department of Astronomy, University of Texas, Austin, TX, USA

[3]Theoretical Astrophysics, California Institute of Technology, Pasadena, CA, USA

2. Simulation

We used the adaptive mesh-refinement (AMR) ASC FLASH code (Fryxell et al. 2000) in 2D cylindrical coordinates to simulate a region 20 Mpc across, with 1 kpc spatial resolution. Gravity is contributed by two rigid, spherical DM halo profiles, whose centers free-fall toward each other from rest at separation 4.6 Mpc. The density associated with the gravitational force is a Navarro-Frenk-White (NFW) profile $\rho = \rho_0 (r/r_s)^{-1}(1 + r/r_s)^{-2}$ for the subcluster, and a cored profile $\rho = \rho_0 (1 + r/r_s)^{-3}$ for the main cluster, to imitate a non-head-on merger. The masses $(M_{500}/10^{15} M_\odot)$ and radii r_{500} and r_s (Mpc) are $(1.25, 1.5, 0.5)$ and $(0.25, 0.5, 0.16)$ for the main cluster and subcluster, respectively.

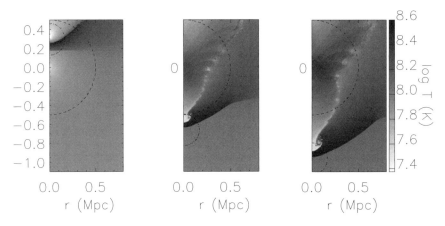

Figure 1. Time evolution of temperature. The larger CDM halo is centered at cylindrical coordinate $(r, z) = (0, 0)$; smaller halo is at $z = 0.29$, -0.66 and -0.96 Mpc, from left to right respectively. Circles indicate the scale radii r_s of the halos.

3. Results

After the subhalo "bullet" passes through the center of primary halo, the subhalo gas trails the subhalo DM, led by a bow shock and contact discontinuity (cold front), as seen in Figure 1. The opening angle and radius of curvature of the bow shock are sensitive to simulation details, but both are larger than expected for steady-state bow shocks of hard spheres with constant velocity in a uniform medium. The wings of the contact discontinuity are Kelvin-Helmholtz unstable. During pericenter passage, the shock and contact discontinuities are slower than the relative velocity of the two CDM halos, due to ram pressure force acting on the gas. Later, the halos climb out of the gravitational potential well and decelerate, but the shock and contact discontinuity do not decelerate (Fig. 2a). At an observed separation D = 720 kpc, the velocity of the shock is 4800 km s^{-1}, consistent with the 4740 km s^{-1} inferred from 1E 0657-56 data. The relative velocity of the simulated halos then is much less, 4050 km s^{-1} (16% smaller).

The X-ray surface brightness and temperature profiles of the simulation viewed transverse to collision axis (Fig. 3) agree with those of 1E 0657-56.

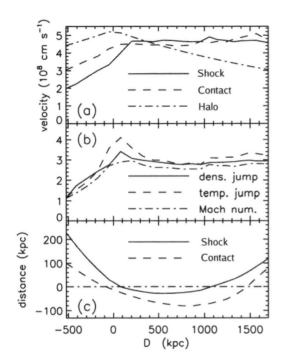

Figure 2. (a) The velocity of the shock $V_{\rm shock}$ (solid line), the contact discontinuity $V_{\rm contact}$ (dashed line), relative to the preshock gas upstream, and the relative velocity of the CDM halos V_{12} (dash-dot line), all as functions of time. The distance D between the two CDM halos is used as the coordinate of time. (b) The density and temperature jump at the shock and the shock Mach number. (c) The position of the shock (solid line) and the contact discontinuity (dashed line) relative to the subhalo center. Figures 2 and 3 are taken from Milosavljević et al. (2007).

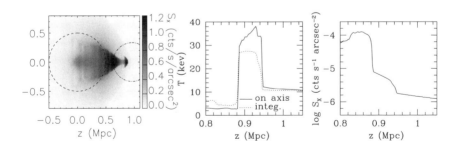

Figure 3. (Left:) Surface brightness map of the simulated cluster in the $0.8 - 4\,{\rm keV}$ band. The circles indicate the scale radii r_s of the CDM halos. (Middle:) Temperature profile along collision axis (solid) and emissivity-weighted temperature, averaged over the (transverse) line of sight (dotted). (Right:) Surface brightness profile along collision axis.

Figure 4 shows the thermal X-ray spectrum of the simulated merging cluster. The observed spectrum of 1E 0657-56 has stronger emission above 30 keV, suggesting some additional non-thermal emission (Petrosian et al. 2006).

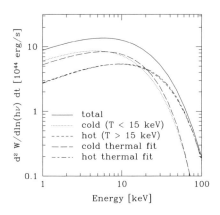

Figure 4. Thermal bremsstrahlung emission spectrum from simulated merging cluster: the contribution from cold gas below 15 keV (dotted) and hot gas above 15 keV (short-dash), total spectrum (solid), which is the sum of cold and hot components, and single-temperature fit to cold (long-dashed) and hot (mixed dash) components. The emission from the hot shock-heated gas ($T > 15$ keV) dominates the spectrum above 20 keV and is fitted well by a single temperature profile with $T = 20$ keV.

4. Discussion and Conclusions

Our simulations of merging galaxy clusters show that the halo collision velocity need not be the same as the intergalactic gas shock velocity. Applying our correction to the speed of the subhalo, the probability that such a massive cluster in ourCDM universe has such a high-speed subhalo increases by 3 orders of magnitude relative to previous estimates, but is still small ($\sim 10^{-4}$). Springel & Farrar (2007) independently reached a similar conclusion, although their relative halo velocity is much smaller, $2600\,\mathrm{km\,s^{-1}}$. They assumed perfectly head-on collision between a cuspy main halo and a more concentrated subhalo than ours, which would make the subhalo decelerate more before reaching the observed separation. More recent simulations by Mastropietro & Burkert (2007) show that the difference between the shock velocity and the halo relative velocity is smaller if the initial halo collision velocity is larger and if the collision is not perfectly head-on.

References

Farrar, G. R., & Rosen, R. A. 2007, Physical Review Letters, 98, 171302
Fryxell, B., et al. 2000, ApJS, 131, 273
Hayashi, E., & White, S. D. M. 2006, MNRAS, 370, L38
Markevitch, M. 2006, The X-ray Universe 2005, 604, 723
Markevitch, M., & Vikhlinin, A. 2007, Phys. Rep., 443, 1
Mastropietro, C., & Burkert, A. 2007, astro-ph/0711.0967
Milosavljević, M., Koda, J., Nagai, D., Nakar, E., & Shapiro, P. R. 2007, ApJ, 661, L131
Petrosian, V., Madejski, G., & Luli, K. 2006, ApJ, 652, 948
Springel, V., & Farrar, G. R. 2007, MNRAS, 380, 911

Beryllium Abundances in Solar Mass Stars

Julie A. Krugler[1] and Ann M. Boesgaard[2]

[1] *Department of Physics and Astronomy & Joint Institute for Nuclear Astrophysics, Michigan State University, East Lansing, MI, USA*
[2] *Institute for Astronomy, University of Hawaii, Honolulu, HI, USA*

Abstract. Light element abundance analysis allows for a deeper understanding of the chemical composition of a star beneath its surface. Beryllium provides a probe down to 3.5×10^6 K, where it fuses with protons. In this study, Be abundances were determined for 52 F and G dwarfs selected from a sample of local thin disc stars. These stars were selected by mass to range from 0.9 to 1.1 $M_\odot$. They have effective temperatures from 5600 to 6400 K, and their metallicities [Fe/H] = -0.65 to $+0.11$. The data were taken with the Keck HIRES instrument and the Gecko spectrograph on the Canada France Hawaii Telescope. The abundances were calculated via spectral synthesis and were analyzed to investigate the Be abundance as a function of age, temperature, metallicity, and its relation to the lithium abundance for this narrow mass range. Be is found to decrease linearly with metallicity down to [Fe/H] ~ -4.0 with slope 0.86 ± 0.02. The relation of the Be abundance to effective temperature is dependent upon metallicity, but when metallicity effects are taken into account, there is a spread ~ 1.2 dex. We find a 1.5 dex spread in A(Be) when plotted against age, with the largest spread occurring from 6-8 Gyr. The relation with Li is found to be linear with slope 0.36 ± 0.06 for the temperature regime of 5900-6300 K.

1. Introduction

The study of the light elements reveals the chemistry of a star beneath its surface. Lithium, Be, and B can all be used as probes to understand surface mixing down to temperatures of 2.5×10^6, 3.5×10^6, and 5×10^6 K respectively. Boron provides the deepest look into the interior of a star, but the B I, B II, and B III lines appear only in the "satellite" UV so it is impossible to observe without going above the atmosphere. Lithium is markedly easier to observe given that its resonance lines occur in the red portion of the spectrum at 6708 Å, but there are multiple sites for the creation and destruction of Li. Beryllium provides a cleaner picture as it has one mechanism of production (cosmic ray spallation) and the resonance lines of Be II are found at 3130.421 and 3131.065 Å, which is still observable with ground-based telescopes. However, the Be II features are near the atmospheric cutoff ($\sim$3000 Å), making them more difficult to observe than Li.

In recent years, there have been many studies regarding the light elements. Lambert & Reddy (2004; hereafter LR04) aimed to compile a database of Li abundances for 451 F and G thin disc stars. The stars were taken from three different surveys of Li in field stars: Reddy et al. (2003), Chen et al. (2001), and Balachandran (1990). The stars are located in the Solar neighborhood and have measured Hipparcos parallaxes. HR diagrams were created for these field stars using the parallaxes to calculate M_V. The majority of these stars were selected to be slightly evolved away from the zero age main sequence in order to calculate evolutionary ages using the Girardi et al. (2000) isochrones. The corresponding evolutionary tracks were used to determine masses.

Given that the LR04 data have a complete set of stellar parameters, along with Li abundances, they provide a good base from which to study the Be abundances. In order to elicit information regarding Be and Li abundances, we placed a one solar mass constraint on the LR04 data in our star selection, selecting 156 stars, of which we have data for 52. This study aims to ameliorate the gap in Be information and to understand better the nature of the processes which affect Be at $1\,M_\odot$.

2. Observations

A total of 52 high resolution spectra were taken over several nights using the HIRES instrument on the Keck I 10 m telescope and using the Gecko instrument on the CFHT. Both Keck and the CFHT are located on Mauna Kea at 14,000 feet elevation, which is above ~40% of the Earth's atmosphere, allowing additional throughput in the ultraviolet region.

3. Analysis and Results

We determined the abundance of Be by fitting the Be II resonance lines at 3130.421 and 3131.065 Å using the spectral synthesis software MOOG (Sneden 1973). Due to the considerable blending around the 3130.421 Å line, the stronger of the doublet, the fit of the 3131.065 Å line gave a more reliably-determined Be abundance. The line list was adopted largely from Kurucz (1993) to best fit the Be region. The Be abundance is calculated in terms of A(Be).

3.1. Trends with [Fe/H]

The stars of this sample are members of the thin disc and generally have considerably higher metallicities than those of the thick disc and halo populations. These stars are of metallicity [Fe/H] < −0.65, with two super solar metallicity stars. However, when the Be abundance is plotted against [Fe/H] in Figure 1, the metal-rich stars of this survey match the trend well with the halo stars sampled in Boesgaard et al. (in preparation). The most metal-rich star (at [Fe/H] = 0.11) in our sample is HD 199960 whose Be abundance is equivalent to that of the meteoritic value of 1.42 dex (Grevesse & Sauval 1998). This value is also the highest Be abundance in this sample, supporting the idea that A(Be) increases with [Fe/H].

3.2. Trends with Age

In order to parametrize the relation of the Be abundance with age, we placed the mass constraint at $1 \pm 0.1\,M_\odot$. Of our 52 stars, all but one are older than the 4.5 Gyr Sun. This fits well with the larger sample of LR04 stars with masses between 0.9 and $1.1\,M_\odot$. The full LR04 sample includes 451 stars, with 156

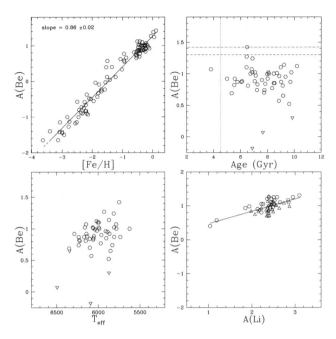

Figure 1. Be abundances from this study along with Boesgaard et al. (in preparation). The upper left panel shows that as [Fe/H] increases, A(Be) increases with a slope of 0.86 ± 0.02, shown as the dashed line. Be is plotted as a function of age (upper right). The dotted line represents the age of the Sun. Only one star is younger than the Sun. The dashed line is placed at the solar Be abundance and the dashed-dotted line represents the meteoritic Be abundance. As mentioned in the text, most of the LR04 stars are somewhat evolved off the zero age main-sequence and of the 156 stars in the mass range 0.9 to 1.1 $M_\odot$ only 6 (or 4%) are < 5 Gyr in age. Given that these stars are older than the sun and have lower Be abundances, there is evidence for slow mixing during the main sequence stage of these stars. The lower left panel shows the distribution of A(Be) with T_{eff} in this study. The upper envelope for detection is present, where higher Be abundances are detectable for cooler stars. A(Be) is plotted against A(Li) (lower right) with the Boesgaard et al. (2004) data. The triangles represent the stars from this study in the T_{eff} range 5900-6300 K; the hexagons represent the sample from Boesgaard et al. The Be and Li relation is more apparent out to larger metallicity scales and masses. The slope is 0.36 ± 0.06. This is consistent with the slope of 0.36 ± 0.05 given by Boesgaard et al.

within our mass constraints and of those 156 stars, only 22% have ages < 6 Gyr and 4% with ages < 5 Gyr. Figure 1 plots A(Be) against the age. The spread in the Be abundance is nearly 1.5 dex for the entire range of ages. It should be noted that only one star (HD 199960) is super solar in A(Be) with the rest being below the solar A(Be) value of ~ 1.3 dex (e.g., Boesgaard et al. 2004). These stars, which have evolved away from the zero age main sequence, have continued to lose Be via slow mixing.

3.3. Trends with Temperature

This solar-mass sample was selected from a survey of F and G stars in the local thin disc. This resulted in a temperature range for our stars from 5600 to 6400 K. When plotted against temperature, an upper envelope for the Be abundance becomes apparent. Higher Be abundances are present in the cooler stars, with reduced detectability for Be in the hotter stars of our sample. This wide spread is seen in the cooler stars (i.e., $T_{\rm eff} \leq 5900$ K). Figure 1 is a plot of the Be abundance against $T_{\rm eff}$. Given the metallicity dependency of Be, the correlation between [Fe/H] and A(Be) must also be considered when evaluating the temperature relation, which could substantially affect the Be-$T_{\rm eff}$ relation.

3.4. Trends with Lithium Abundance

Understanding the correlation between Be and Li is key to understanding the nature of light element depletion and slow mixing in these stars. The range in detectable Be for our data is 0.8 dex and 1.0 dex in Li. This large spread is greatest in Be at 6-8 Gyr. The median Be abundance is ∼1.0 which is low compared to the solar value of ∼1.3. This may be attributed to depletion of Be during main-sequence evolution before these stars evolved away from the main sequence. The median lithium abundance is ∼2.4, which is also below the meteoritic value of 3.3 dex. Figure 1 shows the plot of A(Be) versus A(Li).

Boesgaard et al. (2004) found a linear relationship between A(Li) and A(Be) for stars in the $T_{\rm eff}$ ranges of 5900-6300 K and 6300-6650 K with slopes of 0.365 ± 0.036 and 0.404 ± 0.034 respectively. We find a similar linear relationship in our data combined with the data of Boesgaard et al. for stars in the temperature range of $5900 < T_{\rm eff} < 6300$ K with a slope of 0.36 ± 0.06 versus 0.365 ± 0.04 in the Boesgaard et al. paper.

Acknowledgments. This research was conducted through the Research Experiences for Undergraduate (REU) program at the University of Hawaii's Institute for Astronomy and was funded by the NSF.

References

Balachandran, S. 1990, ApJ, 354, 310
Boesgaard, A. M., Armengaud, E., King, J. R., Deliyannis, C. P., & Stephens, A. 2004, ApJ, 613, 1202
Chen, Y. Q., Nissen, P. E., Benoni, T., & Zhao, G. 2001, A&A, 371, 943
Girardi, L., Bressan, A., Bertelli, G., & Chiosi, C. 2000, A&AS, 141, 371
Grevesse, N., & Sauval, A. J. 1998, Space Science Reviews, 85, 161
Lambert, D. L., & Reddy, B. E. 2004, MNRAS, 349, 757
Reddy, B. E., & Lambert, D. L. 2005, AJ, 129, 2831
Sneden, C. 1973, PhD Thesis, The University of Texas at Austin

Characterizing Barred Galaxies in the Abell 901/902 Supercluster from STAGES

I. Marinova,[1] S. Jogee,[1] D. Bacon,[2] M. Balogh,[3] M. Barden,[4]
F. D. Barazza,[5] E. F. Bell,[6] A. Böhm,[7] J. A. R. Caldwell,[1] M. E. Gray,[8]
B. Häußler,[8] C. Heymans,[9,10] K. Jahnke,[6] E. van Kampen,[4]
S. Koposov,[6] K. Lane,[8] D. H. McIntosh,[11] K. Meisenheimer,[6]
C. Y. Peng,[12,13] H.-W. Rix,[6] S. F. Sánchez,[14] A. Taylor,[15] L. Wisotzki,[7]
C. Wolf,[16] and X. Zheng[17]

[1] *Dept. of Astron., Univ. of Texas, Austin and Fort Davis, Texas, USA*
[2] *Inst. of Cosmology and Gravitation, Univ, of Portsmouth, GBR*
[3] *Department of Physics and Astronomy, Univ. Of Waterloo, CAN*
[4] *Inst. of Astro & Particle Phys., Univ. of Innsbruck, Innsbruck, AUT*
[5] *EPFL, Sauverny, SUI*
[6] *Max-Planck-Institut für Astronomie, Heidelberg, GER*
[7] *Astrophysikalisches Insitut Potsdam, Potsdam, GER*
[8] *School of Physics & Astron., Univ. of Nottingham, Nottingham, GBR*
[9] *Dept. of Physics & Astron., Univ. of Brit. Col., Vancouver, CAN*
[10] *Institut d'Astrophysique de Paris, Paris, FRA*
[11] *Dept. of Astronomy, Univ. of Massachusetts, Amherst, MA, USA*
[12] *NRC Herzberg Institute of Astrophysics, Victoria, CAN*
[13] *Space Telescope Science Institute, Baltimore, MD, USA*
[14] *Centro Hispano Aleman de Calar Alto, Almeria, ESP*
[15] *The Scottish Universities Physics Alliance (SUPA), Institute for Astronomy, University of Edinburgh, Edinburgh, GBR*
[16] *Dept. of Astrophysics, University of Oxford, Oxford, GBR*
[17] *Purple Mountain Observatory, Nanjing, CHN*

Abstract. In dense clusters, higher densities at early epochs as well as physical processes such as ram pressure stripping and tidal interactions become important and can have direct consequences for the evolution of bars and their host disks. To study bars and disks as a function of environment, we are using the STAGES ACS *HST* survey of the Abell 901/902 supercluster, along with earlier field studies based the SDSS and the Ohio State University Bright Spiral Galaxy Survey (OSUBSGS). We explore the limitations of traditional methods for characterizing the bar fraction and particularly highlight uncertainties in disk galaxy selection in cluster environments. We present an alternative approach for exploring the proportion of bars, and investigate the their properties as a function of host galaxy color, Sérsic index, stellar mass, star formation rate (SFR), specific SFR, and morphology.

1. Introduction

The most important internal driver of disk galaxy evolution are stellar bars, because they efficiently redistribute angular momentum between the disk and dark matter halo (e.g., Combes & Sanders 1981; Weinberg 1985; Athanassoula

2002). To put bars in a cosmological context, we must determine what effects environment has on bar and galaxy evolution. The bar fraction and properties in clusters relative to that found in field galaxies depend on several factors, such as the epoch of bar formation, the higher densities in clusters at early times leading to earlier collapse of dark matter halos, and the relative importance of processes such as ram pressure stripping, galaxy tidal interactions, mergers, and galaxy harassment. For example, tidal interactions can induce a bar in a dynamically cold disk, but they may also heat the disk, making it less unstable to bar formation. Previous studies have found opposing results for the bar fraction in isolated galaxies and those that are perturbed or in clusters (van den Bergh 2002; Varela et al. 2004). We use our large supercluster sample of galaxies ($\sim$2000 over M_V -15.5 to -24.0) from the STAGES survey of the Abell 901/902 supercluster ($z \sim 0.165$) to investigate the fraction and properties of bars and their host disks in a dense environment.

2. STAGES Data and Sample

To study galaxies in a dense cluster environment, we use the Space Telescope A901/902 Galaxy Evolution Survey (STAGES; Gray et al., in preparation). The Abell 901/902 supercluster ($z \sim 0.165$, number of galaxies per unit area $N = 250 \, \mathrm{Mpc}^{-2}$) consists of three clusters: A901a, A901b, and A902 with an average core separation of $\sim$ 1 Mpc. The A901 clusters show irregular X-ray morphologies, suggesting that they are not yet relaxed (Ebeling et al. 1996). The STAGES survey includes high resolution HST ACS F606W images (PSF $\sim 0.1''$, corresponding to $\sim 300 \, \mathrm{pc}$ at $z \sim 0.165$[1]) of the Abell 901/902 supercluster, along with spectrophotometric redshifts of accuracy $\delta_z/(1+z) \sim 0.02$ down to $R_{\mathrm{Vega}}= 24$ from the COMBO-17 survey (Wolf et al 2004). Multi-wavelength coverage is available for this field from $GALEX$, $Spitzer$, and XMM-Newton. Dark matter maps for the supercluster have been constructed using gravitational lensing by Gray et al. (2002) and Heymans et al. (2008, MNRAS submitted). The cluster sample contains 798 bright, $M_V \leq -18.0$, galaxies.

3. Method for Identification and Characterization of Bars

To identify and characterize the properties of bars, we employ the widely used method of fitting ellipses to the galaxy isophotes out to sky level with the iraf task 'ELLIPSE' (e.g., Friedli et al. 1996; Jogee et al. 1999; Knapen et al. 2000; Marinova & Jogee 2007). We generate plots of the surface brightness (SB), ellipticity (e), and position angle (PA) as a function of radius for each galaxy. We also plot overlays of the fitted ellipses onto the galaxy image. We use both the overlays and radial plots to identify bars. A galaxy is identified as barred if (a) the e profile rises to a global maximum while the PA stays constant and (b) after the global max, the e drops and the PA changes characterizing the disk region. To ensure reliable morphological classification, we exclude all galaxies with outer $e > 0.5$ (i $> 60°$). With our PSF of $\sim 0.1''$, ($\sim 300 \, \mathrm{pc}$ at $z \sim 0.165$), we cannot

---

[1]We assume a flat cosmology with $\Omega_M = 1 - \Omega_\Lambda = 0.3$ and $H_0 = 70 \, \mathrm{km \, s^{-1} \, Mpc^{-1}}$.

reliably detect bars with diameter smaller than ∼1.8 kpc. However, such small bars are usually nuclear bars, whereas we focus only on primary bars, which have diameters greater than 2 kpc. When working in the rest-frame optical, bars heavily obscured by dust and SF will be missed, while such bars can be detected in the rest-frame NIR (Eskridge et al. 2000; Knapen et al. 2000; Marinova & Jogee 2007). Furthermore, partially obscured bars can be missed in ellipse-fits because dust and SF along the bar can cause the PA to vary marginally more than the 10° allowed by the constant PA criterion. In fact, studies of nearby galaxies suggest that the bar fraction increases by a factor of ∼1.3 in the NIR, compared to the optical (Marinova & Jogee 2007). As we do not have rest-frame NIR images for Abell 901/902, we resort to a second method of select bars: visual classification. The visual classification method tends to capture partially obscured bars somewhat better than ellipse fits, because visual classification takes into account, not only the stellar light in the bar, but also secondary signatures, such as the shape of dust lanes, the overall morphology of the disk, and spiral arms.

4. Preliminary Analysis

All bar studies to date carried out in field samples define the bar fraction $f_{\rm bar}$ as the ratio (number of barred *disks*/ total number of *disks*). An accurate determination of $f_{\rm bar}$ therefore hinges on an accurate way to identify disk galaxies. In field samples, both locally and at intermediate redshifts, two techniques are widely used to identify disks: Sérsic cuts (n <2.5) based on single component fits, and luminosity-color cuts to isolate blue cloud galaxies from the red sequence. These methods are limited even in field samples: the Sérsic cut can miss bright disks with prominent bulges (where n >2.5) and the luminosity-color selection misses bright disks with red colors (caused by old stellar populations or dust-reddening).

In dense cluster environments, where disk galaxies can be red, and where the luminosity function is dominated by faint dwarf galaxies, it becomes even harder to identify disks via either method. We illustrate these uncertainties in disk selection in the supercluster as follows. The barred galaxies identified in § 3, from ellipse fit and visual classification, are plotted on the color-luminosity (Fig 1a) and Sérsic-luminosity (Fig 1b) planes. Because bars are disk signatures, we can use the strongly barred galaxies missed by the two methods as a lower limit on their failure to select disk galaxies. For bright galaxies, we find that 46% (45/97) and 40% (39/97) of disks with prominent, visually-identified bars are missed, respectively, by the blue cloud color-luminosity cut and Sérsic cut. It is clear that the uncertainties in disk selection will cause a large and dominant error in the optical bar fraction $f_{\rm bar}$ in clusters. We therefore adopt the following approach in the A901/902 supercluster (1) We define a new quantity $P_{\rm bar}$ as the proportion of *all galaxies* (rather than disk galaxies), which are barred. Thus, $P_{\rm bar}$ is not as heavily affected as $f_{\rm bar}$ by the uncertainties in disk selection. (2) We explore how $P_{\rm bar}$ and bar properties vary as a function of galaxy properties, such as color (Fig 1a), stellar mass (Fig 1c,d), SFR (Fig 1c), specific SFR (Fig 1d), and bulge-to-disk ratio. (3) When comparing the frequency of bars in clusters to that in the field, we can only use $f_{\rm bar}$, since no measurements of $P_{\rm bar}$ exist in the field. We estimate $f_{\rm bar}$ by selecting disks via visual classification, rather

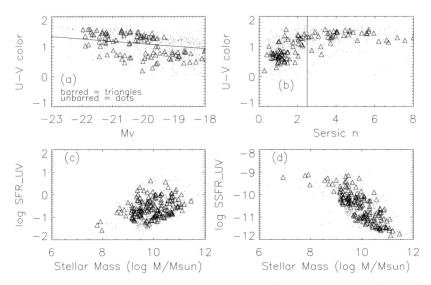

Figure 1. (a) Bright galaxies in rest-frame U-V vs. M_V plane. Those with visually-identified strong bars are marked as triangles. Many lie on the red sequence and would be missed by assuming disks lie only in the blue cloud. (b) Rest-frame U-V color vs. Sérsic n plane. Many disks with prominent bars would be missed by a Sérsic cut $n < 2.5$. (c) SFR vs. stellar mass. (d) Specific SFR vs. stellar mass. Values for red sequence galaxies are upper limits.

than from color-luminosity or Sérsic cuts. Disk are identified visually based on features such as spiral arms and stellar bars, or the presence of a bulge+disk from the light distribution.

Acknowledgments. I.M. and S.J. acknowledge support from NSF grant AST 06-07748, NASA LTSA grant NAG5-13063, as well as HST G0-10395.

References

Athanassoula, E. 2002, ApJ, 569, L83
Combes, F., & Sanders, R. H. 1981, A&A, 96, 164
Ebeling, H., et al. 1996, MNRAS, 281, 799
Eskridge, P. B., et al. 2002, ApJS, 143, 73
Friedli, D., Wozniak, H., Rieke, M., Martinet, L., & Bratschi, P. 1996, A&AS, 118, 461
Gray, M. E., et al. 2002, ApJ, 568, 141
Jogee, S., Kenney, J. D. P., & Smith, B. J. 1999, ApJ, 526, 665
Knapen, J. H., Shlosman, I., & Peletier, R. F. 2000, ApJ, 529, 93
Marinova, I., & Jogee, S. 2007, ApJ, 659, 1176
Van den Bergh, S. 2002, AJ, 124, 782
Varela, J., et al. 2004, A&A, 420, 873
Weinberg, M. D. 1985, MNRAS, 213, 451
Wolf, C., et al. 2004, A&A, 421, 913

New Horizons in Astronomy: Frank N. Bash Symposium 2007
ASP Conference Series, Vol. 393, © 2008
A. Frebel, J. R. Maund, J. Shen, and M. H. Siegel, eds.

Exploring the Impact of Galaxy Interactions over Seven Billion Years with CAS

Sarah H. Miller, [1] S. Jogee, [1] C. Conselice, [2] K. Penner, [1] E. Bell, [3]
X. Zheng, [4] C. Papovich, [5] R. Skelton, [3] R. Somerville, [3] H.-W. Rix, [3]
F. Barazza, [6] M. Barden, [7] A. Borch, [3] S. Beckwith, [8] J. Caldwell, [9]
B. Häußler, [2] C. Heymans, [10,11] K. Jahnke, [3] D. McIntosh, [12]
K. Meisenheimer, [3] C. Peng, [13,14] A. Robaina, [3] S. Sánchez, [15]
L. Wisotzki, [16] and C. Wolf [17]

[1] *Dept. of Astronomy, University of Texas at Austin, Austin, TX, USA*
[2] *School of Physics & Astron., Univ. of Nottingham, Nottingham, GBR*
[3] *Max Planck Institute for Astronomy, Heidelberg, GER*
[4] *Purple Mountain Observatory, Nanjing, CHN*
[5] *Dept. of Astronomy, University of Arizona, Tucson, AZ, USA*
[6] *Laboratory of Astroph., Federal Inst. of Technology, Lausanne, SUI*
[7] *Inst. of Astro & Particle Physics, Univ. of Innsbruck, AUT*
[8] *Dept. of Physics & Astron., J. Hopkins Univ., Baltimore, MD, USA*
[9] *McDonald Observatory, University of Texas, Austin, TX, USA*
[10] *Dept. of Physics & Astron., Univ. of Brit. Col., Vancouver, CAN*
[11] *Institut d' Astrophysique de Paris, Paris, FRA*
[12] *Dept. of Astronomy, Univ. of Massachusetts, Amherst, MA, USA*
[13] *NRC Herzberg Institute of Astrophysics, Victoria, CAN*
[14] *Space Telescope Science Institute, Baltimore, MD, USA*
[15] *Centro Hispano Aleman de Calar Alto, Almeria, ESP*
[16] *Astrophysikalisches Insitut Potsdam, Potsdam, GER*
[17] *Dept. of Astrophysics, University of Oxford, Oxford, GBR*

Abstract. We explore galaxy assembly over the last seven billion years by characterizing "normal" galaxies along the Hubble sequence, against strongly disturbed merging/interacting galaxies with the widely used CAS system of concentration (C), asymmetry (A), and 'clumpiness' (S) parameters, as well as visual classification. We analyze Hubble Space Telescope (HST) ACS images of ~4000 intermediate and high mass ($M/M_\odot > 10^9$) galaxies from the GEMS survey, one of the largest HST surveys conducted to date in two filters. We explore the effectiveness of the CAS criteria [$A > S$ and $A > 0.35$] in separating normal and strongly disturbed galaxies at different redshifts, and quantify the recovery and contamination rate. We also compare the average star formation rate and the cosmic star formation rate density as a function of redshift between normal and interacting systems identified by CAS.

1. Introduction

Galaxy mergers and interactions are believed to have a profound impact on the structural evolution and star formation activity of galaxies. The recent advent of large space-based surveys with thousands of galaxies, such as GEMS,

236 Miller et al.

GOODS, COSMOS, and AEGIS, has led to the use of different methods for identifying interacting/merging galaxies, such as detailed visual classification, and quantitative codes like CAS (Conselice et al. 2000). However, detailed assessment of the results from different methods has been altogether lacking.

2. CAS

We quantify the concentration, asymmetry, and high-spatial frequency clumpiness of the sample galaxies using the CAS concentration (C) (Bershady et al. 2000), asymmetry (A) and clumpiness (S) indices (Conselice et al. 2000). The concentration index (C) is the ratio of the two radii containing 80% and 20% of the total flux, which is then normalized logarithmically. The asymmetry index (A) is found by rotating the original image of the galaxy by 180 degrees, and then subtracting this image from the original. The residual flux of this subtracted image is then normalized by the original galaxy's flux. The clumpiness index (S) is computed by reducing the original galaxy's effective resolution to create a new image that is smoothed so that the high-frequency structure has been washed out. The original image is then subtracted from the smoothed image to produce a residual map, which then contains only the high frequency part of the original galaxy's light distribution. The flux of this light is then summed and divided by the sum of the original galaxy's flux to obtain the S index. The CAS criteria for identifying distorted/interacting systems ($A > S$ and $A > 0.35$) are empirically derived from nearby galaxies (Conselice et al. 2003).

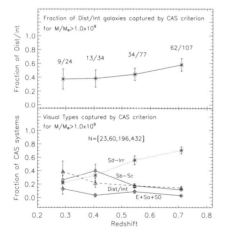
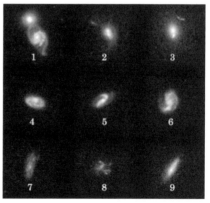

Figure 1. On the left (1a), the top panel shows, by redshift, the fraction of visually-classified, strongly distorted/interacting systems that the CAS criteria also identify as distorted/interacting. The bottom panel shows which visually-classified systems CAS is identifying as distorted/interacting. Notice the large fraction of Sd-Irr type galaxies as redshift increases, in agreement with Jogee et al. (2008). On the right (1b), galaxies 1-3: Visually classified as strongly distorted/interacting, but CAS criteria identify them as normal. Galaxies 4-9: Visually classified as normal types, but CAS criteria identify them as distorted/interacting.

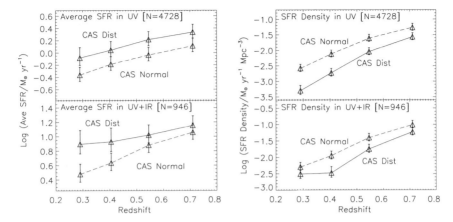

Figure 2. The average star formation rate (*Left*) and the cosmic star formation rate density (*Right*) as functions of redshift between the systems that the CAS criteria [$A > S$ and $A > 0.35$] identify as normal and identify as distorted/interacting systems. According to the CAS criteria, the severe decline in the cosmic SFR density we observe between $z \sim 0.2$–0.8 (Madau et al. 1996) is driven primarily by the shutdown of star formation in normal undisturbed systems rather than distorted/interacting systems.

3. Conclusions

The CAS criteria ($A > S$ and $A > 0.35$) captures 38% to 58% (Figure 1a) of the galaxies visually classified as interacting. This is not unexpected, given that on average, the $A > 0.35$ criterion is only satisfied for one third of the major merger timescale in N-body simulations (Conselice 2006).

We inspect the morphologically distorted galaxies, which the CAS merger criteria fail to capture and find the following. The criteria do not pick up systems where the externally-triggered tidal features (e.g., light bridges between galaxies, tidal tails, arcs, shells, ripples) and tidal debris (e.g., small accreted satellite in the main disk of a galaxy), contribute less than 35% of the total light (e.g., Figure 1b, cases 1 and 3). Also, in galaxies with close double nuclei where the center is assumed to be between the two nuclei, the resulting low A value will prevent the system from satisfying the CAS criteria (e.g., Figure 1b, case 2).

The CAS criteria suffer from contamination by relatively undistorted, so-called 'normal' galaxies. The contamination rate (i.e. fraction of 'normal' galaxies misidentified by CAS as distorted/interacting) increases with redshift (Figure 1a). This is due to several factors:

- S (clumpiness) and A (asymmetry) values are higher in systems with increased star formation. Small-scale asymmetries due to stochastic star formation can be separated from truly interacting systems by visual classification. However, CAS (designed in rest-frame optical) may identify high star-forming systems as interacting (e.g., Figure 1b, cases 4 and 6) at bluer/UV wavelengths as in $z \sim 0.6$–0.8 (Figure 1a).

- Galaxies whose outer parts look irregular (e.g., Figure 1b, cases 7 and 8), whether intrinsically or due to cosmological surface brightness dimming, can be picked by the CAS criteria.
- In systems without a clear center (e.g. dusty bright galaxies, local and distant Irrs, and distant Sds that are dimmed to resemble Irrs) the A (asymmetry) parameter can be high (e.g., Figure 1b cases 4 and 8).
- In edge-on systems and compact systems, where the light profile is steep, small centering inaccuracies can lead to large A (asymmetry) values. (e.g., Figure 1b case 9)

In Jogee et al. (2008), we report the following from visual classification. The average SFR of strongly disturbed/interacting systems is only modestly enhanced, by a factor of 2–3, with respect to normal undisturbed galaxies. Contrary to common lore, large order of magnitude enhancements in the SFR are rare in strongly disturbed systems. In fact, such systems contribute below 20% of the cosmic SFR density over $z \sim 0.2$–0.8. Thus, the decline over this regime is driven primarily by a shutdown in the star formation of relatively undisturbed galaxies. Our results are consistent with earlier findings, over a narrower redshift bin ($z \sim 0.65$–0.75), that relatively undisturbed galaxies produce most of the UV (Wolf et al. 2005) and IR (Bell et al. 2005) luminosity density.

A certain element of uncertainty and subjectivity is inherent in visual classification. However, the robustness of the results can be illustrated by an independent analysis based on the CAS system, yielding the same conclusions. In particular, we show that the cosmic SFR density is dominated by systems classified as 'Normal' by CAS, rather than as distorted (Figure 2). The interpretation here is complicated due to the fact that the CAS 'Normal' class includes a small number of distorted galaxies missed by the CAS merger criteria, while the CAS 'Distorted' class is contaminated by a number of Irregular systems. Nonetheless, it is encouraging that the same conclusions are reached with CAS.

Acknowledgments. We acknowledge support from a NASA LTSA grant NAG5-13063, NSF grant AST-0607748, as well as HST grants GO-9500 from STScI, which is operated by AURA, Inc., for NASA, under NAS5-26555.

References

Bell, E. F. et al. 2005, ApJ, 625, 23
Bershady, M. A., Jangren, A., & Conselice, C. J. 2000, AJ, 119, 2645
Conselice, C. J., Bershady, M. A., & Jangren, A. 2000, ApJ, 529, 886
Conselice, C. J., Bershady, M. A., Dickinson, M., & Papovich, C. 2003, AJ, 126, 1183
Conselice, C. J. 2006,ApJ, 638, 686
Jogee, S. et al. 2008, to appear in proceedings of "Formation and Evolution of Galaxy Disks", Rome, Oct. 2008, ed. J. G. Funes, S. Jogee and E. M. Corsini.
Madau, P. et al. 1996, MNRAS, 283, 1388
Wolf, C. et al. 2005, ApJ, 630, 771

Amplitude Limitation in Multi-Periodic Pulsating White Dwarfs

M. H. Montgomery

Department of Astronomy, University of Texas, Austin, TX, USA

Abstract. While previous investigations have focused on the effect of resonant mode coupling, we explore the consequences of a simple model of linear driving and non-linear, *non*-resonant damping of modes in pulsating white dwarfs. For an assumed spectrum of modes, we find that those with the largest growth rates have the largest final amplitudes, and that smaller growth rate modes can become damped and have final amplitudes of zero. This general trend mimics what is observed in these stars.

1. Astrophysical Context

With the exception of the Solar-like pulsators, all the observed modes in pulsating stars are believed to be linearly unstable, i.e., they have amplitudes $A(t)$ which should grow according to the following equation:

$$\frac{dA}{dt} = \gamma A \quad \Rightarrow \quad A(t) = A_0 e^{\gamma t}, \quad \text{where } \gamma > 0. \tag{1}$$

The calculation of the growth rates, γ, while non-trivial, is relatively straightforward. Theory predicts amplitudes which grow without bound, a clearly unphysical situation. What must therefore happen is that some non-linear mechanism must limit growth, e.g.,

$$\frac{dA}{dt} = \gamma A - \eta A^2 \quad \Rightarrow \quad \lim_{t \to \infty} A = \frac{\gamma}{\eta}. \tag{2}$$

Unlike the case of linear theory, however, there is no universal approach leading to the "correct" non-linear terms (e.g., the η coefficient). This situation is exacerbated for multi-mode systems, in which modes may interact with each other. These non-linear effects may well be different for different types of stars, and we must use our physical intuition to uncover the dominant physical processes for each.

The situation is somewhat more encouraging for pulsating white dwarfs. In this case there is already a proposed mechanism for non-linear energy dissipation: turbulence in a shear layer at the base of the convection zone. While this mechanism in its simplest single-mode form (Wu 1998) is at odds with the fact that longer period modes have higher amplitudes in a given star (it predicts the opposite), we show that this difficulty is resolved when we consider many modes simultaneously present in a star.

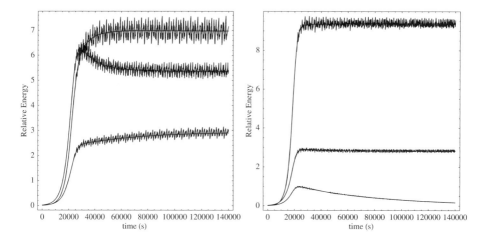

Figure 1. (left panel) The evolution of the amplitudes of modes having growth rates of 5.5, 5.0, and 4.5 $\cdot 10^{-4}$ s^{-1}, with $\eta = 10^{-5}$ s^{-1} for all modes. (right panel) The same as the left panel but with the growth rate of the largest mode increased to $7.0 \cdot 10^{-4}$ s^{-1}. This renders the smallest growth rate mode stable ("damped").

2. The Shear Layer

White dwarfs are g-mode pulsators, with horizontal displacements typically far greater than vertical ones. In general, the displacements have a fair amount of shear. Brickhill (1990) and Goldreich & Wu (1999) showed that within the surface convection zone, the rapid mixing of the convective fluid elements should virtually eliminate shear in the pulsations. Essentially, this shear should be pushed to the base of the convection zone, resulting in a velocity jump there. The energy dissipation rate (per unit area) due to the jump has the form

$$\frac{1}{A}\frac{dE}{dt} = \frac{C_D}{2}\rho\left(\Delta v\right)^3, \qquad (3)$$

where ρ is the density, Δv is the velocity jump, and C_D is a parameter describing the "roughness" of the base of the convection zone (Wu 1998). A naïve application of this assuming that modes do not interact ("single-mode" assumption) led Wu to predict that low period modes should have larger amplitudes than high period ones, despite their lower growth rates. This is contrary to what is observed in these stars (e.g., Winget et al. 1994).

In this preliminary investigation we assume that $\Delta v \sim v$, where v is the velocity of the mode at the base of the convection zone, and we further assume that the rate of change of energy of the ith mode is given by

$$\frac{dE_i}{dt} = \gamma_i \dot{x}_i^2 - \eta \dot{x}_i^2 v_{\text{tot}}^2, \qquad \text{where} \qquad v_{\text{tot}} \equiv \sum_j \dot{x}_j, \qquad (4)$$

and $E_i = \frac{1}{2}a_i^2\omega_i^2$ and x_i are the energy and displacement of the ith mode at the base of the convection zone. Since this is a first exploration, we have simplified

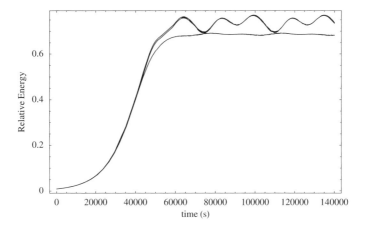

Figure 2. Evolution of an $\ell=1$ triplet with time.

the mathematics by using v_{tot}^2 in the above formula instead of $|v_{\text{tot}}|$; this allows the removal of the absolute value sign and greatly simplifies the time averaging of the equations. It also allows us to recast the equations in terms of the time-averaged energies only, i.e.,

$$\frac{dE_i}{dt} = E_i \left\{ \gamma_i - \eta \left[\frac{3}{2} E_i + \sum_{j \neq i} E_j \right] \right\}. \tag{5}$$

3. The Solutions

In Fig 1 we show the growth of three modes having unequal growth rates. The rapidly oscillating lines are exact solutions of eq 4, and the smooth lines are solutions of eq 5. The agreement between the two sets of calculations is extremely good. As expected, the highest growth rate mode is the one with the largest final amplitude, and similarly for the other modes. The right panel shows that if the disparity between growth rates is large enough (> 50% for this toy problem) then low growth rate modes can become damped. In Fig 2 we show the evolution of an $\ell = 1$ triplet with time. Since the modes have the same radial structure they have identical growth rates. Due to a resonance, the $m = \pm 1$ modes have slightly larger and more time-variable amplitudes than the $m = 0$ mode.

4. A Sample Amplitude Spectrum

Lastly, we consider a final toy model of a pulsating white dwarf. We assume that all modes are $\ell = 1$, $m = 0$ modes, and that the damping coefficient η is the same for each mode (as we have assumed previously). Following Wu, we assume a distribution of growth rates as shown in the left panel of Figure 3. With this set of assumed growth rates, we calculate a final distribution of amplitudes, shown in the right panel of Figure 3. We note that only modes with growth rates near the maximum have non-zero amplitudes. This feature helps explain

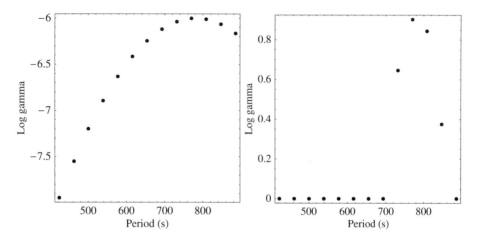

Figure 3. Assumed growth rates (left panel) and final amplitudes (right panel) for an ensemble of modes. All modes are taken to have $\ell=1$, $m=0$.

stars in which a small number of modes are seen, since it is quite likely that such stars have many modes which are linearly unstable to pulsation, but that these modes are stabilized by non-linear dissipation. In a real star, not only would the γ_i be different for each mode, but the damping coefficients η would also likely vary from one mode to another.

Finally, we note that we can resolve the discrepancy between the observations and the predictions of the "single-mode" model (Wu 1998). The "single-mode" model predicts that modes which are linearly unstable (i.e., $\gamma > 0$) should have amplitudes which scale as $\propto \text{Period}^6$, e.g., a 400 s mode should have ~ 60 times the amplitude of an 800 s mode. In the model presented here, the large growth rates of the high period modes result in the low period modes being damped. This result is in better qualitative agreement with the observations of these stars. As a next step, we will investigate the unsimplified form of equation 4, which has $|v_{\text{tot}}|$ instead of v_{tot}^2. This equation has proven to be more difficult to establish a correspondence between solutions of the averaged equations and the instantaneous ones (e.g., as was done in Figure 1), although we expect the solutions to be qualitatively similar to those obtained from equation 5.

Acknowledgments. This research was supported in part by the National Science Foundation under grant AST-0507639.

References

Brickhill, A. J. 1990, MNRAS, 246, 510
Goldreich, P. & Wu, Y. 1999, ApJ, 523, 805
Winget, D. E., Nather, R. E., Clemens, J. C., et al. 1994, ApJ, 430, 839
Wu, Y. 1998, PhD thesis, California Institute of Technology

New Horizons in Astronomy: Frank N. Bash Symposium 2007
ASP Conference Series, Vol. 393, © 2008
A. Frebel, J. R. Maund, J. Shen, and M. H. Siegel, eds.

The Bolocam 1.1 mm Galactic Plane Survey

Miranda K. Nordhaus,[1] Neal J. Evans,[1] James Aguirre,[2] John Bally,[2] Caryn Burnett,[2] Meredith Drosback,[2] Jason Glenn,[2] Glenn Laurent,[2] Richard Chamberlin,[3] Erik Rosolowsky,[4] John Vaillancourt,[5] Josh Walawender,[6] and Jonathan Williams[6]

Abstract. We have mapped approximately 150 square degrees of the Galactic Plane at 1.1 mm with Bolocam on the Caltech Submillimeter Observatory. The survey coverage includes the Galactic Center, most of the molecular ring, and many active star-forming regions such as Cygnus X and the W3/4/5 region. This is the first unbiased survey of emission from cold dust associated with massive star and cluster formation.

1. Introduction

The formation of massive stars in clustered environments is one of the central issues in astronomy (e.g., Lada & Lada 2003). Massive stars affect their environment on Galactic scales via their strong UV radiation, winds, as well as production and pollution of heavy elements. Understanding their formation mechanism through studies of massive dense cores in our own Galaxy will provide critical input into theories of galaxy formation and evolution (e.g., Evans 2005). Understanding clustered star formation will also provide insight into low-mass star formation since many, probably most, low-mass stars also form in such clustered environments.

In contrast to the situation for low-mass star formation, very little is known about the processes in high-mass star formation and our samples of objects in early stages (Massive Young Stellar Objects, or MYSOs) are incomplete (e.g., Hoare et al. 2007). This situation arises partly because most regions of massive star formation lie at large distances and are subsequently highly obscured. In addition, we have lacked complete surveys for their hosts, massive dense cores. Without the discriminator of knowing there is dense gas present, photometric searches for MYSOs have had difficulty achieving complete separation between YSOs and objects such as AGB stars and planetary nebulae (e.g., Lumsden et al. 2002).

---

[1]Department of Astronomy, The University of Texas, Austin, TX, USA

[2]CASA, University of Colorado, Boulder, CO, USA

[3]Caltech Submillimeter Observatory, Hilo, HI, USA

[4]Harvard-Smithsonian Center for Astrophysics, Cambridge, MA, USA

[5]Physics Deptartment, California Institute of Technology, Pasadena, CA, USA

[6]Institute for Astronomy, University of Hawaii, Hilo, HI, USA

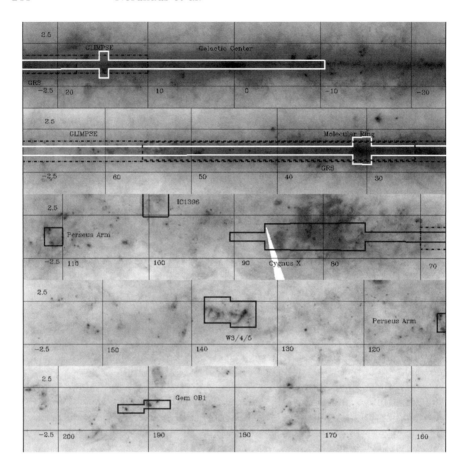

Figure 1. Coverage of te Bolocam 1.1 mm dust continuum survey as of September 2007 superimposed on the IRAS 100μm image of the Galactic Plane. The solid lines (both black and white) show the BGPS coverage. Dashed black lines show the GLIMPSE fields in the northern sky.

The first crucial step in studying massive star formation is identification of massive dense cores, the regions that are currently forming or will form massive stars and clusters. Clusters, and high-mass stars, form from very high-density cloud cores with large column densities and very high extinctions ($A_V > 100$ mag; e.g., Plume et al. 1997). Continuum emission from cold dust provides a reliable, optically thin tracer of the column density and masses on the Rayleigh-Jeans tail. Millimeter and sub-millimeter continuum emission provides a perfect tracer of the cold, dense dust in such cluster forming regions. The Bolocam Galactic Plane Survey has mapped approximately 150 square degrees of the Galactic Plane in dust continuum at 1.1 mm, providing a list of massive dense cores.

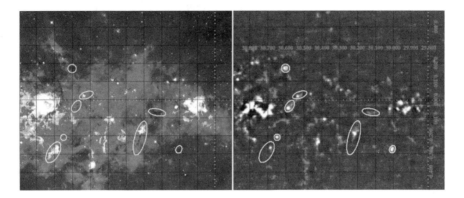

Figure 2. A comparison of the Bolocam map (right) of the $l = 30.5°$ region with the $8\,\mu$m *Spitzer*-GLIMPSE map(left). The white ovals highlight specific objects and regions in each image. Note the bright Bolocam sources that appear as Infrared Dark Clouds (IRDCs) in the $8\,\mu$m image.

As the first large-scale survey of cold molecular dust associated with star and star cluster formation we hope to answer some fundamental questions regarding star formation. Is the initial mass function (IMF) of stars and the cluster mass function (CMF) determined by inital conditions in the regions where they are born or are they the result of dynamical processes such as fragmentation and competitive accretion? Does the clump mass function resemble the IMF? Determination of the clump mass function and physical properties is a critical step in resolving many questions in star formation. With the large area covered by the BGPS we can not only determine the clump mass function, but also study how it varies with environment. We will determine the clump mass function for the Galactic center, the molecular ring, the Solar circle (Cygnus-X), and the outer Galaxy (Perseus arm).

2. The Survey

Between June 2005 and September 2007, the Bolocam Galactic Plane Survey (BGPS) observed approximately 150 square degrees of the Galactic Plane at $\lambda = 1.1$ mm with Bolocam on the Caltech Submillimeter Observatory (CSO). Continuous survey coverage extends from $l = 350°$ to $l = 87°$ with $|b| \leq 0.5°$ and includes other outer Galaxy regions such as the W3/4/5 region, Gem OB1, the Perseus Arm, and IC1396. Survey coverage is shown superimposed on IRAS 100 μm image in Figure 1. Coverage of the *Spitzer* Legacy project Galactic Legacy Infrared Mid-Plane Survey Extraordinare (GLIMPSE; Benjamin et al. 2003) is marked with the dash-dot line.

Bolocam (Glenn et al. 2003) is a 144-element bolometer array with approximately 115 working detectors. The detectors are arranged in a hexagonal grid, each with a 31 $''$ beam, resulting in 7.1$'$ instantaneous field-of-view in the focal plane. Bolocam utilizes a 45 GHz bandwidth filter centered at 268 GHz ($\lambda = 1.1\,mm$) designed to exclude the J=2-1 transition of CO at 230 GHz.

We have identified over 8,000 massive star-forming clumps in the BGPS survey with 1.1 mm fluxes above about 30 mJy/beam. Comparison with *Spitzer* and other IR data indicates that many clumps are forming massive stars and/or clusters. Some clumps are pre-stellar infrared dark clouds (IRDCs) in the earliest phases of star formation while others are more evolved and associated with compact or mature HII regions. Figure 2 shows the GLIMPSE 8 μm image and Bolocam 1.1 mm image of $l = 30.5°$. White ovals mark the same regions in each image. Such a side-by-side comparison clearly shows the range in type of sources we detect at 1.1 mm.

3. Methods and Data Reduction

Each observation consists of raster-scanning the array across either a $1° \times 1°$ or $3° \times 1°$ area at a rate of $120''$ per second. An equal number of parallel and orthogonal scans (relative to the Galactic plane) were obtained to remove striping effects caused by the scanning motion. Individual observations were aligned to each other using the brightest sources in each field. The combined maps are then registered and aligned to coordinates of masers or other sources with well-known positions. The raw data were reduced using the methods described in Laurent et al. (2005) and Enoch et al. (2006).

Upon successful completion of the data reduction and core extraction, the images and source lists will be made publicly available via the Infrared Processing and Analysis Center (IPAC) at Caltech.

References

Benjamin, R., et al., 2003, PASP, 115, 953
Enoch, M. L., et al., 2006, ApJ, 638, 293
Evans, N. J., 2005, IAU Symposium 227, Massive Star Birth: A Crossroads of Astrophysics, 443
Glenn, J., et al. 2003, in Proc. SPIE, Vol. 4855, Millimeter and Submillimeter Detectors for Astronomy, 30
Hoare, M. G., Kurtz, S. E., Lizano, S., Keto, E., & Hofner, P., 2007, Protostars and Planets V, 181
Lada, C. J., & Lada, E. A. 2003, ARAA, 41, 57
Laurent, G. T., et al. 2005, ApJ, 623, 742
Lumsden, S. L., Hoare, M. G., Oudmaijer, R. D., & Richards, D., 2002, MNRAS, 336, 621
Plume, R., Jaffe, D. T., Evans, N. J., Martin-Pintado, J., 1997, ApJ, 476, 730

The Brightest Serendipitous X-ray Sources in ChaMPlane

Kyle Penner,[1] Maureen van den Berg,[2] JaeSub Hong,[2] Silas Laycock,[3] Ping Zhao,[2] and Jonathan Grindlay[2]

[1] *Department of Astronomy, University of Texas, Austin, TX, USA*
[2] *Harvard–Smithsonian Center for Astrophysics, Cambridge, MA, USA*
[3] *Gemini Observatory, Hilo, HI, USA*

Abstract. The *Chandra* Multiwavelength Plane (ChaMPlane) Survey is a comprehensive effort to constrain the population of accretion-powered and coronal low-luminosity X-ray sources ($L_X \lesssim 10^{33}$ erg s^{-1}) in the Galaxy. ChaMPlane incorporates X-ray, optical, and infrared observations of fields in the Galactic Plane imaged with *Chandra* in the past six years. We present the results of a population study of the brightest X-ray sources in ChaMPlane. We use X-ray spectral fitting, X-ray lightcurve analysis, and optical photometry of candidate counterparts to determine the properties of 21 sources. Our sample includes a previously unreported quiescent low-mass X-ray binary or cataclysmic variable ($R = 20.9$) and ten stellar sources ($12.5 \lesssim R \lesssim 15$), including one flare star ($R = 17.3$). We find that quantile analysis, a new technique developed for constraining the X-ray spectral properties of low-count sources, is largely consistent with spectral fitting.

1. Introduction

Accretion powered binaries with compact objects include white dwarfs in cataclysmic variables (CVs) and neutron stars/black holes in quiescent low-mass X-ray binaries (qLMXBs). Space densities of CVs and qLMXBs are poorly constrained physical quantities due to small number statistics and systematic selection effects (Warner 1995). The *Chandra* Multiwavelength Plane (ChaMPlane) Survey aims to constrain the Galactic population of low-luminosity ($L_X \lesssim 10^{33}$ erg s^{-1}) accretion-powered X-ray binaries using serendipitous sources detected in six years of deep ($\gtrsim 20$ ks) *Chandra* observations. The current database includes ∼15000 X-ray sources from 122 fields. ChaMPlane will use *Chandra*'s sensitivity to detect CVs and qLMXBs beyond ∼ 1.2 kpc, the approximate limit of previous surveys; this future sample will allow us to study the spatial and luminosity distributions of CVs and qLMXBs on Galactic scales.

ChaMPlane utilizes deep optical imaging (to $R = 24$) to identify candidate source counterparts for follow-up studies. Each of the 122 *Chandra* fields is imaged in *VRI* and $H\alpha$ with the Mosaic cameras on the KPNO- and CTIO-4m telescopes. Ongoing efforts for complete source classification include optical spectroscopy and infrared imaging of heavily reddened Galactic Bulge regions. A second major goal of ChaMPlane is to study populations of stellar coronal sources. By combining observations of coronal sources with spectral identifications from ChaMPlane's optical survey, we will be able to constrain when stars develop coronae.

In this study, we determine the properties of the brightest serendipitous sources using X-ray spectral fitting, X-ray lightcurve analysis, and optical photometry of candidate counterparts (§2.). We also test the predictions of quantile analysis for the first time (§3.).

2. The bright sample

Grindlay et al. (2005) details the selection criteria for the entire ChaMPlane Survey. For this study, we apply further criteria, including a requirement that the number of net source counts in the broad (B_X, .3 – 8 keV) band $\gtrsim 250$, to confidently model and fit X-ray spectra. We exclude the inner Bulge ($l > 358°$ and $l < 2°$) to study a less crowded environment.

Our bright sample is composed of 21 serendipitous X-ray sources. Ten have candidate optical counterparts with $R \lesssim 24$; five of these have optical spectra in our current spectral database. Optical counterpart matching is in progress for three remaining sources.

2.1. The CV/qLMXB candidate

One X-ray source is well fit (reduced $\chi^2 = .94$) by the absorbed powerlaw model, with $\Gamma = 1.01^{+.14}_{-.13}$ and $N_H = .22^{+.12}_{-.11} \times 10^{22}$ cm^{-2}. The unabsorbed flux in B_X is $5.1 \pm 1.2 \times 10^{-13}$ erg cm^{-2} s^{-1}. Assuming the Galactic dust distribution of Drimmel & Spergel (2001), the N_H implies a distance of 1.3 kpc. The unabsorbed luminosity in B_X is thus 10^{32} erg s^{-1}, consistent with luminosities of both qLMXBs and CVs. The X-ray lightcurve (Fig. 1) appears periodic; it deviates from a constant count rate to 1% significance. Furthermore, the candidate optical counterpart of this source has a large Hα excess ($H\alpha - R = -.48 \pm .13$) and is bluer than many objects in the surrounding field. The ratio of unabsorbed X-ray flux to unabsorbed R flux (i.e., unabsorbed F_X/F_R) is 28.3; for comparison, most stellar coronal sources have unabsorbed $F_X/F_R \sim 10^{-3}$. This source is most likely a CV or qLMXB. Further work will help determine the precise nature of the compact object.

2.2. The remaining sample

Combining X-ray spectral fits with candidate optical counterpart photometry and spectroscopy in the same way as in §2.1., we characterize the remainder of our sample. Our preliminary results indicate the sample consists of the CV/qLMXB candidate; ten stellar sources, one of which shows an X-ray flare during the *Chandra* observation; three probable active galactic nuclei (AGN); one candidate young stellar object (YSO); and six unclassified sources. These six include four sources with optical counterparts too faint for ChaMPlane; if we assume a limiting absorbed $R \simeq 24$, two of these four sources have unabsorbed $F_X/F_R > 1$, values too high for the sources to be stellar coronae. Unabsorbed F_X/F_R for the sample is shown in Fig. 2; the ratios range from $\sim 10^{-3}$ for stellar coronal sources to $\simeq 30$ for the CV/qLMXB candidate.

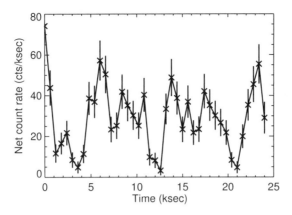

Figure 1. Background subtracted X-ray lightcurve for the CV/qLMXB candidate, binned in 600 second intervals. A K-S test on the photon arrival times shows a significant (p = .009) departure from a constant count rate.

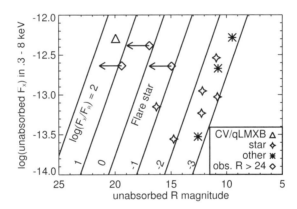

Figure 2. Unabsorbed X-ray flux against unabsorbed R magnitude for the sample. Also shown are lines of constant $\log(F_X/F_R)$. Several X-ray sources have spectra fit well using multiple models; only models with a strictly optimal reduced χ^2 are plotted. Sources for which the optical counterpart is too faint are shown with upper limits to the unabsorbed R magnitude.

3. Quantile analysis

Most X-ray sources in ChaMPlane's catalog do not have enough counts for confident spectral fitting. Quantile analysis, detailed in Hong et al. (2004), is a method of constraining the absorption and spectral parameters of low-count sources (see Fig. 3). We perform quantile analysis for the 21 bright sources using both powerlaw and thermal bremsstrahlung models. For the first time, we test the predictions of quantile analysis against spectral parameters from fitting. We find that, of the 13 sources where the X-ray spectrum is best fit by either a powerlaw or thermal bremsstrahlung model, only 2 have quantile values outside the 1σ results from spectral fitting. All sources have quantile values within the 2σ results from spectral fitting.

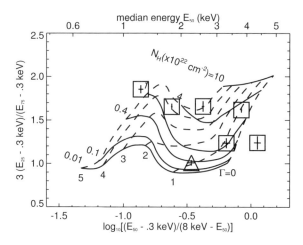

Figure 3. Quantile color-color diagram (QCCD) for the absorbed powerlaw model. Soft, unobscured sources are to the bottom-left, and hard, obscured sources are to the top-right; spectral parameters (here, Γ) and N_H are constrained by comparison with model grids. Plotted on this diagram are a few sources from our bright sample; the CV/qLMXB candidate is marked with a triangle. E_{50} is the median energy; E_{25} and E_{75} are the quartile energies.

4. Conclusions

Our population study of the brightest serendipitous X-ray sources in the ChaMPlane Survey has uncovered a previously unreported CV/qLMXB, ten stellar sources (including a flare star), three possible AGN, and one candidate YSO.

We perform quantile analysis on our sample and find that it predicts spectral parameters and absorption mostly to within the 1σ errors of X-ray spectral fitting.

We plan to further study the CV/qLMXB candidate and determine more properties (e.g. spectral types of stellar objects) of our bright X-ray sample.

Acknowledgments. We gratefully acknowledge the support of NSF grant 9731923. This work was carried out as part of the Research Experience for Undergraduates program at the Harvard-Smithsonian Center for Astrophysics.

References

Drimmel, R., & Spergel, D. N. 2001, ApJ, 556, 181
Grindlay, J. E., et al. 2005, ApJ, 635, 920
Hong, J., Schlegel, E. M., & Grindlay, J. E. 2004, ApJ, 614, 508
Warner, B. 1995, Cataclysmic Variable Stars (Cambridge: Cambridge University Press)

An Investigation of the Canis Major Over-density

W. Lee Powell Jr, Ronald J. Wilhelm, and Kenneth Carrell

Texas Tech University, Department of Physics, Lubbock, TX, USA

Abstract. In this paper we attempt to explain the origin of the Canis Major Over-density (CMa) discovered by Martin et al. (2004a). We present new photometry for various CMa fields obtained at McDonald Observatory and CTIO. We obtained spectra at McDonald Observatory of blue stars selected from the photometry. Based upon their kinematics and their preliminary abundances, we tentatively suggest that the blue plume (BP) present in the CMa fields is a feature of the Galactic disk and likely did not originate from an accreted dwarf galaxy.

1. Introduction

Martin et al. (2004a) discovered an over-density of M-giant stars in Canis Major using 2MASS colors. Their claim was that the CMa is the remnant of a disrupted dwarf galaxy. In later papers, Martin et al. (2004b, 2005) describe the kinematics of the CMa using red clump stars (RC), finding a radial velocity of $109\,\mathrm{km\,s}^{-1}$ with a low dispersion ($13\,\mathrm{km\,s}^{-1}$). They claim that the narrow radial velocity dispersion suggests that CMa is a coherent group of stars moving at similar speeds. Momany et al. (2004, 2006), however, claim that the CMa can be completely explained as due to the warp of the galactic disk. Lopez-Corredoira (2006) also invokes a warp model to explain the origin of the CMa. Moitinho et al. (2006) claim that the CMa and the Monoceros ring can both be explained using their model of the Norma-Cygnus spiral arm.

Further evidence for the dwarf galaxy remnant origin of the CMa is given by Martinez-Delgado et al. (2005) who show that an intermediate age population turnoff (the blue plume, BP) is present near $(B-V)=0.2$, bluer than the thick disk turnoff. The authors also find evidence for an older turnoff near $B = 19.5$. Recent papers by Butler et al. (2007) and De Jong et al. (2007) have analyzed the presence of this BP of young main-sequence stars in the CMa region. De Jong et al. used CMD-fitting to analyze the young and old main-sequence populations in the CMa region. Due to degeneracy between distance and metallicity, they cannot say with confidence whether the young and old populations are co-spatial. There could be a metal-rich young population at a distance of around 9 kpc, or a metal-poor young population co-spatial with the old population at 7.5 kpc. Spectroscopic observations of these young stars can resolve this question.

It is clear that if the BP is associated with a merging dwarf galaxy and is the progenitor of the red clump stars identified by Martin et al., then the BP stars should have the same kinematic and spatial signature as the RC stars. On the other hand, if the BP shows no evidence of being a dynamically cold population with a velocity centered on the RC peak, then the hypothesis that the BP has a common origin with the CMa will be compromised. There is currently no published kinematic data for the BP stars.

2. Methods

Initial photometry was obtained on the McDonald Observatory (McD) 0.8-m telescope to examine the stellar populations in CMa. Preliminary, low-resolution spectra (R∼1250) were obtained at McD using Lage Cassegrain Spectrograph (LCS) on the 2.7m telescope. Further photometry was obtained using the 0.9-m telescope at CTIO 24-29 November 2006. Sixteen $13'5$ fields were imaged in UBV under photometric conditions. The data were reduced using standard techniques, including flat field and bias corrections, using quadproc in IRAF. All photometric measurements were made using aperture photometry methods. Landolt standard fields were observed to calibrate the photometry. Typical residuals in the solution were 1% in V and B, and a few percent in U. We constructed CMDs and color-color plots (CC) from the resulting data. Since the BP is purported to be a young main sequence (De Jong et al. 2007), we fit Schmidt-Kaler (1982) ZAMS curves to the CC plots to measure the effect of differential reddening across the fields. Using the photometry, we identified over 700 BP candidate stars. We used the LCS on the McD 2.7m telescope to obtain spectra of a small sample (8 stars) of the brighter candidates as an initial test.

3. Results

Figure 1 shows the CMD and CC for a representative field - NC9. The stars observed spectroscopically were taken from NC5, centered on (l,b)=(238.72, -7.818), and NC9, centered at (l,b)=(238.70, -7.374). On the CMD and the CC, the stars with spectra are marked with error bars. The stars were selected from the top of the BP to facilitate observations at the 2.7m. There are three S-K ZAMS curves plotted. The solid line is reddened to a E(B-V) of 0.20, the dotted line to 0.33, and the dashed line to 0.50. Stars matching 0.20, 0.33, and 0.50 are plotted on the CMD and CC as diamonds, triangles, and squares, respectively. We are confident we are improving on the SDF dust value (Schlegel et al. 1998) using values obtained from the CC. Table 1 details our spectroscopic results. Figure 2 plots the radial velocity of the stars against distance. The large circle indicates the velocity of the RC found in Martin et al. . The stars marked with diamonds are from field NC5 and those marked with triangles are from NC9. The dark triangle marks a star that has an abundance of [Fe/H]=-1.39. All of the other stars are more metal rich than the Sun.

4. Conclusions

None of the stars in our spectroscopic sample have velocities consistent with the distance and radial velocity found by Martin et al. for the RC stars. The results of De Jong et al. indicate that if the BP is metal rich, it is likely to be associated with the Norma-Cygnus (NC) spiral arm which lies in the background of the CMa. Three of our spectra indicate stars located at around 10 kpc with

Table 1. Spectroscopic Results

Star	V_o	σ	$v_{\rm rad}$	dist(pc)	σ	log g	σ	$T_{\rm eff}$
9 76	14.939	0.026	−36.1	4535	797	3.98	0.44	8779
9 1041	14.622	0.040	11.7	6969	1009	3.83	0.37	9872
9 1156	14.950	0.013	−22.1	16292	1432	3.12	0.26	9875
9 1421	14.785	0.012	−34.1	5746	810	4.4	0.2	10500
5 1888	14.657	0.012	75.3	11271	991	4.33	0.2	15500
5 3092	14.967	0.011	84.8	5754	554	4.22	0.2	9764
5 2670	15.322	0.014	61.5	10680	1001	4.56	0.1	11500
5 2374	15.454	0.011	63.1	10073	1053	3.95	0.18	9840

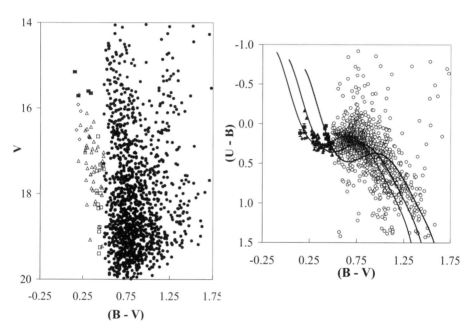

Figure 1. Color-magnitude (left) and color-color (right) diagrams of the NC9 field. The solid lines are the ZAMS at reddening values of 0.20, 0.33 ad 0.50. Error bars indicate spectroscopic targets.

a mean radial velocity of 67 km s^{-1} which is consistent with the NC arm. The remaining five spectra are harder to reconcile but the observed stars are from the top of the BP and we anticipate contamination from foreground stars that have no connection to the BP.

In light of De Jong et al., the fact that seven of our eight stars are metal rich is important. They used main sequence fitting that depends upon the metallicity of the stars. Our metal rich stars would suggest the BP is part of the NC arm. The disk has been shown to exhibit a trend toward decreasing iron abundance and increasing α-abundance with increasing radius (Carraro et al. 2007), which

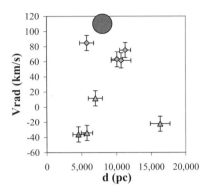

Figure 2. Radial velocity against distance. The large dot is the Martin et al. result.

would explain this result. We are still evaluating the interstellar Calcium, so this result is preliminary. Of further significance is the fact that dwarf galaxies have been shown to exhibit lowered α-abundances (Pritzl et al. 2005) which clearly is not indicated here. We tentatively conclude that the BP is unlikely to be part of an accreted dwarf galaxy and that it is most consistent with sub-structure in the disk of the Milky Way.

References

Butler, D. J., Martinez-Delgado, D., Rix, H. W., Penarrubia, J., & De Jong, J. T. A 2007, AJ, 133, 2274
Carraro, G., Geisler, D., Villanova, S., Frinchaboy, P., & Majewski, S. 2007, A&A in press (astro-ph/0709.2126)
De Long, J. T. A., Butler, D. J., Rix, H.W., Dolphin, A.E., & Martinez-Delgado, D. 2007, AJ, 662, 259
Lopez-Corredoira, M. 2006, MNRAS, 369, 1911
Martin, N. F., Ibata, R. A., Bellazzini, M., Irwin, M. J., Lewis, G.F., & Dehnen, W. 2004a, MNRAS, 348, 12
Martin, N. F. et al. 2004b, MNRAS, 355, L33
Martin, N. F. etl al. 2005, MNRAS, 362, 906
Martinez-Delgado, D., Butler, D. J., Rix, H.-W., Franco, Y. I., & Penarrubia, J. 2005, ApJ, 633, 205
Moitinho, A. et al. 2006, MNRAS, 368, L77
Momany, Y., Zaggia, R.S., Bonifacio, P., Piotto, G., De Angeli, F., Bedin, L.R., & Carraro, G. 2004, A&A, 421, L29
Momany, Y. et al. 2006, A&A, 451, 515
Pritzl, B. J., Venn, K. A., & Irwin, M. 2005, AJ, 130, 2140
Schlegel, D. J., Finkbeiner, D. P., & Davis, M. 1998, ApJ, 500, 525
Schmidt-Kaler, Th. 1982, in Landolt-Bornstein Numerical Data and Functional Relationships in Science and Technology: New Series, Vol. 2(b), 14

Signatures of Granulation in the Spectra of K-Dwarfs

I. Ramírez,[1] C. Allende Prieto,[1] D. L. Lambert,[1] and M. Asplund[2]

[1] *Department of Astronomy, University of Texas, Austin, TX, USA*
[2] *Max-Planck-Institute for Astrophysics, Garching, Germany*

Abstract. Very high resolution ($R > 150,000$) spectra of a small sample of nearby K-dwarfs have been acquired to measure the line asymmetries and central wavelength shifts caused by convective motions present in stellar photospheres. This phenomenon of granulation is modeled by 3D hydrodynamical simulations but they need to be confronted with accurate observations to test their realism before they are used in stellar abundance studies. We find that the line profiles computed with a 3D model agree reasonably well with the observations. The line bisectors and central wavelength shifts on K-dwarf spectra have a maximum amplitude of only about $200\,\mathrm{m\,s^{-1}}$ and we have been able to resolve these granulation effects with a very careful observing strategy. By computing a number of iron lines with 1D and 3D models (assuming local thermodynamic equilibrium), we find that the impact of 3D-LTE effects on classical iron abundance determinations is negligible.

1. Context

K-dwarfs are ideal for Galactic chemical evolution studies because they are not biased by stellar death (lifetimes of K-dwarfs are greater than the age of the Galaxy). These cool dwarf stars have convective envelopes and, therefore, they experience granulation (small-scale convective motions associated with temperature and density fluctuations) in their photospheres. However, classical model atmospheres of K-dwarfs (e.g., ATLAS, MARCS) are static and homogeneous, i.e., incompatible with granulation, yet they are extensively used in stellar abundance work. Granulation affects line profiles and line strengths, which are the basis for chemical abundance determinations from stellar spectra. Severe inconsistencies in K-dwarf abundance studies for key elements such as Fe and O have been reported recently in the literature (e.g., Morel & Micela 2004; Schuler et al. 2006; Ramírez, Allende Prieto, & Lambert 2007), which may originate in the inadequacy of standard spectral line-formation calculations that use classical model atmospheres.

We aim at detecting and quantifying the signatures of granulation in very high resolution spectra of K-dwarfs. The Doppler shifts introduced by the convective motions result in asymmetric absorption line-profiles whose central wavelengths are shifted with respect to their rest values (e.g., Allende Prieto et al. 2002; Dravins, Lindegren, & Nordlund 1981; Dravins 1987; Gray 1982, 2005). In contrast, classical model atmospheres predict non-shifted and perfectly symmetric lines. Furthermore, we use a state-of-the-art three-dimensional radiative-hydrodynamical model atmosphere to explore the impact of granulation on standard abundance studies of K-dwarfs. In this paper, we present preliminary results from our study.

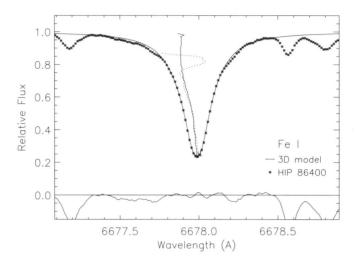

Figure 1. Fe I line profile observed in the spectrum of HIP 86400 (stars). The synthetic 3D line-profile is shown with the solid line. The observed and predicted line-bisectors (midpoints of the horizontal line segments between the wings of the line) are shown within the spectral line and have been expanded for clarity. The maximum amplitude of this bisector is only about $200\,\mathrm{m\,s^{-1}}$. Note that the observed bisector (dotted line) becomes severely affected by small blends as one approaches the continuum level. Residuals (observation−model), expanded by a factor of 2, are shown at the bottom.

2. Observations and modeling

A small sample of bright K-dwarfs has been observed with the 2dcoudé spectrograph (Tull et al. 1995) on the 2.7-m Telescope at McDonald Observatory. The wavelength coverage of these data is complete in the interval $\lambda\lambda = 5580 - 7800$ Å and the spectral resolution ($R = \lambda/\Delta\lambda$) ranges from 150,000 to 210,000. Very high signal-to-noise ratios ($S/N > 300$) were achieved by carefully coadding several exposures of the same object. Observations from the Hobby-Eberly Telescope are also being used in this study. Details on the data reduction and post-reduction processing will be given in a forthcoming publication (Ramírez et al. 2008a). Similar high quality data for a small sample of stars across the HR diagram will be analyzed in a future study (Ramírez et al. 2008b).

A three-dimensional radiative-hydrodynamical model atmosphere of parameters $T_\mathrm{eff} = 4820$ K, $\log g = 4.5$, and [Fe/H]=0 was computed using the prescription described in Stein & Nordlund (1998, see also Asplund et al. 2000). Several absorption lines were calculated using the 3D model and have been used to determine theoretical line-bisectors and central wavelength shifts (see Figures 1 and 2). In particular, in Figure 2 we show the relation between central wavelength shift and equivalent width; we used 10 Fe I lines of different wavelength

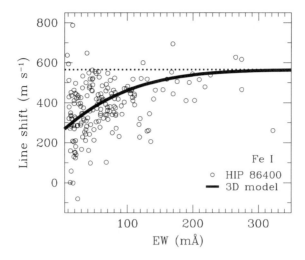

Figure 2. Central wavelength shifts for about 180 Fe I lines with well determined rest wavelengths, as measured in our high resolution spectrum of HIP 86400 (circles) and predicted by the 3D model (solid line). A K-S test shows that the distribution of points around the theoretical line corresponds to the expected random scatter with a 90% confidence level.

and EP values to construct this trend, which has an intrinsic line-to-line scatter of only $10\,\mathrm{m\,s^{-1}}$. Details will be given in Ramírez et al. (2008a).

3. Results

By looking at a few iron lines in our observed spectra we find that the theoretical line profiles predicted by the 3D model are reasonably consistent with the observations. Figure 1 shows an example of this. Furthermore, the central wavelength shifts predicted by the 3D model are in remarkable agreement with the observations, as it is shown in Figure 2 for the case of HIP 86400, our sample star with parameters closest to that of the 3D model. Although the scatter is large, the standard deviation of that distribution ($\sim 130\,\mathrm{m\,s^{-1}}$) is fully explained by errors in the dispersion solution ($\sim 25\,\mathrm{m\,s^{-1}}$), merging of spectral orders and setups ($\sim 50\,\mathrm{m\,s^{-1}}$), uncertain laboratory wavelengths ($<\sim 75\,\mathrm{m\,s^{-1}}$), and finite signal-to-noise ratios ($\sim 90\,\mathrm{m\,s^{-1}}$).

These results validate our 3D model atmosphere and allow us to use it confidently in stellar abundance determinations. Using the small set of Fe I and Fe II lines computed with the 3D model and comparing their strengths with those predicted by a standard spectrum synthesis procedure using a classical 1D Kurucz model, we find that the impact of 3D effects on the abundance of Fe is very small (see Figure 3) and it is unlikely that 3D effects will solve the problems reported by the studies cited in Section 1.

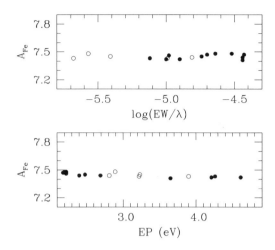

Figure 3. Here, the abundance of Fe was determined using a standard technique, i.e., ATLAS 1D model atmosphere (e.g., Kurucz 1979) and the spectrum synthesis code MOOG (Sneden 1973) with the equivalent widths predicted by the 3D model. Abundances from Fe I (filled circles) and Fe II (open circles) lines are shown as a function of reduced equivalent width (REW = $\log(EW/\lambda)$) and excitation potential (EP). The equivalent widths predicted by the 3D model were computed with a unique Fe abundance ($A_{Fe} = 7.45$). When used in the 1D analysis, the resulting abundance was $A_{Fe} = 7.45 \pm 0.02$ for Fe I lines and $A_{Fe} = 7.44 \pm 0.02$ for Fe II lines. The fact that the abundances from Fe I and Fe II lines are consistent (within 0.01 dex) and no significant trends with EP or REW are introduced in the 1D analysis suggests that the impact of 3D effects on the Fe abundance determination in K-dwarfs is negligible.

Acknowledgments. This research was supported in part by the Robert A. Welch Foundation of Houston, Texas. C. A. P.'s research is funded by NASA (NAG5-13057 and NAG5-13147).

References

Allende Prieto, C., Asplund, M., García López, R. J., & Lambert, D. L. 2002, ApJ, 567, 544
Asplund, M., Nordlund, Å, Trampedach, R., et al. 2000, A&A, 359, 729
Dravins, D., Lindegren, L., & Nordlund, Å. 1981, A&A, 96, 345
Dravins, D. 1987, A&A, 172, 211
Gray, D. F. 1982, ApJ, 255, 200
Gray, D. F. 2005, PASP, 117, 711
Kurucz, R. L. 1979, ApJS, 40, 1
Morel, T. & Micela, G. 2004, A&A, 423, 677
Ramirez, I., Allende Prieto, C., & Lambert, D. L. 2007, A&A, 465, 271
Schuler, S. C., King, J. R., Terndrup, D. M., et al. 2006, ApJ, 636, 432
Sneden, C. 1973, PhD Thesis, The University of Texas at Austin
Stein, N. F. & Nordlund, A. 1998, ApJ, 499, 914
Tull, R. G., MacQueen, P. J., Sneden, C., & Lambert, D. L. 1995, PASP, 107, 251

The Hobby-Eberly Telescope (HET) Delivers the First Ground-Based Detection of an Exoplanetary Atmosphere

Seth Redfield,[1,2] Michael Endl,[1] William D. Cochran,[1] and Lars Koesterke,[1,3]

Abstract. We present the first ground-based detection of absorption in the optical transmission spectrum caused by the atmosphere of the transiting extrasolar planet around HD189733. We combine multiple in-transit observations taken by the 9.2 m Hobby-Eberly Telescope (HET) High Resolution Spectrograph (HRS) and compare directly with out-of-transit observations. We detect a significant excess of absorption in the Na I doublet for in-transit observations, relative to the out-of-transit observations. A strong control line is also analyzed, and shows no variation between in- and out-of-transit observations. We also present empirical Monte Carlo simulations, where our large sample of out-of-transit observations are used to estimate the systematic errors involved in our analysis. With no currently operating high resolution space-based optical or ultraviolet spectrographs, ground-based facilities provide the only opportunity to probe the extrasolar atmospheric signatures in the visible waveband. Measurements, and even upper limits, of atmospheric lines should provide important constraints on models of giant exoplanet atmospheres, including for example, cloud cover altitudes, atomic and molecular composition, and temperature profiles. Full details can be found in Redfield et al. (2008).

1. Introduction

Transiting extrasolar planets provide unique opportunities to measure physical properties of exoplanets and their atmospheres that are not accessible for inclined systems. Physical characteristics of a transiting hot-Jupiter (or hot-Neptune) atmosphere can be probed by the resonance line absorption of starlight transmitted through the extended exoplanetary atmosphere during a transit. Charbonneau et al. (2002) used medium resolution spectra ($R \equiv \lambda/\Delta\lambda \sim 5{,}540$) from the Space Telescope Imaging Spectrograph (STIS) onboard the *Hubble Space Telescope* (*HST*) to detect absorption in Na I caused by the planetary atmosphere of HD209458b during transit. The measured Na I absorption was lower than model predictions by a factor of ~ 3, but adjustments to cloud height, metallicity, rainout of condensates, distribution of stellar flux, and photoionization of sodium could account for the discrepancy between the models and observations (Barman 2007; Fortney et al. 2003).

---

[1]Department of Astronomy and McDonald Observatory, University of Texas, Austin, TX, USA

[2]Hubble Fellow

[3]Texas Advanced Computing Center, University of Texas, Austin, TX, USA

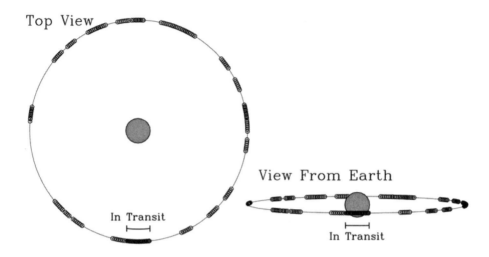

Figure 1. HD189733b is shown to scale relative to the limb-darkened host star HD189733. A planet is indicated for each exposure taken with the HET, and those taken during transit are shaded and labeled. The observations were taken over the course of a year, in which time HD189733b orbited HD1897333 ∼165 times (2.2 day orbital period; Pont et al. 2007). HD189733 experienced ∼30 rotations (12.0 day stellar rotational period; Henry & Winn 2008).

Previous ground-based attempts to detect absorption in the optical transmission spectra of HD209458 and other bright transiting systems have measured only upper limits, likely due to limited observations of only a single effective transit (e.g., Bundy & Marcy 2000; Moutou et al. 2001; Winn et al. 2004; Narita et al. 2005; Bozorgnia et al. 2006). In addition, contaminants, such as telluric lines and, in some cases, the use of an iodine cell, hindered or prevented measurements of several spectral features, most notably the Na I lines.

We describe a program to observe many transits of the bright system, HD189733, to look for absorption in the transmission spectrum. Our observations were taken over the course of a year (11 Aug 2006 to 11 Aug 2007) using the 9.2 m Hobby-Eberly Telescope (HET; Ramsey et al. 1998) High Resolution Spectrograph (HRS; Tull 1998), with a spectral range of 5000-9000 Å, and a resolving power of $R \sim 60{,}000$. No iodine cell was used. We obtained spectra during 11 in-transit visits, along with 25 out-of-transit visits, see Figure 1.

2. Detection of Transmission Absorption

The left plot of Figure 2 shows a clear detection of excess absorption in the in-transit spectrum ($\mathcal{F}_\mathrm{in}$) compared to the out-of-transit spectrum ($\mathcal{F}_\mathrm{out}$) is seen in both Na I lines. Using the same "narrow" spectral region, 5887-5899 Å, defined by Charbonneau et al. (2002), we measure a deficit of $(-67.2 \pm 7.2) \times 10^{-5}$, relative to the adjacent spectrum, which is ∼3× stronger than detected for HD209458. We were able to measure a random error of 7.2×10^{-5} comparable to that achieved by the STIS/*HST* observations, (5.7×10^{-5}).

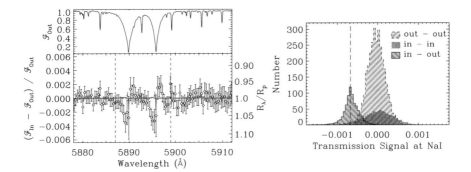

Figure 2. *Top left:* Spectrum of HD189733 near the Na I doublet that is predicted to show excess absorption in the transmission spectrum due to the atmosphere of a transiting exoplanet. *Bottom left:* The difference of the relative flux of the in-transit template ($\mathcal{F}_{\rm in}$) and the out-of-transit template ($\mathcal{F}_{\rm out}$). Absorption is clearly detected in both Na I lines. Also shown in gray is the contribution of differential limb darkening, which is minimal. *Right:* Distributions of the empirical Monte Carlo analysis, which indicates the stability of the observations and the contribution of systematic errors.

3. Empirical Monte Carlo

In order to assess the stability of our observations and contributions of systematic errors due to data reduction (e.g., continuum placement) and astrophysical conditions (e.g., stellar variability), we constructed an empirical Monte Carlo analysis. It involves randomly selecting among the hundreds of individual ten-minute exposures a test sample of "in-transit" and "out-of-transit" exposures and measuring any signal in the transmission spectrum. We repeat this exercise many thousands of times. Three different scenarios are explored: (1) an "out-out" comparison, where a subset of out-of-transit exposures is selected and compared to the rest of the out-of-transit exposures; (2) an "in-in" comparison, where a subset of in-transit exposures is compared to the rest of the in-transit exposures; and (3) an "in-out" comparison, where an increasing number of in-transit exposures are removed from the in-transit template and compared to the out-of-transit template. The distributions of transmission spectrum measurements, for all realizations of the three scenarios, are shown in Figure 2.

Both the "in-in" and "out-out" distributions are centered on zero net signal, indicating that there is no significant difference between the out-of-transit exposures, and likewise, there is no significant difference between the various in-transit exposures. Since we have fewer in-transit exposures in order to do a "in-in" comparison, the S/N is greatly reduced, and the spread in the distribution is much wider than the "out-out" distribution.

The width of the "out-out" distribution provides a powerful diagnostic of the systematic errors involved in our analysis. These realizations measure the contribution of both our reduction systematics, as well as any astrophysical systematics which are not correlated with the exoplanet orbital period. For example, if changing distributions of starspots could mimic an absorption feature in the transmission spectrum, we would expect to see a significant fraction of our "out-out" realizations provide such measurements. The standard deviation of our "out-out" distribution is adopted as the appropriate 1σ error for our transmission spectrum measurements since it encompasses both random and systematic errors. Our Na I absorption detection is $(-67.2 \pm 20.7) \times 10^{-5}$.

4. Conclusions and Future Work

We present the first ground-based detection of absorption in the transmission spectrum of a transiting extrasolar planet (for a full discussion see Redfield et al. 2008). The detection is based on the first high resolution ($R \sim 60,000$) transmission spectrum, with the Na I doublet fully resolved. The Na I absorption for HD189733b, measured in a "narrow" spectral band that encompasses both lines, is $(-67.2 \pm 20.7) \times 10^{-5}$. The error includes both random and systematic contributions. The amount of absorption in Na I due to HD189733b is $\sim 3\times$ larger than detected for HD209458b by Charbonneau et al. (2002), and indicates that the two exoplanets may have significantly different atmospheric properties. The excess absorption appears in a blue shifted component, which may be a result of high speed exoplanetary winds flowing from the hot dayside (e.g., Cho et al. 2003; Cooper & Showman 2005). Future work will include the analysis of other important optical lines (e.g., K I, Hα), detailed analysis of the absorption profile and comparison with atmospheric models, measurements of spectral indicators of stellar variability, and observations of other transiting systems.

Acknowledgments. S. R. would like to acknowledge support provided by the Hubble Fellowship grant HST-HF-01190.01 awarded by the STScI, which is operated by the AURA, Inc., for NASA, under contract NAS 5-26555. The Hobby-Eberly Telescope (HET) is a joint project of the University of Texas at Austin, the Pennsylvania State University, Stanford University, Ludwig-Maximilians-Universität München, and Georg-August-Universität Göttingen. The HET is named in honor of its principal benefactors, William P. Hobby and Robert E. Eberly.

References

Barman, T. 2007, ApJ, 661, L191
Bozorgnia, N., et al. 2006, PASP, 118, 1252
Bundy, K. A., & Marcy, G. W. 2000, PASP, 112, 1421
Charbonneau, D., Brown, T. M., Noyes, R. W., & Gilliland, R. L. 2002, ApJ, 568, 377
Cho, J. Y.-K., Menou, K., Hansen, B. M. S., & Seager, S. 2003, ApJ, 587, L117
Cooper, C. S., & Showman, A. P. 2005, ApJ, 629, L45
Fortney, J. J., et al. 2003, ApJ, 589, 615
Henry, G. W., & Winn, J. N. 2008, AJ, 135, 68
Moutou, C., et al. 2001, A&A, 371, 260
Narita, N., et al. 2005, PASJ, 57, 471
Pont, F., et al. 2007, A&A, 476, 1347
Ramsey, L. W., et al. 1998, in Proc. SPIE Vol. 3352, Advanced Technology Optical/IR Telescopes VI, ed. L. M. Stepp, 34
Redfield, S., Endl, M., Cochran, W. D., Koesterke, L. 2008, ApJ, 673, L87
Tull, R. G. 1998, in Proc. SPIE Vol. 3355, Optical Astronomical Instrumentation, ed. S. D'Odorico, 387
Winn, J. N., et al. 2004, PASJ, 56, 655

New Horizons in Astronomy: Frank N. Bash Symposium 2007
ASP Conference Series, Vol. 393, © 2008
A. Frebel, J. R. Maund, J. Shen, and M. H. Siegel, eds.

Europium, Samarium, and Neodymium Isotopic Fractions in Metal-Poor Stars

Ian U. Roederer,[1] James E. Lawler,[2] Christopher Sneden,[1] John J. Cowan,[3] Jennifer Sobeck,[4] and Catherine A. Pilachowski[5]

Abstract. We have derived isotopic fractions of europium, samarium, and neodymium in two metal-poor giants with differing neutron-capture nucleosynthetic histories. These isotopic fractions were measured from new high resolution ($R \sim 120,000$), high signal-to-noise ($S/N \sim 160$–1000) spectra obtained with the 2dCoudé spectrograph of McDonald Observatory's 2.7 m Smith telescope. Synthetic spectra were generated using recent high-precision laboratory measurements of hyperfine and isotopic subcomponents of several transitions of these elements and matched quantitatively to the observed spectra. We interpret our isotopic fractions by the nucleosynthesis predictions of the stellar model, which reproduces s-process nucleosynthesis from the physical conditions expected in low-mass, thermally-pulsing stars on the AGB, and the classical method, which approximates s-process nucleosynthesis by a steady neutron flux impinging upon Fe-peak seed nuclei. Our Eu isotopic fraction in HD 175305 is consistent with an r-process origin by the classical method and is consistent with either an r- or an s-process origin by the stellar model. Our Sm isotopic fraction in HD 175305 suggests a predominantly r-process origin, and our Sm isotopic fraction in HD 196944 is consistent with an s-process origin. The Nd isotopic fractions, while consistent with either r-process or s-process origins, have very little ability to distinguish between *any* physical values for the isotopic fraction in either star. This study for the first time extends the n-capture origin of multiple rare earths in metal-poor stars from elemental abundances to the isotopic level, strengthening the r-process interpretation for HD 175305 and the s-process interpretation for HD 196944.

Nuclides with $Z > 30$ are produced by stars through either the rapid (r)-process or the slow (s)-process. Two methods can be used to decompose the Solar System (S. S.) isotopic abundances into their constituent s- and r-process origins—the "standard" or "classical" method (e.g., Clayton et al. 1961; Cowan et al. 2006) and the "stellar" model (Arlandini et al. 1999). Isotopic abundances should be more fundamental indicators of neutron (n)-capture nucleosynthesis than elemental abundances because they can directly confront r-process and s-process predictions without the smearing effect of multiple isotopes. The wavelength of a spectral line is split by two effects, hyperfine structure (HFS) and isotope shifts, and the isotopic fractions can be measured by detailed compar-

---

[1] Department of Astronomy, University of Texas, Austin, TX, USA

[2] Department of Physics, University of Wisconsin, Madison, WI, USA

[3] Department of Physics and Astronomy, University of Oklahoma, Norman, OK, USA

[4] European Southern Observatory, Garching bei München, Germany

[5] Department of Astronomy, Indiana University, Bloomington, IN, USA

isons of an observed absorption line profile to synthetic spectra of these line substructures. A handful of previous studies have quantitatively examined the isotopic abundances of Ba and Eu in metal-poor stars, and recently Sm has been qualitatively investigated as well; no measurements of the Nd isotopic fraction have been made outside of the S. S., and no study has yet attempted to measure isotopic fractions of multiple n-capture elements in the same star.

We chose two bright, well-studied, metal-poor giants for our analysis, HD 175305 and HD 196944. The elemental abundance distributions of HD 175305 and HD 196944 suggest an r-process and an s-process origin, respectively. We acquired new observations of these stars using the 2dCoudé cross-dispersed échelle spectrograph on the 2.7 m Harlan J. Smith Telescope at the W. J. McDonald Observatory. Our data have FWHM resolving powers $R \sim 120,000$, covering $4120 \leq \lambda \leq 5890$Å in HD 175305 and $4200 \leq \lambda \leq 6640$Å in HD 196944. Image processing, order extraction, and wavelength calibration were performed using standard IRAF packages. Our final S/N values (per pixel) range from ~ 160 at 4130Å to ~ 330 at 5100Å in HD 175305 and ~ 160 at 4200Å to ~ 400 at 5000Å in HD 196944.

We re-derive atmospheric parameters for our stars that are in good agreement previous studies ($T_{\rm eff}/\log(g)/[{\rm Fe/H}]/v_t = 4870/2.15/-1.60/1.2$ for HD 175305 and $5170/1.80/-2.46/1.7$ for HD 196944). We collect the isotopes of Nd (7 naturally-occurring isotopes), Sm (7 isotopes), and Eu (2 isotopes) into the isotopic fractions shown in Table 1. Eu HFS and isotope shifts are taken from previous studies (Lawler et al. 2001; Ivans et al. 2006). We compute the full HFS and isotope shift patterns for Nd and Sm using the recent laboratory measurements (Nd: Den Hartog et al. 2003; Rosner et al. 2005) and (Sm: Masterman et al. 2003; Lawler et al. 2006; Lundqvist et al. 2007). Synthetic spectra are computed with the LTE spectrum analysis code MOOG (Sneden 1973).

Table 1. Predicted and Measured Isotopic Fractions

Species and Isotopic Fraction	Stellar f^s	f^r	Classical f^s	f^r	S. S.	HD 175305	HD 196944
Nd: $f_{142+144} =$	0.67	0.32	0.69	0.27	0.510	$0.21^{+0.56}_{-0.21}$	$0.36^{+0.64}_{-0.36}$
Sm: $f_{152+154} =$	0.21	0.64	0.20	0.64	0.495	0.51 ± 0.08	0.35 ± 0.14
Eu: $f_{151} =$	0.54	0.47	0.04	0.47	0.478	0.50 ± 0.04	

The isotopic fractions can be measured by using three methods. First, the shape of the synthetic line profile can be compared with the observations using a χ^2 fitting algorithm; this method proved most useful for the Eu and Sm lines. Second, the absolute wavelengths of the observed and synthetic spectra can by matched using wavelength standard Fe I (Learner & Thorne 1988; Nave et al. 1994) and Ti II (Pickering et al. 2001, 2002) lines, and then the peak wavelength of the rare-earth line of interest can be measured. When matching multiple wavelength standards in the same échelle order, the scatter in our measured wavelength offsets necessary to ensure the matching of each line is ~ 5 mÅ, much greater than anticipated and not precise enough to enable the isotopic fraction to be measured. The third method matches the observed and synthetic spectra of each line by using the point in the synthetic spectrum that is insensitive to the s- or r-process mix; this method is the only way we have been able to reliably assess the Nd isotopic fraction.

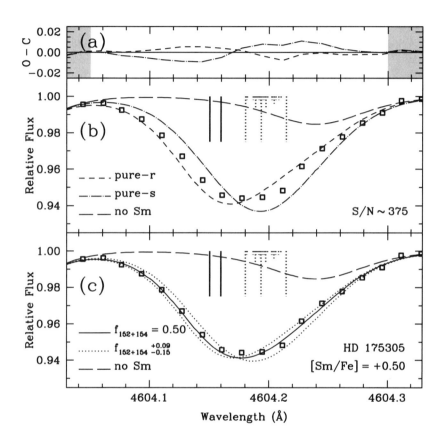

Figure 1. Our synthesis of the Sm II 4604Å line in HD 175305. The observed spectrum is indicated by open squares. In panel **a**, we show an $(O-C)$ plot for the pure-s- and pure-r-process syntheses shown in panel **b**. The unshaded region indicates the wavelengths over which we measure the isotopic fraction using the χ^2 algorithm. In panel **b**, we show syntheses for the pure-s- and pure-r-process syntheses. In panel **c**, we show the synthesis of our best fit and the 3σ uncertainties. The vertical sticks represent the positions and relative strengths of the individual hyperfine components; solid sticks represent the two heaviest isotopes, while dashed sticks represent the lighter five isotopes. Our measured Sm isotopic fraction suggests an r-process origin.

In HD 175305, our Sm isotopic fraction $f_{152+154} = 0.51 \pm 0.08$, measured from profile fits to the 4424 and 4604Å lines (see Figure 1), suggests an r-process origin. It is tempting to surmise that our Nd isotopic fraction, $f_{142+144} = 0.21^{+0.56}_{-0.21}$, derived from three lines, is suggestive of an r-process origin, but the large uncertainty cannot exclude an s-process origin. Our Eu isotopic fraction, $f_{151} = 0.50 \pm 0.04$, derived from only the 4129Å line and shown to be a reasonable match for the 4205 and 4435Å lines, implies an r-process origin by the classical method predictions but excludes neither a pure-s- nor a pure-r-process origin by the stellar model predictions. Our isotopic fractions in HD 175305 suggest a nucleosynthetic history primarily dominated by the r-process. The agreement between the elemental and isotopic abundance distributions in HD 175305 add to the preponderance of evidence (e.g., the growing number of r-process-enhanced stars that conform to the scaled-S. S. r-process elemental abundance distribution) that supports the hypothesis of a universal r-process mechanism for elements with $Z \geq 56$.

In HD 196944, our measured Sm isotopic fraction, $f_{152+154} = 0.35 \pm 0.14$, is suggestive of an s-process origin. Our Nd isotopic fraction, $f_{142+144} = 0.36^{+0.64}_{-0.36}$ can exclude no possible values of the isotopic mix. The interpretation for the Sm is consistent with the s-process elemental abundance signatures and radial velocity variations observed for this star (Lucatello et al. 2005). Our best measurements for each species are summarized in Table 1 for both HD 175305 and HD 196944.

Acknowledgments. Funding for this project has been generously provided by the U. S. National Science Foundation (grant AST 03-07279 to J. J. C., grant AST 05-06324 to J. E. L., and grants AST 03-07495 and AST 06-07708 to C. S.) and by the Sigma Xi *Grants-in-Aid-of-Research* program.

References

Arlandini, C., Käppeler, F., Wisshak, K., Gallino, R., Lugaro, M., Busso, M., & Straniero, O. 1999, ApJ, 525, 886
Clayton, D. D., Fowler, W. A., Hull, T. E., & Zimmermann, B. A. 1961, Ann. Phys., 12, 331
Cowan, J. J., Lawler, J. E., Sneden, C., Den Hartog, E. A., & Collier, J. 2006, in Proc. NASA LAW, ed. V. Kwong, (Washington: NASA), 82
Den Hartog, E. A., Lawler, J. E., Sneden, C., & Cowan, J. J. 2003, ApJS, 148, 543
Ivans, I. I., Simmerer, J., Sneden, C., Lawler, J. E., Cowan, J. J., Gallino, R., & Bisterzo, S. 2006, ApJ, 645, 613
Lawler, J. E., Wickliffe, M. E., den Hartog, E. A., & Sneden, C. 2001, ApJ, 563, 1075
Lawler, J. E., Den Hartog, E. A., Sneden, C., & Cowan, J. J. 2006, ApJS, 162, 227
Learner, R. C. M., & Thorne, A. P. 1988, J. Opt. Soc. Am. B, 5, 2045
Lucatello, S., Tsangarides, S., Beers, T. C., Carretta, E., Gratton, R. G., & Ryan, S. G. 2005, ApJ, 625, 825
Lundqvist, M., Wahlgren, G. M., & Hill, V. 2007, A&A, 463, 693
Masterman, D., Rosner, S. D., Scholl, T. J., Sharikova, A., & Holt, R. A. 2003, Can. J. Phys., 81, 1389
Nave, G., Johansson, S., Learner, R. C. M., Thorne, A. P., & Brault, J. W. 1994, ApJS, 94, 221
Pickering, J. C., Thorne, A. P., & Perez, R. 2001, ApJS, 132, 403
Pickering, J. C., Thorne, A. P., & Perez, R. 2002, ApJS, 138, 247
Rosner, S. D., Masterman, D., Scholl, T. J., & Holt, R. A. 2005, Can. J. Phys., 83, 841
Sneden, C. A. 1973, Ph.D. Thesis, Univ. Texas at Austin

New Horizons in Astronomy: Frank N. Bash Symposium 2007
ASP Conference Series, Vol. 393, © 2008
A. Frebel, J. R. Maund, J. Shen, and M. H. Siegel, eds.

Coordinated Galaxy Growth on Inner and Outer Disk Scales: Analysis of an Unusually Resonant S0 Galaxy and Its Companion

Tara A. Scarborough[1] and Sheila J. Kannappan[2]

Abstract. We present an unusual S0 galaxy with narrow, luminous, very blue inner and outer resonance rings along with a strong bar. Its dwarf companion at $\Delta X \sim 30\,\mathrm{kpc}$ and $\Delta V \lesssim 30\,\mathrm{km/s}$ is also bright blue. We perform stellar population analysis of *ugriz* photometry to test the hypothesis of coeval interaction-triggered starbursts in both systems across a wide range of scales. We also investigate the possibility of nuclear activity. The rare combination of resonances in the primary galaxy may reflect a short-lived evolutionary state that is key to understanding coordinated bulge and disk growth, possibly as part of a transformation from a classical S0 to an S0 with a pseudobulge, or a later type galaxy.

1. Introduction

Drawn from a recently identified class of E/S0s with colors typical of spirals at the same mass (Kannappan, Guie, & Baker 2008), CGCG065-002 is neither spiral nor typical S0. It may be the only known system with *narrow, intense blue rings* at both the inner 4:1 and outer Lindblad resonances (private communication; R. Buta 2006, J. Kormendy 2006). This structure could reflect either a short-lived state experienced by many galaxies or a truly unique state. We propose that the observed structure represents a passing phase that is not uncommon because the rings match canonical resonances. Furthermore, the rings appear unstable (Figure 1): we see possible break-up of the inner ring, and the outer ring appears intrinsically elongated and vulnerable to shear (deep imaging to enable deprojection is pending). The formation and subsequent broadening or dispersal of rings may contribute to the growth of both outer disks and inner-disk "pseudobulges" — rotationally supported, flattened bulges found in many galaxies (Kormendy & Kennicutt 2004). Although the central stellar velocity dispersion of our galaxy (120 km/s from SDSS; Adelman-McCarthy, et al. 2007) places it on the dynamically hot, i.e., classical-bulge, side of the stellar mass vs. velocity dispersion relation, any growth on the scale of the inner ring could help to build a larger pseudobulge over the smaller classical bulge (as seen by Erwin 2007). Here we investigate whether a companion interaction may have caused the direct onset of coupled star formation on both inner and outer disk scales in CGCG065-002. A very blue dwarf companion is located $\sim$30 kpc south of the primary, within $\sim$30 km/s in radial velocity (Fig 2). We consider if the starburst ages in the rings and companion are consistent with interaction-triggered star formation. We also test for nuclear activity in either galaxy.

---

[1]Department of Astronomy, University of Texas, Austin, TX, USA

[2]Department of Physics and Astronomy, University of North Carolina, Chapel Hill, NC, USA

 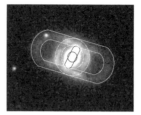

Figure 1. SDSS g-band images of the primary show strong resonance structures and possible break-up of the inner ring. Left: Brightness adjusted to show outer structures. Right: Brightness adjusted to show inner structures. White and black lines indicate the component positions determined by eye and optimized iteratively in all five bands.

2. Component Starburst Analysis

Similar starburst ages in the companion and in the two rings of the primary may indicate a common star-formation trigger. As input data for determining starburst ages, we measured *ugriz* magnitudes for the inner and outer rings, bar, bulge, and companion (Figure 1). SDSS images were finely sampled in order to induce a higher accuracy in the magnitude calculations, but the difference amounted to less than 0.02 mag for a grid 10 times finer than the original. Background subtraction was included in the measurement using three large blank areas in the image. We checked our algorithm for computing magnitudes by reproducing the SDSS fiber magnitude calculations; measurements in all bands were within 0.03 mag of SDSS values. For component colors, the systematic errors estimated by varying the size of the components are larger than the formal errors. For example, the systematic error in u-r ranges from $\lesssim 0.1$ mag (bar, companion) to $\sim$0.4-0.5 mag (both rings), with the bulge intermediate.

We fit the *ugriz* photometry for each component using a grid of Bruzual-Charlot (2003) models, as described in Kannappan & Gawiser (2007). Briefly, each model combines a pair of single bursts, with pairs spanning mass ratios 0%:100%, 2%:98%, 4%:96% ... 100%:0%. The single bursts have a broad range of age, dust, and metallicity values. Binning the fitted likelihoods of the models over age, Figure 2 reveals that the inner ring and companion have similar, young stellar-population ages within broad uncertainties while the bar and classical bulge appear older. The outer ring is also consistent with having a young (sub)population, but when we exclude the overlap with the inner ring, the age distribution favors ages more intermediate between the bar/bulge and inner ring/companion (Figure 2). In this case the age is still uncertain, however, due to the low S/N of the outer ring's u-band flux. We obtain similar results when we vary the input using larger or smaller component outlines, except that expanding the circle that defines the bulge in Figure 1 broadens its age distribution towards younger ages, consistent with a pseudobulge growth scenario.

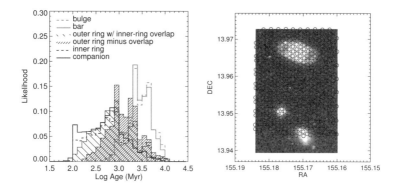

Figure 2. Left: Likelihood of mass-weighted stellar population age for the primary galaxy components and the companion. Right: VIRUS-P IFU fiber positions overlaid on the primary galaxy plus companion. Another set of positions covers the companion completely (not shown). Spatially resolved spectroscopy will enable refinement of the analysis in the left panel (pending).

In summary, the companion and the inner ring of the primary have likely experienced coordinated starbursts, while the outer ring contains a coeval or slightly older starburst population. We are working on reducing the uncertainties in this analysis by combining the *ugriz* data from SDSS with IFU spectroscopy from the VIRUS-P spectrograph (taken April 2007 at McDonald Observatory by T.A.S. and the VIRUS-P team), shown schematically in Figure 2.

3. Nuclear Activity Analysis

Fits to Hα and [NII] in the SDSS spectrum of the primary show moderately broad lines (FWHM ~ 230 km/s after quadrature subtraction of $\sigma \sim 70$ km/s instrumental broadening) and high [NII]/Hα, possibly suggesting an active nucleus. Alternatively, the broad line widths in the primary may indicate a rapidly rising rotation curve, and the high [NII]/Hα may indicate absorption. A detailed line ratio analysis using diagnostic plots of AGN and star-forming activity (Figure 3) suggests that the primary may have either a star-forming or a composite nucleus. A more conclusive result may require higher spatial-resolution spectroscopy, to minimize dilution of the nuclear spectrum by the bulge region. For the companion, Gaussian fits to Hα and [NII] show narrow line widths and an [NII]/Hα ratio consistent with star formation, with no evidence for AGN activity. Diagnostic plots (not shown) confirm this result.

4. Discussion

We believe that CGCG065-002 is the only known galaxy with narrow, intense blue rings at both the 4:1 and outer Lindblad resonances. This galaxy may be an anomaly or may represent a short-lived state that is not rare. The starburst ages in the inner ring and in the companion are consistent with a coeval starburst a few hundred Myr old, possibly involving the outer ring as well. The

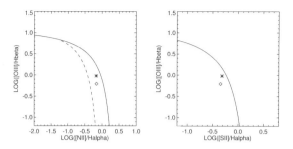

Figure 3. Diagnostic plots of emission line ratios for CGCG065-002. Asterisks show ratios measured by gaussian+continuum fitting, and diamonds show ratios measured directly from the spectrum, with the continuum level adjusted by eye to compensate for absorption. Dividing lines are taken from Kewley et al. (2006). To the right of the solid line are active LINER/AGN nuclei. To the left of the solid line are predominantly star-forming nuclei, but galaxies between the dashed and solid lines in the left panel may have composite nuclei.

proximity of the primary and companion ($\sim$30 kpc on the sky) are also consistent with an interaction occurring a few hundred Myr ago, assuming a velocity $\sim$100 km/s. New VIRUS-P IFU spectroscopy will allow for a more accurate analysis of the stellar populations, as these data can be used to model absorption line structure in each galaxy component separately. Additional data will also be needed to test conclusively for low-level nuclear activity in the primary, which is neither confirmed nor ruled out by emission line ratios measured from SDSS spectroscopy. With current and ongoing analysis of this system, we will learn more about whether and how galaxy interactions may link star formation on outer and inner disk scales, perhaps enabling coevolution of disk and (pseudo)bulge components.

Acknowledgments. We are grateful to B. Wills and S. Jogee for their scientific feedback. We thank J. Adams and the VIRUS-P team for their collaboration. This research was partially supported by an NSF Astronomy & Astrophysics Postdoctoral Fellowship to S. J. K. under award AST-0401547.

References

Adelman-McCarthy, J. et al. 2007, ApJS, 172, 634
Bruzual, G., & Charlot, S. 2003, MNRAS, 344, 1000
Erwin, P. 2007, astro-ph/0709.3793
Kannappan, S. J., Guie, J. M., & Baker, A. J. 2008, AJ, submitted
Kannappan, S. J., & Gawiser, E. 2007, ApJ, 657, L5
Kewley, L. J., Groves, B., Kauffmann, G., &, Heckman, T. 2006, MNRAS 372, 961
Kormendy, J., & Kennicutt, R. C. 2004, ARA&A, 42, 603
Oke, J.B. 1990, AJ, 99, 1621

The HST/ACS Survey of Galactic Globular Clusters: First Results

Michael Siegel

McDonald Observatory, University of Texas, Austin, TX, USA

Ata Sarajedini

Department of Astronomy, University of Florida, Gainesville, FL, USA

Brian Chaboyer, Aaron Dotter

Department of Physics and Astronomy, Dartmouth College, Hanover, NH, USA

Steven Majewski, David Nidever

Department of Astronomy, University of Virginia, Charlottesville, VA, USA

Abstract. The HST/ACS Survey of Galactic Globular Clusters has obtained F606W and F814W images of 65 clusters with the Advanced Camera for Surveys. Our scientific programs exploit both the precise ACS photometry and the proper motions derived by combining our data with the HST archives. Our initial photometric results have included: deep photometry of clusters never before surveyed with HST; measurement of differential reddening in the foreground of Palomar 2; precise measurement of the properties of the Sagittarius dSph galaxy and its associated globular clusters; examination of the NGC 1851 cluster; and the derivation of relative ages for our complete sample. The data exceed expectations and are revealing unprecedented detail in globular cluster CMDs.

Future endeavors will explore main sequence luminosity functions, derive absolute ages and measure relative and absolute proper motions. Among our secondary science goals will be a much clearer picture of the chemodynamical history of the Milky Way and a more precise measure of the Galactic potential.

1. Overview

The globular clusters of the Milky Way are astrophysical laboratories that provide insight into stellar and chemical evolution, cluster dynamics, and luminosity/mass functions. Their ensemble properties provide insight into the gravitational potential and chemodynamical history of the Milky Way.

We have begun a systematic survey of 65 globular clusters using the Hubble Space Telescope Advanced Camera for Surveys. The survey consists of a uniform deep ($V \sim 26$) imaging sample for our target clusters consisting of a single orbit each in F606W and F814W. Our new photometric reductions methods (Anderson et al. submitted) provide extraordinarily precise photometry from the images. With tens or hundreds of thousands of well-measured stars in each of 65 clusters spanning a range of properties, we are addressing issues in stellar evolution, main sequence (MS) luminosity functions, horizontal branch morphology, binary fractions, relative and absolute ages, relative and absolute proper motions, density distributions and MS distancing.

To date our published results have included:

• Color-magnitude diagrams of nine clusters previously unobserved with HST. This included the revision of distances and reddenings with MS-fitting, the derivation of a younger age for the Palomar 1 cluster and the analysis of differential reddening in the Palomar 2 cluster.

• The detection of a doubling in the subgiant branch of the NGC 1851 globular cluster (Milone et al. 2007) – a feature only detailed in our extremely precise deep sample.

• A definitive detection of young populations in the field of the M54 globular cluster, which resides in the core of the tidally disrupted Sagittarius (Sgr) dSph galaxy (Siegel et al. 2007, §2).

Final photometric and astrometric catalogs from this treasury program will be made available to the public, allowing programs to be developed beyond our current endeavors. In this contribution, we highlight two recent results focused upon the Sgr dwarf – the star formation history of the Sgr core and the three-dimensional shape of the Sgr's orbit.

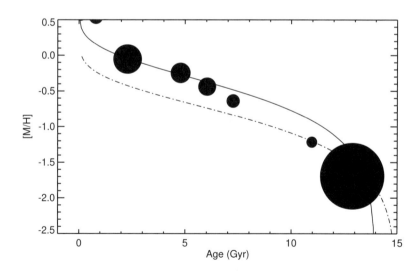

Figure 1. The age-metallicity relationship (AMR) of the M54+Sgr field. Point size indicates population strength. The dashed line is the closed-box AMR of Layden & Sarajedini (2000); the solid line is that of Siegel et al. (2007).

2. The Star Formation History of the Sagittarius dSph Galaxy

As detailed in Siegel et al. (2007), we have used the unprecedented deep and precise photometry afforded by the HST/ACS data to constrain the star formation history of the Sgr core, within which the M54 globular cluster resides. Through StarFISH population synthesis (Harris & Zaritsky 2001), we reproduced the observed color-magnitude diagram (see Figure 1 of Siegel et al. 2007).

Figure 1 shows the derived star-formation history of the M54 field. The large old metal-poor population of the globular cluster dominates the field, as expected. However, the young (1.75 Gyr) population of Sgr is plainly visible as is a wide range of the intermediate populations that are known to dominate the larger Sgr tidal stream.

3. The Three-Dimensional Shape of the Sagittarius Tidal Stream

Constraining the distance and three-dimensional shape of the Sgr tidal debris stream is of critical importance to models of its disruption (Law et al. 2005). The Sgr debris stream, however, is both diffuse and faint, making measurement of its three-dimensional shape difficult.

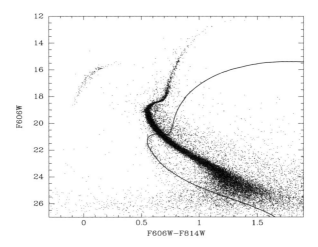

Figure 2. Measurement of Sgr distance. The dominant sequence is the foreground NGC 6681 globular cluster. The faint sequence is Sgr, overlayed with the isochrones fit in Siegel et al. (2007) and shifted to $(m - M) = 17.4$.

Figure 2 demonstrates the method we are using to provide critical distance measures along the tidal stream. The bulge globular cluster NGC 6681 lies in the foreground of the diffuse Sgr stream. Below the main sequence of the primary cluster is a faint diffuse sequence from Sgr. Our revised and improved isochrones, applied to 25,000 NGC 6681 cluster members, allows for the precise measurement of NGC 6681's extinction, distance, abundance and age. We can then use the well-measured foreground cluster to get a relative distance measure to the diffuse Sgr population, utilizing the isochrones derived from the our study of the Sgr core (§2).

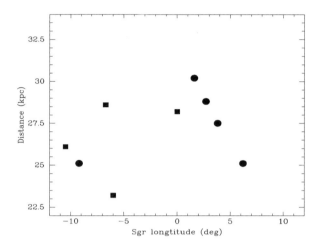

Figure 3. The latitude-distance relation of the Sgr member clusters (squares) and Sgr debris in the background of bulge clusters (circles). The curve of Sgr toward the Galactic Center can be discerned.

Figure 3 shows the spatial location and distance of Sgr's four member clusters and the background features identified in five bulge clusters. The three-dimensional shape of Sgr can be clearly discerned. The stream arcs back along the line of sight with decreasing longitude, consistent with the known orbital shape of the Sgr stream.

Acknowledgments. This work is based on observations with the NASA/ESA *Hubble Space Telescope*, obtained at the Space Telescope Science Institute, which is operated by AURA, Inc., under NASA contract NAS 5-26555, under program GO-10775 (PI: Sarajedini).

References

Dotter, A. et al. 2007, AJ, 134, 376
Harris, J., & Zaritsky, D. 2001, ApJS, 136, 25
Layden, A. C., & Sarajedini, A. 2000, AJ, 119, 1760
Law, D. R., Johnston, K. V., & Majewski, S. R. 2005, ApJ, 619, 80
Milone, A. P., et al. 2007, astro-ph/0709.3762
Sarajedini, A., et al. 2007, AJ, 133, 1658
Siegel, M. H., et al. 2007, ApJ, 667, L57

The Impact of Cosmic Rays on Population III Star Formation

Athena Stacy and Volker Bromm

Department of Astronomy, University of Texas, Austin, TX, USA

Abstract. We explore the implications of a possible cosmic ray (CR) background generated during the first supernova explosions that end the brief lives of massive Population III stars. We show that such a CR background could have significantly influenced the cooling and collapse of primordial gas clouds in minihalos around redshifts of z $\sim$ 15 − 20, provided the CR flux was sufficient to yield an ionization rate greater than about 10^{-19} s^{-1} near the center of the minihalo. The presence of CRs with energies less than approximately 10^7 eV would indirectly enhance the molecular cooling in these regions, and we estimate that the resulting lower temperatures in these minihalos would yield a characteristic stellar mass as low as $\sim 10\,M_\odot$. CRs have a less pronounced effect on the cooling and collapse of primordial gas clouds inside more massive dark matter halos with virial masses greater than approximately $10^8\,M_\odot$ at the later stages of cosmological structure formation around z $\sim$ 10 − 15. In these clouds, even without the CR flux, the molecular abundance is already sufficient for cooling to the floor set by the temperature of the cosmic microwave background.

1. Cosmic Rays in the high-z Universe

Though CR effects will be examined for a range of star formation rates, the typical Pop III star formation rate is taken to be that found in Bromm & Loeb (2006) at z $\sim$ 15, which is approximately $\Psi_* \simeq 2 \times 10^{-2}\,M_\odot$ yr^{-1} Mpc^{-3} in a comoving volume. We assume that those Pop III stars whose masses lie in the pair instability supernova (PISN) range (140 − 260 $M_\odot$) have an average mass of 200 $M_\odot$ (e.g., Heger et al. 2003). CRs are assumed to have been generated in the PISNe that may have marked the death of Pop III stars. The CRs are accelerated in the SN shock wave through the first-order Fermi process. This yields a differential energy spectrum in terms of CR number density per energy (e.g., Longair 1994):

$$\frac{\mathrm{d}n_{\mathrm{CR}}}{\mathrm{d}\epsilon} = \frac{n_{\mathrm{norm}}}{\epsilon_{\mathrm{min}}} \left(\frac{\epsilon}{\epsilon_{\mathrm{min}}}\right)^x \tag{1}$$

$$n_{\mathrm{norm}} = \frac{U_{\mathrm{CR}}}{\epsilon_{\mathrm{min}} \ln\left(\frac{\epsilon_{\mathrm{max}}}{\epsilon_{\mathrm{min}}}\right)} \approx \frac{1}{10} \frac{U_{\mathrm{CR}}}{\epsilon_{\mathrm{min}}} \tag{2}$$

Here, $x = -2$, ϵ is the CR kinetic energy, ϵ_{min} is the minimum kinetic energy, and n_{CR} is the CR number density. The CR energy density U_{CR} is given by

$$U_{\mathrm{CR}}(z) \approx p_{\mathrm{CR}} E_{\mathrm{SN}} f_{\mathrm{PISN}} \Psi_*(z) t_H(z)(1+z)^3 \tag{3}$$

where p_{CR} is the fraction of SN explosion energy, E_{SN}, that goes into CR energy, and f_{PISN} is the number of PISNe that occur for every solar mass unit of star-

2. Evolution of Primordial Clouds

forming material. Here we take the above star-formation rate to be constant over a Hubble time t_H.

We examine the evolution of primordial gas by adapting the one-zone models used in Johnson & Bromm (2006; see also Mackey et al. 2003). We solve the comprehensive chemical reaction network for all the species included in Johnson & Bromm (2006), and consider cooling due to H, H_2, and HD. The temperature of the CMB sets the lower limit to which the gas can cool radiatively (e.g., Larson 1998). Assuming that CRs with the above energy spectrum impinge upon the primordial gas cloud, we add the respective heating and ionization rates:

$$\Gamma_{\rm CR}(D) = \frac{E_{\rm heat}}{50{\rm eV}} \int_{\epsilon_{\rm min}}^{\epsilon_{\rm max}} \left(\frac{d\epsilon}{dt}\right)_{\rm ion} \frac{dn_{\rm CR}}{d\epsilon} e^{-D/D_{\rm p}} d\epsilon \qquad (4)$$

$$\zeta_{\rm CR}(D) = \frac{\Gamma_{\rm CR}}{n_{\rm H} E_{\rm heat}} \qquad (5)$$

where $\left(\frac{d\epsilon}{dt}\right)_{\rm ion}$ is the rate of CR energy loss due to ionization, $D_{\rm p}$ is the maximum depth into a cloud a CR can reach before losing all its energy to ionization, D is the distance to the central region of the cloud, $n_{\rm H}$ is the density of hydrogen, and $E_{\rm heat}$ is an efficiency factor dependent upon the ionization fraction of the gas.

Our first case (Figure 1) is a minihalo which undergoes free-fall collapse. The minihalos form during hierarchical mergers at velocities too low to cause any shock or ionization in the minihalo, so the electron-catalysed molecule formation and cooling is insufficient to allow stars less than around 100 $M_\odot$ to form (e.g., Bromm, Coppi, & Larson 1999, 2002; Abel, Bryan, & Norman 2002). CR ionization and heating, however, can potentially alter the thermal and chemical evolution of primordial gas.

As our second case (Figure 2), we consider strong virialization shocks that arise in later stages of structure formation during the assembly of the first dwarf galaxies. The virial velocities of these DM halos are much greater, implying merger velocities that are now high enough to create a shock that can partially ionize the primordial gas (Johnson & Bromm 2006 and references therein). The post-shock evolution is taken to be roughly isobaric (e.g., Shapiro & Kang 1987; Yamada & Nish 1998).

3. Conclusion

We have investigated the effect of CRs on the thermal and chemical evolution of primordial gas clouds in two important sites for Pop III star formation: minihalos and higher-mass halos undergoing strong virialization shocks (i.e., the first dwarf galaxies). The presence of CRs has a negligible impact on the evolution in the latter case (Figure 2), since the primordial gas has sufficient molecular abundance to be able to cool to the CMB floor even without the help of CR ionization. The direct CR heating is also unimportant unless extremely high

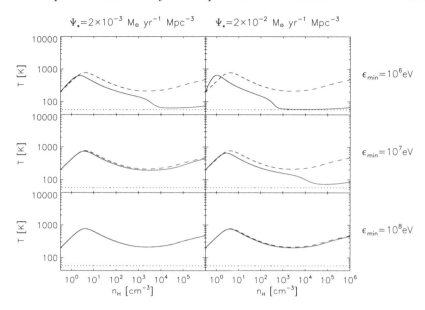

Figure 1. Thermal evolution of primordial gas clouds undergoing free-fall collapse inside minihalos at $z = 20$ for various combinations of Pop III star formation rate and $\epsilon_{\min}$. The solid lines show the evolution when the effect of CRs is included, while the dashed lines depict the standard evolution where only cooling due to H_2 and HD are considered. The dotted line represents the CMB temperature floor at $z = 20$. (Adopted from Stacy & Bromm 2007)

star formation rates are assumed (upper right panel of Figure 2). Thus, the CR emission from the deaths of previously formed Pop III stars will not initiate any feedback in the first dwarf galaxies.

The impact of CRs on gas in minihalos, which would typically have formed before the more massive dwarf galaxy systems, could have been much more pronounced, given a sufficiently low $\epsilon_{\min}$ and Pop III star formation rates that are not too low (Figure 1). In each case, the additional molecular cooling induced by the CRs allows the gas in a minihalo to cool to lower temperatures. As is evident in Figure 1, given a strong enough energy density, CRs can serve to lower the minimum temperature that a collapsing cloud inside a minihalo is able to reach. This could have important implications for the fragmentation scale of such gas clouds. We estimate the possible change in fragmentation scale due to CRs by assuming that the immediate progenitor of a protostar will have a mass approximately given by the Bonnor-Ebert (BE) mass, $M_{\rm BE} \propto T_{\rm f}^{3/2} n_{\rm f}^{-1/2}$, where $n_{\rm f}$ and $T_{\rm f}$ are the density and temperature of the primordial gas at the point when fragmentation occurs. Evaluating $M_{\rm BE}$ shows that, for CR ionization rates above a critical value of $\sim 10^{-19}$ s^{-1}, the BE mass decreases by an order of magnitude down to around 10 M$_\odot$. This is the mass scale of what has been termed 'Pop II.5' stars (e.g., Mackey et al. 2003; Johnson & Bromm 2006). We thus find that CR effects might significantly reduce the mass scale of metal-free stars.

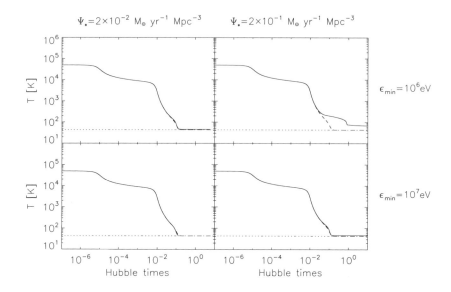

Figure 2. Thermal evolution of primordial gas clouds experiencing virialization shocks during the assembly of the first dwarf galaxies at $z = 15$. We adopt the same manner of presentation, and the same convention for the lines, as in Figure 1. Adopted from Stacy & Bromm 2007.

References

Abel, T., Bryan, G. L., & Norman, M. L. 2002, Science, 295, 93
Bromm, V., & Loeb, A. 2006, ApJ, 642, 382
Bromm, V., Coppi, P. S., & Larson, R. B. 1999, ApJ, 527, L5
Bromm, V., Coppi, P. S., & Larson, R. B. 2002, ApJ, 564, 23
Heger, A., Fryer, C. L., Woosley, S. E., Langer, N., & Hartmann, D. H. 2003, ApJ, 591, 288
Johnson, J. L., & Bromm, V. 2006, MNRAS, 366, 247
Larson, R. B. 1998, MNRAS, 301, 569
Longair, M. S. 1994, High Energy Astrophysics Vol. 2 (Cambridge: Cambridge Univ. Press)
Mackey, J., Bromm V., & Hernquist, L. 2003, ApJ, 586, 1
Shapiro, P., & Kang, H. 1987, ApJ, 381, 32
Stacy, A., & Bromm, V. 2007, MNRAS, 382, 229
Yamada, M., & Nishi, R. 1998, ApJ, 505, 148

Constraining Galaxy Evolution With Bulge-Disk-Bar Decomposition

Tim Weinzirl,[1] Shardha Jogee,[1] and Fabio D. Barazza[2]

Abstract.
Structural decomposition of galaxies into bulge, disk, and bar components is important to address a number of scientific problems. Measuring bulge, disk, and bar structural parameters will set constraints on the violent and secular processes of galaxy assembly and recurrent bar formation and dissolution models. It can also help to quantify the fraction and properties of bulgeless galaxies (those systems having no bulge or only a relatively insignificant disky-pseudobulges), which defy galaxy formation paradigms requiring almost every disk galaxy to have a classical bulge at its core.

We demonstrate a proof of concept and show early results of our ongoing three-component bulge-disk-bar decomposition of NIR images for samples spanning different environments (field and cluster). In contrast to most early studies, which only attempt two-component bulge-disk decomposition, we fit three components using GALFIT: a bulge, a disk, and a bar. We show that it is important to include the bar component, as this can significantly lower the bulge-to-total luminosity ratio (B/T), in many cases by a factor of two or more, thus effectively changing the Hubble type of a galaxy from early to late.

1. Introduction

The formation of galaxies is a classic problem in astrophysics. Contemporary galaxy formation models combine the well-established Lambda-Cold Dark Matter (LCDM) cosmology, which describes behavior of dark matter on very large scales, with baryonic physics to model galaxy formation. In the early Universe, pockets of dark matter decoupled from the Hubble flow, collapsed into virialized halos, and then clustered hierarchically into larger structures. Meanwhile, gas aggregated in the interiors of the halos to form rotating disks, which are the building blocks of galaxies (Navarro & Steinmetz 2002). Such disks were destroyed during mergers of their parent halos, leaving behind classical de Vaucouleurs bulges. Spiral disk galaxies formed subsequently as gaseous disks accreted around spheroids (Burkert & Naab 2004).

Troubling inconsistencies exist between real galaxies and LCDM models of galaxy formation. One issue is the angular momentum problem; simulated galaxy disks have smaller scalelengths and, therefore, less specific angular momentum than their counterparts in nature (D'Onghia & Burkert 2006). A second problem is the severe under prediction in the fraction of galaxies with low bulge-to-total mass ratio (B/T <0.2) and of so-called bulgeless galaxies, which lack a

---

[1]Department of Astronomy, University of Texas, Austin, TX, USA

[2]Laboratoire d'Astrophysique, École Polytechnique Fédérale de Lausanne (EPFL), Switzerland

classical bulge. Simulated spiral galaxies feature prominent classical bulges in their cores. Such predictions are in glaring contradiction with emerging observations that suggest 15-20% of disk galaxies out to $z\sim0.3$ are bulgeless (Kautsch et al. 2006; Barazza et al. 2007)

There are many unanswered questions about the assembly of bulges, the distribution of B/T, and the properties of so-called bulgeless galaxies with low B/T. How do properties, such as disk scalelengths, mass, kinematics, colors, and star formation histories vary across galaxies of different B/T, ranging from bulge-dominated systems to quasi-bulgeless systems? Are quasi-bulgeless systems confined to low mass systems with high specific star formation rates, while classical bulges populate high mass systems? How do the fraction, mass function, and structural properties of galaxies with different B/T vary across environments with different large-scale cosmological overdensities? If environment plays a central part in suppressing bulge formation, then differences would be expected in the properties of bulgeless galaxies in different environments, such as field versus dense galaxy clusters. How does the frequency and properties of galaxies with low B/T as a function of redshift over $z = 0.2 - 0.8$ compare to the recently reported merger history of galaxies over this epoch (Jogee et al. 2007)? Answering these questions will help us to understand the reasons behind the apparent failure of LCDM galaxy formation models, and shed light on how galaxies assemble.

Progress is possible by observationally constraining properties of enigmatic bulgeless galaxies. A powerful technique for measuring the structural properties (e.g. scalelengths, Sérsic indexes, B/T) of galaxies is the decomposition of the 2D light distribution into separate structural components with GALFIT (Peng et al. 2002). Most earlier work has only performed 2D bulge-disk decomposition, but because late-type spirals have been shown to have higher optical bar fractions than early-type galaxies (Barazza et al. 2007), it is important to include the bar when analyzing disk-dominated systems. Bars can contain a significant fraction of light, so failure to account for bars could lead to inflated B/T (Laurikainen et al. 2006).

2. Methodology and Samples

We perform three-component decomposition of the 2D galaxy light distribution, while taking the PSF into account, with GALFIT. Since GALFIT utilizes a non-linear least squares algorithm, initial guess parameters are required for each component GALFIT attempts to fit. Accordingly, our decomposition is broken into three separate invocations of GALFIT.

We first perform one and two-component fits to constrain the bulge and disk parameters. The single-component fit models the entire galaxy with only a Sérsic profile. In addition to constraining the bulge parameters and measuring the centroid, the total luminosity is also determined.

A two-component model, consisting of a Sérsic bulge and exponential disk, is generated based on the single-component fit. If GALFIT is allowed to make an unconstrained bulge-disk fit in a strongly barred galaxy, it will often try to fit the bar by artificially stretching the disk along the bar PA. To avoid this, we constrain the fit by fixing the position angle and axis ratio (b/a) of the outer disk to values pre-determined by fitting an ellipse to the outermost disk isophote.

Finally, a three-component bulge-disk-bar fit is performed, using the two-component fit parameters as initial guesses and fixing the disk b/a and PA as before. The bar is modeled with an elongated, low-index Sérsic component whose initial parameters for the effective radius and PA are estimated by eye. All objects are subjected to the three-component fits, regardless of whether they appear by eye to possess a bar. If there is independent evidence for an AGN or nuclear cluster, a point source is fitted as a fourth component.

In order to decide which of the two or three-component fits is better, a number of criteria are used. 1) If the one or two-component residuals show a bar signature that is removed in the three-component residual, then the three component fit is favored; 2) Structural parameters (scalelength, Sérsic index, b/a) of the bar fit must well behaved; 3) Visual evidence of a strong bar in the input images favors the three-component fit.

To address the questions outlined in §1, we are applying this decomposition method to complementary samples spanning different environments (field and cluster): 1) a $z \sim 0$ sample of ~ 200 galaxies with Hubble types S0 to Sm drawn from the OSU Bright Spiral Galaxy Survey (OSUBSG) (Eskridge et al. 2002) and UKIDSS (McLure et al. 2006); 2) a sample of galaxies in the dense environment of the Coma cluster from our ACS Treasury survey (Carter et al. 2007).

3. Preliminary Findings

An example of our method is presented in Figure 1, which illustrates the complete three-step decomposition for NGC 4643. We now summarize our preliminary findings:

1. Luminosity is conserved between the two and three-component fits.

2. Modeling the bar in the three-component fits forces a reshuffling of luminosity. Generally, the bulge declines in luminosity, whereas light can be taken from, or added back, to the disk.

3. Inclusion of the bar can reduce bulge fractional luminosity B/T by a factor of two or more. Larger changes (a factor of 10 or more) may occur for objects with prominent bars.

4. The scalelength of the disk is generally unchanged by including the bar. However in a few cases, the two-component disk structure can be erroneous, as in the case of NGC 4643, shown in Figure 1.

We have provided a proof of concept of our ongoing three-component bulge-disk-bar decomposition with GALFIT. We are optimistic about our on-going work, which will be described in T. Weinzirl et al. 2008 (in prep).

Acknowledgments. TW and SJ acknowledge support from NSF grad AST-0607748, LTSA grant NAG5-13063, and HST-GO-10861 from STScI, which is operated by AURA, Inc., for NASA, under NAS5-26555.

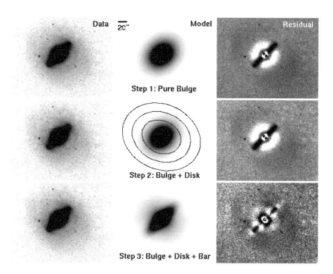

Figure 1. Shown is the complete three-step decomposition for NGC 4643. The residuals for the one and two-component fit show a distinct bar signature. In Step 2, the fitted disk has an unphysically large scalelength (335″) that does not match the galaxy. This fitted disk is hard to see, and ellipses are drawn to show its PA and b/a. In Step 3, the addition of the bar component restores the disk scalelength to a reasonable value. The fit parameters are presented in Table 1.

Table 1. Fit parameters for NGC 4643

Fit		r_e or h (″)	n	b/a	Position Angle	Fractional light
Step 1	Bulge	27.90	4.44	0.80	-51.08	100%
Step 2	Bulge	23.86	4.16	0.80	-51.08	34.6%
	Disk	335.88	1.0	0.84	66.94	65.4%
Step 3	Bulge	5.43	2.53	0.90	60.52	25.0 %
	Disk	48.22	1.0	0.84	66.94	54.1 %
	Bar	21.30	0.62	0.37	-45.84	20.9 %

References

Barazza, F. D., Jogee, S., Marinova, I. 2007, ApJ, accepted, astro-ph/0710.4674
Burkert, A., & Naab, T. 2004, Coevolution of Black Holes and Galaxies, 422
Carter, D. et al. 2007, ApJ, submitted
D'Onghia, E. & Burkert, A. 2006, MNRAS, 372, 1525
Eskridge, P. B. et al. 2002, ApJS, 143, 73
Jogee, S., et al. 2007, to appear in proceedings of "Formation and Evolution of Galaxy Disks", Rome, October 2007, ed. J. G. Funes, S. J. and E. M. Corsini
Kautsch, S. J. et al. 2006, A&A, 445, 765
Laurikainen, E. et al. 2006, AJ, 132, 2634
McLure, R. J. et al. 2006, MNRAS, 372, 357
Navarro, J. F., & Steinmetz, M. 2000, ApJ, 538, 477
Peng C. Y., Ho L. C., Impey, C. D., & Rix, H.-W. 2002, AJ, 124, 266

New Horizons in Astronomy: Frank N. Bash Symposium 2007
ASP Conference Series, Vol. 393, © 2008
A. Frebel, J. R. Maund, J. Shen, and M. H. Siegel, eds.

The Relics of Structure Formation: Extra-Planar Gas and High-Velocity Clouds Around M31

Tobias Westmeier,[1,2] Christian Brüns,[2] and Jürgen Kerp[2]

[1] CSIRO Australia Telescope National Facility, Epping, Australia
[2] Argelander-Institut für Astronomie, Universität Bonn, Bonn, Germany

Abstract. We mapped a large region around M31 in H I with the 100-m radio telescope at Effelsberg to search for high-velocity clouds (HVCs) out to large projected distances in excess of 100 kpc. We argue that tidal stripping may have played an important role in forming several of the 17 detected HVCs, whereas other clouds could be primordial dark-matter satellites.

1. Introduction

Numerical simulations in the framework of cold dark matter (CDM) cosmologies predict a hierarchical formation of gravitationally bound structures in the universe. Larger entities, such as spiral galaxies or galaxy clusters, are expected to form through the merging and accretion of smaller dark-matter halos. One consequence of this scenario is the prediction of hundreds of low-mass dark-matter satellites around galaxies like the Milky Way (Klypin et al. 1999; Moore et al. 1999), whereas only about 20 satellite galaxies of the Milky Way have been discovered so far (Walsh et al. 2007). This discrepancy has been known as the "missing satellites" problem, and several attempts have been made over the past years to solve it.

A promising solution was proposed by Blitz et al. (1999), who suggested that most of the high-velocity clouds (HVCs) observed all over the sky in the 21-cm line emission of neutral hydrogen could be the gaseous counterparts of the "missing" dark-matter satellites in the Local Group. In this picture, most dark-matter satellites would not have formed stars on a grand scale and therefore remained dark at optical wavelengths. In addition, HVCs would have typical distances from the Galaxy of hundreds of kpc and high H I masses of the order of $10^7 \, M_\odot$.

Recent distance measurements (Wakker et al. 2007 and references therein) indicate that the large HVC complexes are typically located only a few kpc above the Galactic plane. For the large population of compact high-velocity clouds (CHVCs; Braun & Burton 1999; de Heij, Braun, & Burton 2002), however, distance measurements are not available. Therefore, the most straightforward way to test the scenario of Blitz et al. (1999) is the study of other galaxies and galaxy groups. Earlier searches for HVCs in nearby galaxy groups (Zwaan 2001; Braun & Burton 2001; Pisano et al. 2004) did not produce any detections and provided an upper distance limit for HVCs of the order of 150 kpc.

The next logical step is the search for HVCs in the direct vicinity of nearby spiral galaxies. The nearest galaxy comparable in mass and size to the Milky

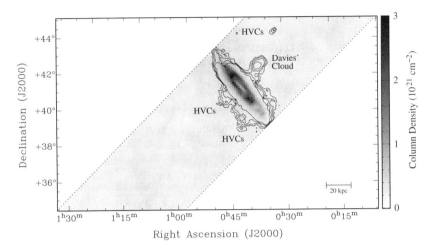

Figure 1. H I column density map of our Effelsberg survey of M31 in the velocity range of $v_{\mathrm{LSR}} = -620\ldots-24$ km s^{-1} (grey-scale image) and $v_{\mathrm{LSR}} = -608\ldots-135$ km s^{-1} (black contours). The survey region is enclosed by dotted lines. The contours correspond to 2, 5, 10, and 20×10^{18} cm^{-2}. Several HVCs and regions of extended extra-planar gas can be seen near M31.

Way is the Andromeda Galaxy (M31). A first comprehensive search for HVCs around M31 was carried out by Thilker et al. (2004) in H I with the Green Bank Telescope. In their field of $7° \times 7°$ they detected several HVCs within about 50 kpc of M31, demonstrating that the Milky Way is not the only galaxy surrounded by HVCs. We carried out a complementary spectroscopic H I survey with the 100-m radio telescope at Effelsberg to trace the population of HVCs out to much larger projected distances in excess of 100 kpc and to search for the expected extended population of CHVCs around M31.

2. Observations and data analysis

Our survey region has a trapezoidal shape and is oriented along the minor axis of M31 to minimize confusion with the disk emission of M31. It extends out to a projected distance of about 140 kpc in the south-eastern direction and about 70 kpc in the north-western direction (Fig. 1), although the azimuthal coverage is limited at larger projected distances. Our $3\,\sigma$ column density detection limit for the warm neutral medium is about 2×10^{18} cm^{-2} for emission filling the $9'$ beam of the Effelsberg telescope. This translates into an H I mass detection limit of about 8×10^4 $M_\odot$ if a distance of 780 kpc (Stanek & Garnavich 1998) is assumed for M31 and its HVC population.

Each H I spectrum was flux-calibrated and corrected for the main beam efficiency of the telescope. Next, the spectral baseline was subtracted through a special two-stage polynomial fitting procedure to the spectral channels free of emission. Afterwards, each spectrum was searched by eye for H I gas separated in phase space from the disk emission of M31. This includes clouds seen in projection against the disk of M31 but at different radial velocities.

3. Results

The total H I column density map of our survey region is shown in Fig. 1. The overlaid contour lines show the column density integrated over a limited velocity range of $v_{\rm LSR} = -608 \ldots -135$ km s^{-1} to avoid confusion with local H I emission from the Galactic disk. Several isolated HVCs as well as regions of extended extra-planar gas can be seen all around M31.

In total we identified 17 individual HVCs near M31 with typical H I masses of a few $10^5 \, M_\odot$ and diameters of the order of 1 kpc. With these parameters the clouds resemble some of the large HVC complexes observed around the Milky Way. Although our Effelsberg survey reaches out to projected distances in excess of 100 kpc, we do not detect any HVCs beyond a projected distance of about 50 kpc from M31. In particular, we do not find an extended population of hundreds of CHVCs (Braun & Burton 1999; de Heij et al. 2002). Assuming that the Galaxy and M31 have similar CHVC populations, we can derive an upper distance limit for CHVCs from their host galaxies of about 60 kpc from our non-detection. Hence, the small angular diameters of CHVCs are not due to their large distances from the Galaxy but simply reflect their small intrinsic sizes.

4. Discussion

In their numerical structure formation simulations of Galaxy-sized dark-matter halos, Kravtsov et al. (2004) made the attempt to include the effects of gas and star formation in the evolution of dark-matter satellites. They concluded that about 2–5 dark-matter satellites with total gas masses of $M_{\rm gas} > 10^6 \, M_\odot$ should exist within a distance of 50 kpc from M31. This is less than the 17 HVCs detected in our Effelsberg survey. However, as indicated by our follow-up high-resolution observations with the WSRT (Westmeier et al. 2005) and suggested by Kravtsov et al. (2004), only a few of the HVCs found near M31 could be primordial dark-matter satellites, whereas others were most likely created during the tidal distortion of satellite galaxies of M31.

Nonetheless, the simulations of Kravtsov et al. (2004) predict many more dark-matter satellites (about 50–100 with $M_{\rm gas} > 10^6 \, M_\odot$) at distances beyond 50 kpc from M31 which have not been detected in our Effelsberg H I survey. Obviously, the direct identification of HVCs with dark-matter satellites fails, and the simulations still predict too many gaseous dark-matter halos. To solve this problem we can look at the structure of the individual halos in more detail. Sternberg et al. (2002) carried out hydrostatic simulations of spherical, dark-matter-dominated CHVCs at a distance of 150 kpc from the Milky Way and compared their results with the parameters of the CHVCs observed on the sky. The basic parameters of their model clouds (such as H I mass, H I peak column density, or diameter) are in excellent agreement with the parameters of the HVCs detected near M31 with the WSRT (Westmeier et al. 2005). In their model, Sternberg et al. (2002) had to introduce an external pressure of

$P/\text{k} \gtrsim 50\,\text{K}\,\text{cm}^{-3}$. This pressure – exerted by the hot, ionised circum-galactic corona – is required to stabilize the clouds in addition to their own gravitational potential. Without this constraint the clouds would become so diffuse that the gas would no longer be sufficiently shielded against the extra-galactic radiation field. Consequently, the HVCs would become mainly ionised and undetectable in the 21-cm line emission of neutral hydrogen.

Therefore, we can construct a consistent picture in which part of the HVCs observed around M31 are the remnants of tidal stripping whereas others could be the gaseous counterparts of primordial dark-matter halos with parameters similar to the CHVC model of Sternberg et al. (2002). Within 50 kpc radius of M31, the number of potential primordial HVCs is in agreement with the predictions made by Kravtsov et al. (2004) based on their ΛCDM structure formation simulations. Beyond this radius we do not detect any HVCs, although many more dark-matter satellites with sufficiently high gas masses should exist according to Kravtsov et al. (2004). These "missing" satellites, however, could be mainly ionised and therefore undetectable in the H I line if the pressure exerted by the circum-galactic corona decreases below the critical value of $P/\text{k} \approx 50\,\text{K}\,\text{cm}^{-3}$ at larger distances from M31. Consequently, we would detect only those dark-matter satellites which are close enough to M31 to be stabilized by the external pressure, but many more highly ionised or pure dark-matter halos could exist at larger distances.

Acknowledgments. This work was supported by the Deutsche Forschungsgemeinschaft under grant number KE 757/4–2. Based on observations with the 100-m telescope of the MPIfR (Max-Planck-Institut für Radioastronomie) at Effelsberg.

References

Blitz, L., Spergel, D. N., Teuben, P. J., Hartmann, D., & Burton, W. B. 1999, ApJ, 514, 818
Braun, R. & Burton, W. B. 1999, A&A, 341, 437
Braun, R. & Burton, W. B. 2001, A&A, 375, 219
de Heij, V., Braun, R., & Burton, W. B. 2002a, A&A, 392, 417
Klypin, A., Kravtsov, A. V., Valenzuela, O., & Prada, F. 1999, ApJ, 522, 82
Kravtsov, A. V., Gnedin, O. Y., & Klypin, A. A. 2004, ApJ, 609, 482
Moore, B., Ghigna, S., Governato, F., Lake, G., Quinn, T., Stadel, J., & Tozzi, P. 1999, ApJ, 524, L19
Pisano, D. J., Barnes, D. G., Gibson, B. K., Staveley-Smith, L., Freeman, K. C., & Kilborn, V. A. 2004, ApJ, 610, L17
Stanek, K. Z. & Garnavich, P. M. 1998, ApJ, 503, 131
Sternberg, A., McKee, C. F., & Wolfire, M. G. 2002, ApJS, 143, 419
Thilker, D. A., Braun, R., Walterbos, R. A. M., Corbelli, E., Lockman, F. J., Murphy, E., & Maddalena, R. 2004, ApJ, 601, L39
Wakker, B. P., York, D. G., Howk, J. C., Barentine, J. C., Wilhelm, R., Peletier, R. F., van Woerden, H., Beers, T. C., Ivezić, Ž., Richter, P., & Schwarz, U. J. 2007, ApJ, 670, L113
Walsh, S. M., Jerjen, H., & Willman, B. 2007, ApJ, 662, L83
Westmeier, T., Braun, R., & Thilker, D. A. 2005, A&A, 436, 101
Zwaan, M. A. 2001, MNRAS, 325, 1142

Embers of the Dark Ages: Recombination Radiation from the First Stars

Amelia C. Wilson, Volker Bromm, and Jarrett L. Johnson

Department of Astronomy, University of Texas, Austin, TX, USA

Abstract. The first stars in the Universe (the so-called Population III, or Pop III) formed a few hundred million years after the Big Bang, at the end of the Cosmic Dark Ages. Arising from tiny fluctuations in the primordial cosmic matter, their presence forever changed the Universe. Some of these stars died as energetic supernovae, whereas others, depending on their mass, directly collapsed into massive black holes. We here consider the latter case, and in particular ask: How does the nebula of ionized, primordial gas that surrounds the remnant black hole recombine after the Population III star's nuclear engine turns off, thus allowing the gas to cool? And: What kind of recombination radiation is emitted during this process? Specifically, we derive the line luminosities for Ly_α, H_α, and He II $\lambda 1640$ from the nebula left behind by a $100\,M_\odot$ star at a redshift of $z \sim 23$, using sophisticated computer simulations to construct snapshots of the ionized gas as it evolves with time. We also make predictions for the line fluxes that could be observed at $z = 0$ with future telescopes. The James Webb Space Telescope will possibly see far enough into the depths of time and space to make the first observations of these crucial first sources of light.

1. Introduction

The first stars owe their existence to tiny quantum ripples in the Universe which were present before inflation, evident in the Cosmic Microwave Background Radiation (Spergel et al. 2007). The formation of these massive stars marked a significant milestone in building up the large-scale structure we see around us today by reionizing the early Universe. After inflation, gravity was able to organize the baryonic matter into the filamentary structure. In the densest regions, a Pop III star consisting only of hydrogen and helium, with mass of $\sim 100\,M_\odot$, would be born (e.g., Bromm & Larson 2004). The high temperatures of the stars ($\sim 10^5$ K) created ultraviolet radiation that would ionize the surrounding gas. As more and more of these stars began to form, these ionized bubbles would merge and the entire Universe would start to become completely reionized, disrupting the initial simplicity and thus setting the Universe on the course to complexity.

2. Methodology

2.1. Simulation

Using Smoothed Particle Hydrodynamics (SPH), we simulated a Pop III star at $z < 23$, just after the central engine turns off. We took several "snapshots" of the remnant HII bubble as the gas is allowed to cool and recombine. From these snapshots, we derived estimates of the luminosities of three characteristic

Pop III recombination lines: Ly$_\alpha$ at 1216 Å, H$_\alpha$ at 6340 Å, and He II at 1640 Å as the bubble evolves with time over a period of $\sim$ 100 million years. We chose only those SPH particles with an electron fraction above a threshold of 10^{-3}, to ensure that we determined the ionized bubble and only calculated the fluxes in the ionized region. Then, we find the electron, proton, and nuclei densities of each selected SPH particle.

2.2. Recombination Coefficients and Calculations

To calculate Ly$_\alpha$ luminosity, we chose the case A recombination coefficient (see Osterbrock & Ferland 2006 for an explanation of cases A and B recombination coefficients). We used case B for H$_\alpha$ and the He II λ1640 line to account for the optical thickness of the bubble to these lines, in which the atoms can destroy the photons by absorbing them and re-emitting them at different wavelengths. Case A is a good approximation for Ly$_\alpha$ because at this stage in the Universe, no significant destroyers of this line exist (no interstellar dust, and 2-photon emission is rare enough to be ignored). Instead of being destroyed, the Ly$_\alpha$ photons will be repeatedly scattered, until they finally escape the bubble. If this process takes too long, the expansion of the Universe will actually be significant enough to create a mismatch between the photon energies and the absorbers, and the bubble becomes optically thin to the redshifted Ly$_\alpha$, and then these photons will freely escape.

We found the luminosity (eq. 1) for each SPH particle in the bubble (as determined by the electron fraction), and summed over all the selected SPH particles to find the total luminosity of each line in each snapshot. Luminosity is given by

$$L = \sum_{i=1}^{N_{SPH}} \alpha_i(T) n_e n_n \Delta V_i, \qquad (1)$$

where the sum is over all the selected SPH particles. $\alpha_i(T)$ is the recombination coefficient, n_e is the electron density in the selected SPH particle, n_n is the nuclei density in the selected SPH particle, and ΔV is the volume element in the SPH particle, defined as:

$$\Delta V = \frac{M_{SPH}}{(n_n M_H)}, \qquad (2)$$

where M_{SPH} is the SPH particle mass, n_n is the nuclei density and M_H is the mass of hydrogen (Osterbrock & Ferland 2006)

We did this for each "snapshot" in the sequence to understand how the bubble cools and what the nature of its emission is as it cools. To determine an accurate estimate for the flux of such an object as observed from a redshift of $z = 0$, we needed to determine the luminosity distance, D_L, which is given by

$$D_L = \frac{c(1+z)}{H_o} \int_0^z \frac{dz'}{\sqrt{\Omega_m(1+z')^3 + \Omega_\Lambda}}. \qquad (3)$$

The WMAP parameters are used: $H_o = 73 \text{ km s}^{-1}\text{Mpc}^{-1}$, $\Omega_m = 0.24$, and $\Omega_\lambda = 0.76$ (Spergel et al. 2007). The flux is thus given by

$$F = \frac{L}{4\pi D_L^2}, \qquad (4)$$

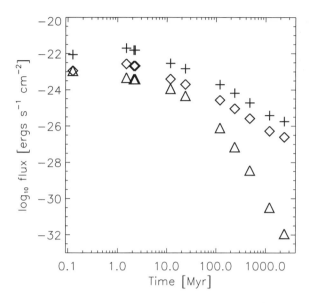

Figure 1. Observed flux of the HII region at $z = 0$, just after the star dies at $z < 23$ and is allowed to cool and recombine. Ly$_\alpha$, H$_\alpha$, and He II $\lambda 1640$ are plotted by the plus, diamond, and triangle symbols, respectively. The time dilation caused by the expanding Universe would make this process last ~ 1000 million years to an observer at $z = 0$, if they could observe it that long. These fluxes are below the threshold for detection by JWST (sensitivity $\sim 10^{-19}$ ergs s^{-1} cm^{-2}).

where L is the luminosity (see eq. 1), and D_L is the luminosity distance (see eq. 3.

2.3. Special Considerations

We needed to consider the fact that the simulated snapshots were not co-moving, so the redshift actually went down with respect to the "observer." F is a function of L and D_L, and D_L is itself a function of z. We calculate D_L for the redshift of every snapshot. We calculated the D_L for the same redshift that the snapshots and their respective luminosities were at. The physical significance of this is that a telescope looking further and further back in time (at larger and larger z) will see these different "snapshots" of the evolution. For the interpretation of these snapshots, we must estimate the correct flux for each.

3. Results

If the first stars died as massive black holes (MBH), the surrounding remnant HII regions would be allowed to cool and recombine. This recombination radiation is what we investigated here, to determine how - and if it is even possible - for the James Webb Space Telescope (JWST) to see evidence of the smallest scale structure formation and the first sources of light.

We find that the angular diameter of a HII region right after it formed at a redshift of $z \sim 23$ is not too small - about $4''$. Unfortunately, JWST's sensitivity is $10^{-19}\,\mathrm{ergs\,s^{-1}\,cm^{-2}}$, which is above the brightest Ly$_\alpha$ line, produced at the very end of the star's life (at the beginning of our calculations). JWST's sensitivity would only enable the detection of a young remnant HII region's early Ly$_\alpha$ line up to a luminosity distance of $\sim 9100\,\mathrm{Mpc}$ ($z \sim 1.3$). It is extremely unlikely that a Pop III remnant formed at this low redshift. Perhaps if a cluster of such Pop III remnants existed at $z \sim 23$ (10^3 of these regions for Ly$_\alpha$, and 10^4 for H$_\alpha$ and He II $\lambda 1640$), JWST could detect it. However, the likelihood of such a large cluster forming is extremely small. Therefore the most feasible possibility is to search for HII regions at $z \sim 10$. Here, the flux will be approximately a factor of ten larger, and so the requirement for detection is a cluster of ~ 100 HII regions. Such a modest number is physically possible, although we found it to be statistically very unlikely.

For each case, as the bubble cools, the ionization fraction and recombination radiation line luminosities drop, so the optimal time to detect a remnant is at the very onset of recombination, right after the MBH forms.

4. Conclusions and Future Outlook

It will not be possible for JWST to see single regions such as the one simulated at $z \sim 23$, but clusters of 10^3 such remnants at $z \sim 10$ will not be out of reach of the JWST (in terms of resolution) if it is statistically possible for such large clusters to form. Characteristic properties such as mass, lifetimes, could be determined, and unanswered questions pertaining to the bottom-up hierarchical model could be addressed. We still require a more sensitive instrument that will be able to detect small groups of these remnants at the beginning of the cosmic renaissance, the ghosts of the first structures that formed at the end of the cosmic dark ages.

Acknowledgments. The first author especially thanks her co-authors Dr. Bromm and J. Johnson for their time and dedication to not only this project, but to her as their student.

References

Bromm, V. & Larson, R. B. 2004, ARAA, 42, 79
Oh, S. P., Haiman, Z., & Rees, M. J. 2001, ApJ, 553, 73
Osterbrock, D. E., Ferland, G. J. 2006, Astrophysics of Gaseous Nebulae and Active Galactic Nuclei, 2nd Ed., Sausalito, CA: University Science Books,
Spergel, D. N. et al., 2007, ApJS, 170, 377

Dougal Mackey and Peter Yoachim discuss their
Russian doll theory of globular clusters.

Kelle Cruz, Eric Mamajek, Anna Frebel and Justyn Maund at the reception.

Jane Rigby, Kurtis Williams and Eric Mamajek. Extrapolating, we find that the next astronomer in this sequence will be 9 feet tall.

Amanda Bauer, Craig Wheeler and Robert Quimby discuss the pending beer supernova.

Kyle Penner and Julie Krugler enjoy the reception buffet.

Enrico Ramirez-Ruiz and Pawan Kumar discuss the next GRB.

Tobias Westmeier and Neal Evans.

Mike Siegel, Kurtis Williams and Robert Tull chat at the reception.

Eichiro Komatsu, Masami Ouchi and Takashi Okamoto say "Cheers!".

At the conference dinner held at La Fonda San Miguel.

McDonald Observatory Director David Lambert delivers the meeting's concluding remarks.

Author Index

Aguirre, J., 243
Allende Prieto, C., 187, 203, 207, 255
Asplund, M., 255

Böhm, A., 211, 231
Bacon, D., 211, 231
Bally, J., 243
Balogh, M., 211, 231, 235
Barazza, F. D., 211, 231, 235, 279
Barden, M., 211, 231, 235
Barentine, J. C., 179
Barnes, S., 79
Beckwith, S., 235
Beers, T. C., 179, 203, 207
Bell, E. F., 211, 231, 235
Bisterzo, S., 207
Boesgaard, A. M., 227
Bromm, V., 215, 275, 287
Brüns, C., 283
Burnett, C., 243

Caldwell, J. A. R., 211, 231, 235
Carrell, K., 251
Chaboyer, B., 271
Chamberlin, R., 243
Chen, A. A., 219
Chen, J., 219
Cieza, L. A., 35
Cochran, W. D., 259
Conselice, C., 235
Couch, S. M., 183
Cowan, J. J., 187, 203, 207, 263
Cruz, K., 51

Davies, L. A., 187
Deen, C. P., 191
Dotter, A., 271
Drosback, M., 243

Emery, J. P., 3

Endl, M., 259
Evans, N. J., 243

Falcon, R. E., 195
Fernandez, E., 199
Frebel, A., 63, 187, 203, 207

Gallino, R., 207
Glenn, J., 243
Gray, M. E., 211, 231
Greif, T. H., 215
Grindlay, J., 247

Häußler, B., 211, 231, 235
Hayakawa, S., 219
Heiderman, A., 211
Heymans, C., 211, 231, 235
Hong, J., 247
Howk, J. C., 179

Ivezić, Ž., 179

Jaffe, D. T., 191
Jahnke, K., 211, 231, 235
Jogee, S., 211, 231, 235, 279
Johnson, J. L., 215, 287

Kahl, D., 219
Kannappan, S. J., 267
Kerp, J., 283
Kim, A., 219
Koda, J., 223
Koesterke, L., 259
Komatsu, E., 199
Koposov, S., 211, 231
Krugler, J. A., 227
Kubono, S., 219

Lambert, D. L., 255
Lane, K., 211, 231
Laurent, G., 243

Lawler, J. E., 263
Laycock, S., 247

Mackey, D., 95
Majewski, S., 271
Mandell, A. M., 19
Marinova, I., 211, 231
McIntosh, D., 211, 231, 235
Meisenheimer, K., 211, 231, 235
Michimasa, S., 219
Miller, S., 235
Milosavljević, M., 183, 223
Montgomery, M. H., 239

Nagai, D., 223
Nakar, E., 223
Nidever, D., 271
Nordhaus, M. K., 243

Okamoto, T., 111
Ouchi, M., 127

Papovich, C., 235
Peletier, R. F., 179
Peng, C. Y., 211, 231, 235
Penner, K., 235, 247
Pilachowski, C. A., 263
Powell, Jr, W. L., 251

Quimby, R. M., 141

Ramírez, I., 255
Redfield, S., 195, 259
Rhee, J., 203, 207
Richter, P., 179
Rigby, J. R., 155
Rix, H.-W, 211, 231, 235
Robaina, A., 235
Roederer, I. U., 203, 207, 263
Rosolowsky, E., 243

Sánchez, F., 211, 231, 235
Sarajedini, A., 271
Scarborough, T. A., 267
Schwarz, U. J., 179
Setoodehnia, K., 219
Shapiro, P. R., 223
Shetrone, M. D., 203, 207
Siegel, M., 271
Skelton, R., 235

Sneden, C., 187, 203, 207, 263
Sobeck, J., 263
Somerville, R., 211, 231, 235
Stacy, A., 275

Taylor, A., 211, 231

Vaillancourt, J., 243
van den Berg, M., 247
van Kampen, E., 211, 231
van Woerden, H., 179

Wakabayashi, Y., 219
Wakker, B. P., 179
Walawender, J., 243
Weinzirl, T., 279
Westmeier, T., 283
Wheeler, J. C., 183
Wilhelm, R., 179, 251
Williams, J., 243
Wilson, A. C., 287
Wisotzki, L., 211, 231, 235
Wolf, C., 211, 231, 235

Yamaguchi, H., 219
York, D. G., 179

Zhao, P., 247
Zheng X., 211, 231, 235

ASTRONOMICAL SOCIETY OF THE PACIFIC (ASP)

An international, nonprofit, scientific and educational organization founded in 1889 established the

ASP CONFERENCE SERIES (ASPCS)

in 1988 to publish books on recent developments in astronomy and astrophysics.

A list of recently published volumes follows. For a complete list of all volumes published please see our web site at
http://www.astrosociety.org/pubs.html

For electronic versions of volumes please see our e-book site at
http://www.aspbooks.org

All book orders or inquiries concerning

ASTRONOMICAL SOCIETY OF THE PACIFIC
CONFERENCE SERIES
(ASPCS)

and

INTERNATIONAL ASTRONOMICAL UNION VOLUMES
(IAU)

should be directed to

Astronomical Society of the Pacific
390 Ashton Avenue
San Francisco   CA   94112-1722   USA

Phone: 800-335-2624 (within USA)
Phone: 415-337-2126
Fax: 415-337-5205

E-mail: service@astrosociety.org
Web Site: http://www.astrosociety.org
E-book site: http://www.aspbooks.org

ASP CONFERENCE SERIES VOLUMES
Published by the Astronomical Society of the Pacific

PUBLISHED: 2003

Vol. CS 300   RADIO ASTRONOMY AT THE FRINGE, A Conference held in honor of Kenneth
I. Kellermann, on the occasion of his 65th Birthday
eds. J. Anton Zensus, Marshall H. Cohen, and Eduardo Ros
ISBN: 1-58381-147-8   e-Book ISBN: 978-1-58381-243-3

Vol. CS 301   MATTER AND ENERGY IN CLUSTERS OF GALAXIES
eds. Stuart Bowyer and Chorng-Yuan Hwang
ISBN: 1-58381-149-4

Vol. CS 302   RADIO PULSARS, In celebration of the contributions of Andrew Lyne,
Dick Manchester and Joe Taylor – A Festschrift honoring their 60th Birthdays
eds. Matthew Bailes, David J. Nice, and Stephen E. Thorsett
ISBN: 1-58381-151-6

Vol. CS 303   SYMBIOTIC STARS PROBING STELLAR EVOLUTION
eds. R. L. M. Corradi, J. Mikołajewska, and T. J. Mahoney
ISBN: 1-58381-152-4

Vol. CS 304   CNO IN THE UNIVERSE
eds. Corinne Charbonnel, Daniel Schaerer, and Georges Meynet
ISBN: 1-58381-153-2

Vol. CS 305   International Conference on MAGNETIC FIELDS IN O, B AND A STARS:
ORIGIN AND CONNECTION TO PULSATION, ROTATION AND MASS LOSS
eds. Luis A. Balona, Huib F. Henrichs, and Rodney Medupe
ISBN: 1-58381-154-0

Vol. CS 306   NEW TECHNOLOGIES IN VLBI
ed. Y. C. Minh
ISBN: 1-58381-155-9

Vol. CS 307   SOLAR POLARIZATION 3
eds. Javier Trujillo Bueno and Jorge Sanchez Almeida
ISBN: 1-58381-156-7

Vol. CS 308   FROM X-RAY BINARIES TO GAMMA-RAY BURSTS
eds. Edward P. J. van den Heuvel, Lex Kaper, Evert Rol, and Ralph A. M. J. Wijers
ISBN: 1-58381-158-3

PUBLISHED: 2004

Vol. CS 309   ASTROPHYSICS OF DUST
eds. Adolf N. Witt, Geoffrey C. Clayton, and Bruce T. Draine
ISBN: 1-58381-159-1   e-Book ISBN: 978-1-58381-244-0

Vol. CS 310   VARIABLE STARS IN THE LOCAL GROUP, IAU Colloquium 193
eds. Donald W. Kurtz and Karen R. Pollard
ISBN: 1-58381-162-1   e-Book ISBN: 978-1-58381-245-7

Vol. CS 311   AGN PHYSICS WITH THE SLOAN DIGITAL SKY SURVEY
eds. Gordon T. Richards and Patrick B. Hall
ISBN: 1-58381-164-8   e-Book ISBN: 978-1-58381-246-4

Vol. CS 312   Third Rome Workshop on GAMMA-RAY BURSTS IN THE AFTERGLOW ERA
eds. Marco Feroci, Filippo Frontera, Nicola Masetti, and Luigi Piro
ISBN: 1-58381-165-6   e-Book ISBN: 978-1-58381-247-1

ASP CONFERENCE SERIES VOLUMES
Published by the Astronomical Society of the Pacific

PUBLISHED: 2004

Vol. CS 313   ASYMMETRICAL PLANETARY NEBULAE III: WINDS, STRUCTURE AND THE THUNDERBIRD
eds. Margaret Meixner, Joel H. Kastner, Bruce Balick, and Noam Soker
ISBN: 1-58381-168-0   e-Book ISBN: 978-1-58381-248-8

Vol. CS 314   ASTRONOMICAL DATA ANALYSIS SOFTWARE AND SYSTEMS XIII
eds. Francois Ochsenbein, Mark G. Allen, and Daniel Egret
ISBN: 1-58381-169-9   e-Book ISBN: 978-1-58381-249-5
ISSN: 1080-7926

Vol. CS 315   MAGNETIC CATACLYSMIC VARIABLES, IAU Colloquium 190
eds. Sonja Vrielmann and Mark Cropper
ISBN: 1-58381-170-2   e-Book ISBN: 978-1-58381-250-1

Vol. CS 316   ORDER AND CHAOS IN STELLAR AND PLANETARY SYSTEMS
eds. Gene G. Byrd, Konstantin V. Kholshevnikov, Aleksandr A. Mylläri, Igor' I. Nikiforov, and Victor V. Orlov
ISBN: 1-58381-172-9   e-Book ISBN: 978-1-58381-251-8

Vol. CS 317   MILKY WAY SURVEYS: THE STRUCTURE AND EVOLUTION OF OUR GALAXY, The 5th Boston University Astrophysics Conference
eds: Dan Clemens, Ronak Shah, and Tereasa Brainerd
ISBN: 1-58381-177-X   e-Book ISBN: 978-1-58381-252-5

Vol. CS 318   SPECTROSCOPICALLY AND SPATIALLY RESOLVING THE COMPONENTS OF CLOSE BINARY STARS
eds. Ronald W. Hilditch, Herman Hensberge, and Krešimir Pavlovski
ISBN: 1-58381-179-6   e-Book ISBN: 978-1-58381-253-2

Vol. CS 319   NASA OFFICE OF SPACE SCIENCE EDUCATION AND PUBLIC OUTREACH CONFERENCE
eds. Carolyn Narasimhan, Bernhard Beck-Winchatz, Isabel Hawkins, and Cassandra Runyon
ISBN: 1-58381-181-8   e-Book ISBN: 978-1-58381-254-9

Vol. CS 320   THE NEUTRAL ISM OF STARBURST GALAXIES
eds. Susanne Aalto, Susanne Hüttemeister, and Alan Pedlar
ISBN: 1-58381-182-6   e-Book ISBN: 978-1-58381-255-6

Vol. CS 321   EXTRASOLAR PLANETS: TODAY AND TOMORROW
eds. J. P. Beaulieu, A. Lecavelier des Etangs, and C. Terquem
ISBN: 1-58381-183-4   e-Book ISBN: 978-1-58381-256-3

Vol. CS 322   THE FORMATION AND EVOLUTION OF MASSIVE YOUNG STAR CLUSTERS
eds. Henny J. G. L. M. Lamers, Linda J. Smith, and Antonella Nota
ISBN: 1-58381-184-2   e-Book ISBN: 978-1-58381-257-0

Vol. CS 323   STAR FORMATION IN THE INTERSTELLAR MEDIUM: In Honor of David Hollenbach, Chris McKee and Frank Shu
eds. D. Johnstone, F. C. Adams, D. N. C. Lin, D. A. Neufeld, and E. C. Ostriker
ISBN: 1-58381-185-0   e-Book ISBN: 978-1-58381-258-7

Vol. CS 324   DEBRIS DISKS AND THE FORMATION OF PLANETS: A Symposium in Memory of Fred Gillett
eds. Larry Caroff, L. Juleen Moon, Dana Backman, and Elizabeth Praton
ISBN: 1-58381-186-9   e-Book ISBN: 978-1-58381-259-4

ASP CONFERENCE SERIES VOLUMES
Published by the Astronomical Society of the Pacific

PUBLISHED: 2004

Vol. CS 325   THE SOLAR-B MISSION AND THE FOREFRONT OF SOLAR PHYSICS, The Fifth Solar-B Science Meeting
eds. Takashi Sakurai and Takashi Sekii
ISBN: 1-58381-187-7   e-Book ISBN: 978-1-58381-260-0

Vol. CS 326   Never Submitted by Editor
ISBN: 1-58381-188-5   LOC#: 2004118371

Vol. CS 327   SATELLITES AND TIDAL STREAMS
eds. F. Prada, D. Martínez Delgado, and T. J. Mahoney
ISBN: 1-58381-190-7   e-Book ISBN: 978-1-58381-261-7

PUBLISHED: 2005

Vol. CS 328   BINARY RADIO PULSARS
eds. F. A. Rasio and I. H. Stairs
ISBN: 1-58381-191-5   e-Book ISBN: 978-1-58381-262-4

Vol. CS 329   NEARBY LARGE-SCALE STRUCTURES AND THE ZONE OF AVOIDANCE
eds. A. P. Fairall and P. A. Woudt
ISBN: 1-58381-192-3   e-Book ISBN: 978-1-58381-263-1

Vol. CS 330   THE ASTROPHYSICS OF CATACLYSMIC VARIABLES AND RELATED OBJECTS
eds. J.-M. Hameury and J.-P. Lasota
ISBN: 1-58381-193-1   e-Book ISBN: 978-1-58381-264-8

Vol. CS 331   EXTRA-PLANAR GAS
ed. Robert Braun
ISBN: 1-58381-194-X   e-Book ISBN: 978-1-58381-265-5

Vol. CS 332   THE FATE OF THE MOST MASSIVE STARS
eds. Roberta M. Humphreys and Krzysztof Z. Stanek
ISBN: 1-58381-195-8   e-Book ISBN: 978-1-58381-266-2

Vol. CS 333   TIDAL EVOLUTION AND OSCILLATIONS IN BINARY STARS: THIRD GRANADA WORKSHOP ON STELLAR STRUCTURE
eds. Antonio Claret, Alvaro Giménez, and Jean-Paul Zahn
ISBN: 1-58381-196-6   e-Book ISBN: 978-1-58381-267-9

Vol. CS 334   14TH EUROPEAN WORKSHOP ON WHITE DWARFS
eds. D. Koester and S. Moehler
ISBN: 1-58381-197-4   e-Book ISBN: 978-1-58381-268-6

Vol. CS 335   THE LIGHT-TIME EFFECT IN ASTROPHYSICS: Causes and Cures of the $O - C$ Diagram
ed. Christiaan Sterken
ISBN: 1-58381-200-8   e-Book ISBN: 978-1-58381-269-3

Vol. CS 336   COSMIC ABUNDANCES as Records of Stellar Evolution and Nucleosynthesis, in honor of Dr. David Lambert
eds. Thomas G. Barnes, III and Frank N. Bash
ISBN: 1-58381-201-6   e-Book ISBN: 978-1-58381-270-9

Vol. CS 337   THE NATURE AND EVOLUTION OF DISKS AROUND HOT STARS
eds. Richard Ignace and Kenneth G. Gayley
ISBN: 1-58381-203-2   e-Book ISBN: 978-1-58381-271-6

ASP CONFERENCE SERIES VOLUMES
Published by the Astronomical Society of the Pacific

PUBLISHED: 2005

Vol. CS 338   ASTROMETRY IN THE AGE OF THE NEXT GENERATION OF LARGE TELESCOPES
eds. P. Kenneth Seidelmann and Alice K. B. Monet
ISBN: 1-58381-205-9   e-Book ISBN: 978-1-58381-272-3

Vol. CS 339   OBSERVING DARK ENERGY
eds. Sidney C. Wolff and Tod R. Lauer
ISBN: 1-58381-206-7   e-Book ISBN: 978-1-58381-273-0

Vol. CS 340   FUTURE DIRECTIONS IN HIGH RESOLUTION ASTRONOMY: A Celebration of the 10th Anniversary of the VLBA
eds. Jonathan D. Romney and Mark J. Reid
ISBN: 1-58381-207-5   e-Book ISBN: 978-1-58381-274-7

Vol. CS 341   CHONDRITES AND THE PROTOPLANETARY DISK
eds. Alexander N. Krot, Edward R. D. Scott, and Bo Reipurth
ISBN: 1-58381-208-3   e-Book ISBN: 978-1-58381-275-4

Vol. CS 342   1604–2004: SUPERNOVAE AS COSMOLOGICAL LIGHTHOUSES
eds. M. Turatto, S. Benetti, L. Zampieri, and W. Shea
ISBN: 1-58381-209-1   e-Book ISBN: 978-1-58381-276-1

Vol. CS 343   ASTRONOMICAL POLARIMETRY: CURRENT STATUS AND FUTURE DIRECTIONS
eds. Andy Adamson, Colin Aspin, Chris J. Davis, and Takuya Fujiyoshi
ISBN: 1-58381-210-5   e-Book ISBN: 978-1-58381-277-8

Vol. CS 344   THE COOL UNIVERSE: OBSERVING COSMIC DAWN
eds. C. Lidman and D. Alloin
ISBN: 1-58381-211-3   e-Book ISBN: 978-1-58381-278-5

Vol. CS 345   FROM CLARK LAKE TO THE LONG WAVELENGTH ARRAY: Bill Erickson's Radio Science
eds. Namir E. Kassim, Mario R. Pérez, William Junor, and Patricia A. Henning
ISBN: 1-58381-213-X   e-Book ISBN: 978-1-58381-279-2

Vol. CS 346   LARGE SCALE STRUCTURES AND THEIR ROLE IN SOLAR ACTIVITY
eds: K. Sankarasubramanian, M. Penn, and A. Pevtsov
ISBN: 1-58381-214-8   e-Book ISBN: 978-1-58381-280-8

Vol. CS 347   ASTRONOMICAL DATA ANALYSIS SOFTWARE AND SYSTEMS XIV
ed: Patrick L. Shopbell, Matthew C. Britton, and Rick Ebert
ISBN: 1-58381-215-6   e-Book ISBN: 978-1-58381-281-5
ISSN: 1080-7926

PUBLISHED: 2006

Vol. CS 348   ASTROPHYSICS IN THE FAR ULTRAVIOLET: FIVE YEARS OF DISCOVERY WITH FUSE
eds. George Sonneborn, H. W. Moos, and B-G Andersson
ISBN: 1-58381-216-4   e-Book ISBN: 978-1-58381-282-2

Vol. CS 349   ASTROPHYSICS OF VARIABLE STARS
eds. Christiaan Sterken and Conny Aerts
ISBN: 1-58381-217-2   e-Book ISBN: 978-1-58381-283-9

Vol. CS 350   BLAZAR VARIABILITY WORKSHOP II: ENTERING THE GLAST ERA
eds. H. R. Miller, K. Marshall, J. R. Webb, and M. F. Aller
ISBN: 1-58381-218-0   e-Book ISBN: 978-1-58381-284-6

ASP CONFERENCE SERIES VOLUMES
Published by the Astronomical Society of the Pacific

PUBLISHED: 2006

Vol. CS 351   ASTRONOMICAL DATA ANALYSIS AND SOFTWARE SYSTEM XV
eds. Carlos Gabriel, Christophe Arviset, Daniel Ponz, and Enrique Solano
ISBN: 1-58381-219-9   e-Book ISBN: 978-1-58381-285-3
ISSN: 1080-7926

Vol. CS 352   NEW HORIZONS IN ASTRONOMY: FRANK N. BASH SYMPOSIUM 2005
eds. Sheila J. Kannappan, Seth Redfield, Jacqueline E. Kessler-Silacci, Martin Landriau, and Niv Drory
ISBN: 1-58381-220-2   e-Book ISBN: 978-1-58381-286-0

Vol. CS 353   STELLAR EVOLUTION AT LOW METALLICITY: MASS LOSS, EXPLOSIONS, COSMOLOGY
eds. H. Lamers, N. Langer, T. Nugis, and K. Annuk
ISBN: 978-1-58381-221-1   e-Book ISBN: 978-1-58381-287-7

Vol. CS 354   SOLAR MHD THEORY AND OBSERVATIONS: A High Spatial Resolution Perspective, in Honor of Robert F. Stein
eds. Han Uitenbroek, John Leibacher, and Robert F. Stein
ISBN: 978-1-58381-222-8   e-Book ISBN: 978-1-58381-288-4

Vol. CS 355   STARS WITH THE B[e] PHENOMENON
eds. Michaela Kraus and Anatoly S. Miroshnichenko
ISBN: 978-1-58381-223-5   e-Book ISBN: 978-1-58381-289-1

Vol. CS 356   REVEALING THE MOLECULAR UNIVERSE: ONE ANTENNA IS NEVER ENOUGH, in Honor of the academic retirement of Jack Welch
eds. D. C. Backer, J. L. Turner, and J. M. Moran
ISBN: 978-1-58381-224-2   e-Book ISBN: 978-1-58381-290-7

Vol. CS 357   THE SPITZER SPACE TELESCOPE: NEW VIEWS OF THE COSMOS
ed. L. Armus and W. T. Reach
ISBN: 978-1-58381-225-9   e-Book ISBN: 978-1-58381-291-4

Vol. CS 358   SOLAR POLARIZATION 4
eds. R. Casini and B. W. Lites
ISBN: 978-1-58381-226-6   e-Book ISBN: 978-1-58381-292-1

Vol. CS 359   NUMERICAL MODELING OF SPACE PLASMA FLOWS: ASTRONUM-2006, 1st IGPP – CalSpace International Conference
eds. Nikolai V. Pogorelov and Gary P. Zank
ISBN: 978-1-58381-227-3   e-Book ISBN: 978-1-58381-293-8

Vol. CS 360   AGN VARIABILITY FROM X-RAYS TO RADIO WAVES
eds. C. Martin Gaskell, Ian M. McHardy, Bradley M. Peterson, and Sergey G. Sergeev
ISBN: 978-1-58381-228-0   e-Book ISBN: 978-1-58381-294-5

PUBLISHED: 2007

Vol. CS 361   ACTIVE OB-STARS: LABORATORIES FOR STELLAR AND CIRCUMSTELLAR PHYSICS
eds. Stanislav Štefl, Stanley P. Owocki, and Atsuo T. Okazaki
ISBN: 978-1-58381-229-7   e-Book ISBN: 978-1-58381-295-2

Vol. CS 362   THE 7TH PACIFIC RIM CONFERENCE ON STELLAR ASTROPHYSICS
eds. Young-Woon Kang, Hee-Won Lee, Kwong Sang Cheng, and Kam-Ching Leung
ISBN: 978-1-58381-230-3   e-Book ISBN: 978-1-58381-296-9

ASP CONFERENCE SERIES VOLUMES
Published by the Astronomical Society of the Pacific

PUBLISHED: 2007

Vol. CS 363   THE NATURE OF V838 MON AND ITS LIGHT ECHO
eds. Romano L. M. Corradi and Ulisse Munari
ISBN: 978-1-58381-231-0   e-Book ISBN: 978-1-58381-297-6

Vol. CS 364   THE FUTURE OF PHOTOMETRIC, SPECTROPHOTOMETRIC AND POLARIMETRIC STANDARDIZATION
ed. Christiaan Sterken
ISBN: 978-1-58381-232-7   e-Book ISBN: 978-1-58381-298-3

Vol CS 365   SINS – SMALL IONIZED AND NEUTRAL STRUCTURES IN THE DIFFUSE INTERSTELLAR MEDIUM
eds. M. Haverkorn and W. M. Goss
ISBN: 978-1-58381-233-4   e-Book ISBN: 978-1-58381-299-0

Vol. CS 366   TRANSITING EXTRASOLAR PLANETS WORKSHOP
eds. C. Afonso, D. Weldrake, and Th. Henning
ISBN: 978-1-58381-234-1   e-Book ISBN: 978-1-58381-300-3

Vol. CS 367   MASSIVE STARS IN INTERACTING BINARIES
eds. Nicole St-Louis and Anthony F. J. Moffat
ISBN: 978-1-58381-235-8   e-Book ISBN: 978-1-58381-301-0

Vol. CS 368   THE PHYSICS OF CHROMOSPHERIC PLASMAS
eds. Petr Heinzel, Ivan Dorotovič, and Robert J. Rutten
ISBN: 978-1-58381-236-5   e-Book ISBN: 978-1-58381-302-7

Vol. CS 369   NEW SOLAR PHYSICS WITH SOLAR-B MISSION, The Sixth Solar-B Science Meeting
eds. K. Shibata, S. Nagata, and T. Sakurai
ISBN: 978-1-58381-237-2   e-Book ISBN: 978-1-58381-303-4

Vol. CS 370   SOLAR AND STELLAR PHYSICS THROUGH ECLIPSES
eds. Osman Demircan, Selim O. Selam, and Berahitdin Albayrak
ISBN: 978-1-58381-238-9   e-Book ISBN: 978-1-58381-304-1

Vol. CS 371   STATISTICAL CHALLENGES IN MODERN ASTRONOMY IV
eds. G. Jogesh Babu and Eric D. Feigelson
ISBN: 978-1-58381-240-2   e-Book ISBN: 978-1-58381-305-8

Vol. CS 372   15th EUROPEAN WORKSHOP ON WHITE DWARFS
eds. Ralf Napiwotzki and Matthew R. Burleigh
ISBN: 978-1-58381-239-6   e-Book ISBN: 978-1-58381-306-5

Vol. CS 373   THE CENTRAL ENGINE OF ACTIVE GALACTIC NUCLEI
eds. Luis C. Ho and Jian-Min Wang
ISBN: 978-1-58381-307-2   e-Book ISBN: 978-1-58381-308-9

Vol. CS 374   FROM STARS TO GALAXIES: BUILDING THE PIECES TO BUILD UP THE UNIVERSE
eds. Antonella Vallenari, Rosaria Tantalo, Laura Portinari, and Alessia Moretti
ISBN: 978-1-58381-309-6   e-Book ISBN: 978-1-58381-310-2

Vol. CS 375   FROM Z-MACHINES TO ALMA: (SUB)MILLIMETER SPECTROSCOPY OF GALAXIES
eds. Andrew J. Baker, Jason Glenn, Andrew I. Harris, Jeffrey G. Mangum, and Min S. Yun
ISBN: 978-1-58381-311-9   e-Book ISBN: 978-1-58381-312-6

ASP CONFERENCE SERIES VOLUMES
Published by the Astronomical Society of the Pacific

PUBLISHED: 2007

Vol. CS 376 ASTRONOMICAL DATA ANALYSIS AND SOFTWARE SYSTEM XVI
eds. Richard A. Shaw, Frank Hill, and David J. Bell
ISBN: 978-1-58381-313-3   e-Book ISBN: 978-1-58381-314-0
ISSN: 1080-7926

Vol. CS 377 LIBRARY AND INFORMATION SERVICES IN ASTRONOMY V:
COMMON CHALLENGES, UNCOMMON SOLUTIONS
eds. Sandra Ricketts, Christina Birdie, and Eva Isaksson
ISBN: 978-1-58381-316-4   e-Book ISBN: 978-1-58381-317-1

Vol. CS 378 WHY GALAXIES CARE ABOUT AGB STARS: THEIR IMPORTANCE AS ACTORS AND PROBES
eds. F. Kerschbaum, C. Charbonnel, and R. F. Wing
ISBN: 978-1-58381-318-8   e-Book ISBN: 978-1-58381-319-5

Vol. CS 379 COSMIC FRONTIERS
eds. Nigel Metcalfe and Tom Shanks
ISBN: 978-1-58381-320-1   e-Book ISBN: 978-1-58381-321-8

Vol. CS 380 AT THE EDGE OF THE UNIVERSE: LATEST RESULTS FROM THE DEEPEST ASTRONOMICAL SURVEYS
eds. José Afonso, Henry C. Ferguson, Bahram Mobasher, and Ray Norris
ISBN: 978-1-58381-323-2   e-Book ISBN: 978-1-58381-324-9

PUBLISHED: 2008

Vol. CS 381 THE SECOND ANNUAL SPITZER SCIENCE CENTER CONFERENCE: INFRARED DIAGNOSTICS OF GALAXY EVOLUTION
eds. Ranga-Ram Chary, Harry I. Teplitz, and Kartik Sheth
ISBN: 978-1-58381-325-6   e-Book ISBN: 978-1-58381-326-3

Vol. CS 382 THE NATIONAL VIRTUAL OBSERVATORY: TOOLS AND TECHNIQUES FOR ASTRONOMICAL RESEARCH
eds. Matthew J. Graham, Michael J. Fitzpatrick, and Thomas A. McGlynn
ISBN: 978-1-58381-327-0   e-Book ISBN: 978-1-58381-328-7

Vol. CS 383 SUBSURFACE AND ATMOSPHERIC INFLUENCES ON SOLAR ACTIVITY
eds. R. Howe, R. W. Komm, K. S. Balasubramaniam, and G. J. D. Petrie
ISBN: 978-1-58381-329-4   e-Book ISBN: 978-1-58381-330-0

Vol. CS 384 14TH CAMBRIDGE WORKSHOP ON COOL STARS, STELLAR SYSTEMS, AND THE SUN
ed. Gerard van Belle
ISBN: 978-1-58381-331-7   e-Book ISBN: 978-1-58381-332-4

Vol. CS 385 NUMERICAL MODELING OF SPACE PLASMA FLOWS: ASTRONUM-2007
eds. Nikolai V. Pogorelov, Edouard Audit, and Gary P. Zank
ISBN: 978-1-58381-333-1   e-Book ISBN: 978-1-58381-334-8

Vol. CS 386 EXTRAGALACTIC JETS: THEORY AND OBSERVATION FROM RADIO TO GAMMA RAY
eds. Travis A. Rector and David S. De Young
ISBN: 978-1-58381-335-5   e-Book ISBN: 978-1-58381-336-2

Vol. CS 387 MASSIVE STAR FORMATION: OBSERVATIONS CONFRONT THEORY
eds. Henrik Beuther, Hendrik Linz, and Thomas Henning
ISBN: 978-1-58381-642-4   e-Book ISBN: 978-1-58381-643-1

ASP CONFERENCE SERIES VOLUMES
Published by the Astronomical Society of the Pacific

PUBLISHED: 2008

Vol. CS 388   MASS LOSS FROM STARS AND THE EVOLUTION OF STELLAR CLUSTERS
eds. Alex de Koter, Linda J. Smith, and Laurens B. F. M. Waters
ISBN: 978-1-58381-644-8   e-Book ISBN: 978-1-58381-645-5

Vol. CS 389   EPO AND A CHANGING WORLD: CREATING LINKAGES AND EXPANDING PARTNERSHIPS
eds. Catharine Garmany, Michael G. Gibbs, and J. Ward Moody
ISBN: 978-1-58381-648-6   e-Book ISBN: 978-1-58381-649-3

Vol. CS 390   PATHWAYS THROUGH AN ECLECTIC UNIVERSE
eds. J. H. Knapen, T. J. Mahoney, and A. Vazdekis
ISBN: 978-1-58381-650-9   e-Book ISBN: 978-1-58381-651-6

Vol. CS 391   HYDROGEN-DEFICIENT STARS
eds. Klaus Werner and Thomas Rauch
ISBN: 978-1-58381-652-3   e-Book ISBN: 978-1-58381-653-0

Vol. CS 392   HOT SUBDWARF STARS AND RELATED OBJECTS
eds. Ulrich Heber, Simon Jeffery, and Ralf Napiwotzki
ISBN: 978-1-58381-654-7   e-Book ISBN: 978-1-58381-655-4

Vol. CS 393   NEW HORIZONS IN ASTRONOMY: FRANK N. BASH SYMPOSIUM 2007
eds. Anna Frebel, Justyn R. Maund, Juntai Shen, and Michael H. Siegel
ISBN: 978-1-58381-656-1   e-Book ISBN: 978-1-58381-657-8